Extraterrestrial Ethics

By

Dr Jensine Andresen

Extraterrestrial Ethics

By Dr Jensine Andresen

This book first published 2023

Ethics International Press Ltd, UK

British Library Cataloguing in Publication Data

A catalogue record for this book is available from the British Library

Print Book ISBN: 978-1-871891-35-5

eBook ISBN: 978-1-871891-36-2

Contents

Preface

This book articulates a path forward to kindness, compassion, justice, and beauty. A primary thesis of the book is that human beings must replace ontological misinformation with an accurate view of the nature of reality to overcome conflict, violence, and suffering, and to survive.

The advanced extraterrestrial intelligence (ETI) in our midst is trying to help humankind see its way forward to a correct ontology by means of pedagogical displays of Unidentified Anomalous Phenomena (UAP). This ETI embodies a deontological ethical approach to existence in contrast to the human tendency to approach existence from a utilitarian ethical perspective. Humankind's strategic and tactical approach has led *Homo sapiens sapiens* to the brink of extinction because it lacks empathy and depth. In contrast, ETI's ontological approach to existence, which is consistent with deontological ethics, facilitates compassion, relationality, and social justice because it views existence from the perspective of the whole.

This book knits together multiple topics including UAP, geopolitics, physics, mathematics, and philosophy to demonstrates the richness of the creative acculturation unfolding now between humankind and ETI. As creative acculturation with ETI continues, human understanding of the nature of reality will advance, since the ETI/UAP around us now clearly operate on the basis of a whole physics. Creative acculturation with ETI also provides humankind with many opportunities to assess its own development as a species. We urgently must demilitarize and de-weaponize ourselves as a species while increasing social justice, which requires thinking ontologically rather than strategically and tactically. We also must expand our awareness and cognition in part by expanding our understanding of geometry and levels of order. Importantly, we must behave and operate from the point of view of the whole rather than in the pursuit of individualistic gain. This is the only sane path forward, and it promises experiences of deep beauty, creativity, and love.

Chapter 1:
Estimate of the Situation

Entangled in an Ontological Ball of Yarn

Schrödinger's cat is out of the bag. An advanced, extraterrestrial intelligence (ETI) is operating many Unidentified Anomalous Phenomena (UAP) in our midst now.

The title of this book, *Extraterrestrial Ethics*, has two meanings—the ethics *of* extraterrestrials, and ethics *for* extraterrestrials. On the first topic—the ethics *of* extraterrestrials—I ask what kind of ethics ETI embodies. Based on the actions of the ETI/UAP around us now, we can conclude that ETI's ethics is based on an ontological understanding of the nature of reality that comprehends the unbroken wholeness of existence. This provides the basis for ETI's deontological ethics. On the second topic—ethics *for* extraterrestrials—I propose that to participate peacefully and constructively with ETI, human beings must learn to perceive, understand, and embody an ontology of wholeness and a deontological rather than utilitarian ethical approach to existence. This is necessary for humankind to create a socially just society and, also, to survive as a species. It also is the same approach that enables human beings to participate constructively in the ongoing and accelerating process of creative acculturation with ETI.

To articulate the nature and meaning of wholeness, I refer to the conceptual framework described by physicist and philosopher David J. Bohm in his groundbreaking interpretation of the quantum theory. Born in Wilkes-Barre, Pennsylvania in 1917, Bohm enriched physics deeply. In 1990, he was elected Fellow in the Royal Society of London. Bohm's central contribution to the field of physics is his articulation of a causal, ontological interpretation of the quantum theory. Bohm articulates his views in terms of mathematics and physics and, also, in terms of linguistics and philosophy. In addition to understanding Bohm's framework conceptually, one also should strive for a direct, felt experience of what Bohm conveys, especially in his discussion of subtlety and levels of order.

Bohm refers to two basic levels of order in his writing—explicate order and implicate order. Beyond this, however, Bohm postulates *a series of levels of order of increasing subtlety*, from implicate to super-implicate, super-super-implicate, etc., potentially to infinity (Bohm 1986a, 96-8). These levels of order are related to Bohm's concept of holomovement, which recognizes the inherent dynamism of the whole. The holomovement is a dynamic continuum comprised of explicate order and subtler, implicate, etc., levels of order. Throughout the book, I often write "implicate, etc." to indicate implicate, super-implicate, super-super-implicate, etc., levels of order, potentially to infinity. Bohm's framework is described in more detail in chapter 5, in case some readers wish to review that material now.

Bohm's framework helps one understand *morphogenesis*, namely the genesis of form, and how the forms we perceive in explicate order arise from subtler levels of order beyond explicate order. Many people mistake explicate order as the entirety of existence, when in fact existence also includes these subtler levels of order—implicate, super-implicate, super-super-implicate, etc., potentially to infinity.

Bohm proposes that the implicate order can be defined by the quantum potential, Q, which is discussed in more detail in chapter 5. Even the quantum potential appears has a geometrical aspect. Fiscaletti (2018, 13) writes:

> Taking account of the geometrodynamic nature of the quantum potential, one can also say that the quantum potential expresses the geometric properties of space which determine the behaviour of the particles and are derived just from a more fundamental geometrodynamic background, namely the implicate order.

Geometry is inherent to explicate order forms, which arise in nature according to geometric principles. For example, numerous forms that manifest around us, including our own embodied forms, are based on fractal geometry. Indeed, one sees the spiraling repetition of fractal geometry in the branching of trees and the patterning of leaves.

Not only are the forms we see and otherwise perceive geometrical in nature, but there is also geometry in our thinking. This becomes clearer given Bohm's explanation of how matter and consciousness are inseparable, also described in chapter 5. To broach the topic here, however, I present an intuitive example. When you shut your eyes and visualize something, or when you dream, this occurs in three dimensions, not in two. Particularly gifted people may be able to visualize things in more than three dimensions—but people naturally are not prone to visualizing things or dreaming in two dimensions, unless they intentionally are trying to visualize something that is two-dimensional. Even then, however, the 'visualization' per se usually, if not always, occurs in three-dimensions in one's mind's-eye, as for example when one 'sees' a piece of paper on which a two-dimensional rendering is made.

When we reason and think tactically and strategically—especially tactically, as for example when military strategists pour over a two-dimensional map of a battlefield—reality is being simplified from three dimensions to two dimensions. Tactical and strategic thinking are emphasized in such situations, which leads to a utilitarian approach to ethics that is based on oversimplifying and reducing a much more complex reality to a flat surface. By increasing the depth of our thinking, however—for example by introducing three-dimensional imaging—we increase our cognitive range in a way that also shifts our ethical outlook. When we include in our awareness and understanding implicate, etc., levels of order, our ethical approach opens even further, moving from a utilitarian emphasis to much more of a deontological one.

Ontological thinking that brings in implicate, etc., levels of order is a very abstract form of thinking that is unfamiliar to most people. In fact, the educational system and other institutions of the so-called modern world socialize us *not* to perceive implicate, super-implicate, super-super-implicate levels of order, though certain conceptual mathematicians, theoretical physicists, and mystics often share the capacity to perceive these realms (Andresen in preparation).

Whereas tactical and strategic thinking based on an oversimplification of reality is consistent with a utilitarian ethical approach, ontological thinking based on perception of subtler levels of order is consistent with a deontological ethical approach. As I describe elsewhere (Andresen in preparation), moving from a tactical and strategic mindset to an ontological involves expanding our understanding of geometry and the relationship of geometry to morphogenesis. Here, there are two 'realms' of geometrical understanding. At the broadest level, there is a fulsome geometry of existence. In Bohm's framework, for example, one can observe that conceptually, the idea of levels of order is geometrical. Then, a narrower focus on explicate order and morphogenesis—i.e., the coming into being of form—illuminates how geometry comes into being in explicate order, since naturally occurring forms around us are naturally geometrical. Accordingly, a central aspect of morphogenesis is the genesis of geometry. Said simply, the genesis of form is the genesis of geometry, as perceived from an explicate order perspective (though there is a caveat here, namely that implicate, etc., levels of order also have a geometrical aspect, for example as regards the quantum potential, Q).

Bohm's implicate order framework has significant implications for human society. With respect to how embodying a perspective based on subtler aspects of order will change human society, Bohm states:

> It's hard to say at first, but it will clearly imply something very different, a different attitude in the sense that we won't give that much primary weight to the external and the mechanistic side—the side of fragmentation and partiality [i.e., explicate order]. Also, it encourages us much more toward a creative attitude, and fundamentally it opens the way to the transformation of the human being because a change of meaning is a change of being. At present we say because of the confused fragmentary meanings we have a confused fragmentary being, both individually and socially. Therefore this opens the way to a whole being, in society and in the individual (Weber 1987).

Awareness and understanding of implicate, etc., levels of order facilitate the shift from tactical and strategic thinking to ontological thinking. As one deepens one's experience of subtler aspects of reality and of wholeness, one also assigns more priority to ethical and spiritual progress in contrast to economic and geopolitical control and domination. Accomplishment then becomes understood in terms of ethical, moral, and ontological/spiritual progress instead of being measured by the self-interested accumulation of money, power, and status in society. This is the morally direction in which we need to move as human beings.

How and Why History Repeats

How human thoughts and actions unfold in history impacts our local, regional, and global environments—and it also impacts the whole. This process occurs by means of a continuous cycle of projection out of implicate order to explicate order and introjection from explicate order back into implicate order. As Bohm puts it, "If we keep on introjecting similar content, it will build up a certain meaning" (Weber 1987; see also Bohm 1989). In fact, *this is the essence of why history repeats.*

It is not mere happenstance that the individual and collective histories of humankind appear to repeat in a cyclical manner. This occurs because *similar content* moves back and forth between levels of order and patterns build up. While people often notice that history seems to be repeating, they often feel powerless to change history because they do not understand the physics behind the repetition—in fact, most people are unaware that physics plays a role in history, though of course physics plays a role in everything. Understanding how physics plays a role in history will help our species avoid negative cycles and, instead, survive and flourish.

Furthermore, by understanding how reality unfolds and why history repeats, one strengthens one's resolve to change society for the better. Embodying a firm, internal locus of control posture, one realizes that each being matters, and everything each being does matters. Individual thoughts and actions *can and do* make a difference. As we become more integrated and centered, more considered in our speech and actions, and more attuned to how patterns unfold energetically to create reality, we *choose* how we participate in the universe. Here, we must recognize that each one of us can shape reality positively, and we must endeavor to do so. Engaging in negative actions and spending time tormented by negative thoughts and emotions reduces the time one has available to contribute positively to the whole.

Physics helps us understand how reality is constituted afresh on a moment-to-moment basis. Since the content we create really matters, we absolutely must create positive patterns rather than negative ones. Existence is never final—it is an ongoing co-participation in the inexhaustible creation of meanings. By manifesting constructive rather than destructive thoughts and actions and having constructive content introjected into the implicate order, similar constructive patterns will be projected back out into explicate order, causing existence as a whole to change for the better. When beings participate together in this process of constructive introjection into the implicate order, human society shifts for the better more quickly. Even some meaningful percentage of human beings participating together in such a constructive process has the power to shift reality positively—as does a single individual impacting reality profoundly. When all beings participate together constructively, the whole improves immediately.

The process of projection, introjection, projection, introjection, etc., between implicate and explicate levels of order has a temporal dimension, too. One's thoughts and actions are not merely important in the moment—they have far-reaching implications for the whole *throughout the entirety of time*—which is eternal. It therefore is vitally important to manifest positive thoughts and actions to the utmost of one's ability at every possible moment and in every possible circumstance. Given that the present and the future are completely entangled, whatever one does literally reverberates in the whole *forever*.

Renée Weber, former Professor of Philosophy at Rutgers University, discusses with Bohm whether his ontological interpretation of the quantum theory implies that human beings contribute to the deeper, inward nature of existence because they contribute to what is re-injected into the implicate order from the explicate order. Bohm states:

> Yet, it [i.e., implicate order] contributes in some way and we could say that we ourselves maybe can bring order to what we're doing, we can play a functional role in producing a higher order than would be possible without us. We do not merely modify it [i.e., implicate order] somewhat, but rather although we make a very tiny change in the whole thing, nevertheless, it may be crucial in allowing this higher order to come to something new to realize its potential.

Weber notes her view that evolution is a creative aspect of nature, and she asks Bohm if he is saying that not only are we being shaped by that evolution, but that we also can help shape it. In reply, Bohm states, "We're part of the movement, there's no separation between us and it, we are part of the way in which that [evolution] shapes itself." Weber then states her understanding that "what is introjected back into the implicate order is not lost, but its essence is in some way preserved," to which Bohm replies, "There may have been something out of which the Big Bang came" (Bohm 1986b, 32-33).

The Pitfalls of Incorrect, Consensus Thinking

Human beings' consensus view of reality is ontologically incorrect. It also is inside out. Bohm uses the metaphor of ripples on the surface of an ocean to help people understand the relationship between explicate and implicate, etc., levels of order. He states, "Matter is like a small ripple on this tremendous ocean of energy, having some relative stability and being manifest" (Bohm 1986b, 28).

Ripples on the ocean is the metaphorical description Bohm uses for explicate order, while the ocean is the metaphorical description Bohm uses to represent implicate order. While most human beings see the explicate order 'ripples,' they completely miss the implicate, etc., levels of the 'ocean' itself. This means that our consensus, historical narrative acknowledges 'parts' of reality but misses the 'whole' of existence itself. One of the most profound mistakes people make is to think that explicate order is the entirety of existence. Today, this ontological misinformation has taken humankind to the point of imminent extinction. It therefore is urgent that we change our perception, broaden our view, and act with expediency for the good of the whole.

The analytical aspect of human cognition parses explicate order manifestations into small parts that it subjects to analysis. But this parsing of reality distracts human beings from the wholeness of reality. It also results in social injustice, since fragmenting reality reinforces an egoistic sense of personal identity according to which individuals rigidly separate themselves from one another. People have forgotten that they are part of a whole, and that they must behave ethically and morally in a manner that supports the whole.

Society today is completely dysfunctional. It is not merely that a few things do not work—*nothing* works. This is because all aspects of society are infected with the same ontological misinformation. However, since all the challenges facing human society today arise from the same ontological misinformation, if we correct our ontology, these myriads of challenges will begin to resolve themselves. That is fortunate, since we simply do not have enough time to solve every issue individually. Approaching the situation that way also would be ineffective, since even if one challenge were resolved, another quickly would take its place unless the underlying ontological misinformation also was remedied. In other words, humankind has one *fundamental* thing to fix—our ontology. Fixing that will help with all the challenges facing humankind.

When misinformation is present at a central position in a system, all manner of chaos ensues (see also Andresen, 2022b, 295-96, 316-17). Bohm (1989, 22-23) discusses how a small amount of misinformation can have enormous, negative consequences for a system. He relates his comments to the concept of meaning:

> Culture *is* meaning, and when we have the wrong meaning within culture, it is like misinformation.
>
> So we can say there's not only information, there's also misinformation. A virus in the DNA molecule could be called a bit

of misinformation, in that it enters the genetic structure and causes the cells to produce more viruses instead of more cells. ...

Society, too, is full of this sort of misinformation. We have various ways of dealing with biological misinformation. The best way is by the immune system which recognizes it and gets rid of it, but we have no such system in society. Misinformation accumulates and society gradually decays. You see, the older society gets, the more chance it has to accumulate all sorts of misinformation and the more it starts to fall apart. The society is blocked because misinformation is held *rigidly*. ...

Behind this sensitive being in us there is some sort of a meaning which has built up throughout the lifetime. It has come from the culture and it produces this fear, resistance and rigidity. There are several ways of approaching this problem. There's the approach of meditation which attempts to deal with this problem at the individual level; and there is the attempt of psychiatry to get at some of this partly by going through childhood experiences. But there is also a third approach according to which a great deal of our anxiety is basically sociocultural in origin and therefore best approached in a dialogue which is in a sociocultural context. ... [Built up meaning is mostly] misinformation at a central position. It's the same as with the DNA molecule. A tiny bit of misinformation in the DNA can cause everything to go wrong. If there is misinformation at a crucial part of the information that is determining a process, the consequences can be very serious. This can be the situation with the mind. Just a little bit of misinformation, say, in our image of ourselves, can distort the whole mind. We are not aware of it, but I think that our mind is so conditioned by such misinformation that it is frightened even to approach the question whether something is wrong with it. ...

We are living in contact with an illusory being which is being projected by the misinformation. We have an illusory view of ourselves which is built up largely in a sociocultural way. Most of our view of ourselves comes from society, doesn't it? For example, the fear of being a fool is a sociocultural consequence, and also the fear of letting go. So a tremendous part of this fear is sociocultural in original. It is sociocultural misinformation that is transmitted subliminally (typographical errors in the PDF corrected).

A significant aspect of misinformation in human society today is ontological in nature. Furthermore, this misinformation exists at a central position in our meaning system as human beings. Until human beings correct their ontology, their minds will remain confused and society will continue to reflect this confusion.

Because they are accustomed to living their lives in accordance with ontological misinformation, many people hold onto that misinformation very rigidly. One sees this across all aspects of society, where the rigidity of institutional stagnation and entrenched bureaucracy limits genuine creativity and dialogue. In this respect, the academy, corporations and other such organizations, and the military are quite similar—they all are institutions that suffer from materialist reductionism, ontological misperception and misunderstanding, and moral confusion. Materialist reductionism depends on the erroneous view that mind and matter are separate—an incorrect idea that one can trace back in Western philosophy to writing of René Descartes (b. 1596). Materialist reductionism is linked to theoretical reductionism according to which the vast preponderance of models and theories proposed in academia are based on the incorrect view that explicate order manifestations comprise the entirety of existence.

In addition to suffering from materialist reductionism, institutions based on militarism and a national security mindset also suffer from identity reductionism, namely the tendency to construe almost all situations as battles between 'oneself' (or one's 'group') and the perceived 'other' (or 'other group')—with 'the other' often being characterized as one's mortal enemy and demonized rhetorically. In war and other instances of violent conflict, such reductionism is taken to the extreme when the false divide between self and other is used to justify an 'us/them' mentality that attempts to rationalize dominating the other to the point of killing and in some instances even genocide.

To perceive and understand the nature of reality correctly, human beings must be willing to let go of the misinformation clutter—including various forms of reductionism—that has accumulated in their minds and, hence, in society. We must disentangle ourselves from the consensus and reductionist views that mind and matter are separate and that explicate order is all that exists. As I state elsewhere (Andresen 2022b, 295-300, 316-18), these two pieces of misinformation have resulted in our current confused and chaotic human society.

In contrast to human beings, the ETI in our midst now is not suffering from the same sort of misinformation that is plaguing human society. This ETI appears highly rationale, very serious, and technologically masterful—while it also displays sagacious whimsy, perspicuous aesthetic sensibilities, and profound kindness. It also is very, very, very, very sentient—by which I mean it is very, very, very, very aware—and not the least bit artificial in this awareness and sentience. To relate with this ETI, human beings must cultivate calmness, consistency, morality, and spiritual depth. They also must expand their capacity for empathy, compassion, and love while manifesting deep-seated kindness and a sense of responsibility for the whole.

Since existence is a seamless whole in which everything is enfolded in everything else, we already are participating with ETI/UAP, and ETI/UAP already are participating with us. To the extent that we can become kinder and gentler participants in this process, things will improve for everyone. Creative acculturation between humankind and ETI has sweeping implications for everything, including geopolitics, economics, energy markets, technology, science, the environment, theology, religion, and everything else in between. But the most profound and positive changes will occur at the levels of human embodiment, consciousness, character, philosophy, ethics, morality, and spirituality.

The Three, Core Postulates of This Book

This book proposes and explains three, interrelated postulates:

<u>One</u>: ETI mostly thinks ontologically while human beings mostly think tactically and strategically. In the context of ethics discourse, ontological thinking is consistent with a deontological approach to ethics, whereas tactical and strategic thinking are consistent with a utilitarian approach to ethics.

<u>Two</u>: To avoid the collapse of human civilization and to survive as a species, *Homo sapiens sapiens* must transform from thinking and acting tactically and strategically and must assume a correct, ontological view with respect to the nature of reality. It also must live in accordance with a deontological ethical framework that arises from a correct, ontological perspective. Using Bohm's terminology, *human beings must transition from focusing only on the explicate order to perceiving and understanding, in addition, implicate, etc., levels of order.*

<u>Three</u>: The path human beings must take to realize an ontological rather than tactical and strategic approach to existence is the same path that will help human beings: realize a society based on social justice; understand ETI/UAP more clearly and participate with ETI/UAP more constructively; and explore the universe in a way that contributes positively to the whole. An accurate ontology will foster an effective, deontological approach to ethics that moves beyond the current, ineffective, status quo that is based primarily on utilitarian ethics. In contrast to a utilitarian approach, which optimizes criteria on behalf of certain individuals and/or groups, deontological ethics supports constructive rather than destructive outcomes for the whole.

The first postulate relates to the ethics *of* extraterrestrials, whereas the second and third postulates relate to ethics *for* extraterrestrials.

First Postulate: ETI Thinks Ontologically, Humans Strategically/Tactically

<u>One</u>: ETI mostly thinks ontologically while human beings mostly think tactically and strategically. In the context of ethics discourse, ontological thinking is consistent with a deontological approach to

ethics, whereas tactical and strategic thinking are consistent with a utilitarian approach to ethics.

The behavior of ETI's UAP demonstrates an understanding of physics that requires a clear and accurate ontology. In contrast, most human beings do not operate according to a correct ontology, relying instead upon tactical and strategic thinking. This is particularly true across the geopolitical space, in which foreign policy and military doctrine demonstrate utilitarian thinking taken to the extreme in bald and ruthless Machiavellianism.

In a deontological ethical approach, the ends do not justify the means. One primarily finds this kind of orientation to existence in certain philosophical and spiritual systems of thought. In contrast, in a utilitarian ethical approach, the ends are seen to justify the means. In contemporary culture, most human beings take this approach to almost everything. This is true whether one lives in a liberal democracy or in countries under autocratic control.

Another commonality in today's human society is that *the types* of strategies and tactics countries use to realize their utilitarian objectives mostly are the same, regardless of what theory of rule predominates in each country. In the world of intelligence gathering, for example, all major nation-states rely on satellite imagery, intercepts, and embedded sources to try to determine what other actors—particularly those perceived as adversaries—will do. The world over, such 'intelligence gathering' is the basis on which utilitarian calculations are made in the realms of foreign policy and military action.

In stark contrast to humankind's utilitarian approach to existence, ETI is working from an ethics understanding that is deontological in nature. ETI's approach is far superior to the human method of doing things, for the simple reason that it is better aligned ontologically with the nature of reality itself. If human beings hope to survive as a species, they must learn from ETI how to think ontologically and how to embody a deontological approach to ethics.

Many indications exist that the ETI in our midst is aware of how much humankind is struggling as a species. ETI's pedagogical demonstrations and precisely choreographed UAP displays seem tailor-made to help humankind move forward ontologically, ethically, morally, and spiritually. With UAP displays and other forms of outreach, ETI literally is pointing the way to a correct ontological view based on perception and understanding of wholeness.

Why is ontology so pivotal? The commonsense answer to this question is correct, namely that when a species does not perceive and understand the nature of reality accurately, it makes numerous errors that accumulate and reinforce one another. Initial errors become amplified until the disruptions they cause become larger and larger, especially when a society is very interconnected, as ours is. On Earth today, misinformation propagates across the entire planet by means of the Internet and via other pathways of complex, global interactions. This means that issues that begin locally often morph very rapidly to create regional and even global disturbances. By means of multiple feedback loops, this process often accelerates until various aspects of different yet interrelated systems collapse. Partial collapses are abundant across the globe, and when enough elements of a larger system collapse, total collapse is not far behind. This is our current predicament on Earth, where we face the possibility of a total collapse of the entire planetary system unless we change our ontological and ethical approach to everything.

The task of transforming our ontological understanding of the nature of reality could not be more urgent. With society on the verge of total collapse (Knorr 2020) and local and regional instances of collapse becoming increasingly common, as for example evidenced by the number of failed and vulnerable states (WPR 2022), we are out of time. While strengthening predicative collapse analysis and engaging in collapse avoidance (Gross 2022, 118) can provide short-term reprieve, these measures alone are insufficient to ensure against the total collapse of human civilization. The only viable path forward for human beings now is to correct our ontology—as rapidly as possible.

Second Postulate: Moving from Tactical/Strategic to Ontological Thinking

> Two: To avoid the collapse of human civilization and to survive as a species, *Homo sapiens sapiens* must transform from thinking and acting tactically and strategically and must assume a correct, ontological view with respect to the nature of reality. It also must live in accordance with a deontological ethical framework that arises from a correct, ontological perspective. Using Bohm's terminology, *human beings must transition from focusing only on the explicate order to perceiving and understanding, in addition, implicate, etc., levels of order.*

To avoid further partial collapses and a total collapse of human society, human beings must transition from a tactical and strategic mindset to an ontological one, which can be facilitated by a correct interpretation of the quantum theory. Since ETI already thinks ontologically and has a much clearer and more accurate understanding of the nature of reality than human beings do, creative acculturation with ETI will help humankind enormously in this endeavor.

The nuclear crisis and climate change are the two most serious challenges facing humankind today. Of the two, the challenge posed by the continued existence of nuclear weapons and other nuclear technologies is the most urgent, simply because in the case of an exchange of nuclear weapons, or even the detonation of a single nuclear weapon, things could spiral completely out of control very quickly. Even though climate change is a broader issue than the situation involving nuclear weapons, and even though climate change is moving rapidly, it does not have the potential to unfold catastrophically in the space of a few hours the way an exchange of nuclear weapons could.

During the current war in Ukraine, threats regarding the use of nuclear weapons have been broached repeatedly. In addition, nuclear power installations have been occupied, which de facto is a weaponization of these installations. That this has occurred indicates that there is an entirely new level of recklessness and chaos emanating from the human mind. Not even during the Cold War was the overt threat of nuclear war invoked in the way it has been of late. When CNN publishes stories with titles such as "The most likely nuclear scenario" (CNN Opinion 2022), one knows something truly is amiss.

In a recent chapter, "Mind of the Matter, Matter of the Mind" (Andresen 2022b), which appears in a recent edited volume, *Extraterrestrial Intelligence: Academic and Societal Implications* (Andresen and Chon Torres 2022), I discuss physics and, also, nuclear weapons and other nuclear technologies. Theoretically, the chapter demonstrates the applicability of Bohm's ontological interpretation of the quantum theory to the elucidation of ETI/UAP. The chapter focuses on two aspects of Bohm's framework in particular, the inseparability of consciousness and matter, and the wholeness of reality. Practically, the chapter focuses on the dangers of nuclear weapons and other nuclear technologies, which I argue must be abolished because fission literally fragments reality—ripping through coherence and creating incoherence in its wake.

The title of the chapter, "Mind of the Matter, Matter of the Mind," also poses an implicit question: What is the matter with the human mind? In this book, I make that question explicit by asking what human beings are doing wrong cognitively to create such a dysfunctional society. My conclusion is that human perceptual and cognitive errors emanate from an incorrect ontology. Accordingly, by perceiving and understanding ontology more clearly and accurately, human beings will gain a better understanding of how history unfolds. This in turn will make many positive outcomes imminently attainable—remediation of the environment, implementation of a socially just society, and constructive, creative acculturation with ETI.

The current situation on Earth, and, increasingly, in space, is very serious. The nuclear crisis and climate change are not the only two issues we face, and because of how interconnected things are, the overall situation is far more serious than many people realize. Failure of any one aspect of the system can impact the entire system precipitously. Although data analysis and modeling in Earth sciences show that Earth is a complex system, these methodologies do not even begin to approach the deeper mathematics (including geometry) and physics of the situation.

At the level of the planetary system of Earth, human beings currently face two types of overlapping challenges—systemic and dynamic. Systemically, human beings are experiencing the widespread militarization of society, according to which almost everything, including language and social media discourse, is being weaponized. Two alarming manifestations of the militarization of human society are the multiple advanced weapons systems operating on Earth and in space (Andresen 2021; forthcoming) and the continued proliferation of weapons of mass destruction (WMDs), including biological weapons, or bioweapons. The proliferation of advanced weapons systems depends upon civilian-military partnerships in which these weapons systems are designed and manufactured. These partnerships create enormous profits for certain elite individuals, with war profiteering being a major reason that largescale military conflicts and a cyclical war economy continue to manifest in explicate order. People want us to believe this is inherent to human nature. It is not. Much of the conflict around us now is manufactured and then choreographed by people who profit from it, and who also profit from the 'reconstruction' that occurs afterwards.

Also systemically, our global energy infrastructure is both unsustainable and unstable. We rely on nonrenewable resources such as oil and natural gas, which in turn rely on an unjust and viciously exploitative petrodollar system. Many regions of the world also continue to rely on the dirty burning of coal. In addition, technologies such as nuclear power are far too risky to be prudent. Economically, we live in an increasingly financialized economy in which poverty results from systemic injustices caused by the heavily manipulated yet still unstable global central banking system. This system puts society through unnecessary, repetitive cycles of 'boom' and 'bust' deceptively referred to as 'the business cycle.' It also results in unnecessary and stultifying periods of stagflation and hyperinflation. Run by central bankers, financiers, and other supranational individuals and groups who set agendas that impact virtually all eight billion people on the planet, the global central banking system is completely unjust (KI 2023; see also Werner 2003; Totten 2003; Oswald 2014; Andresen forthcoming).

At a broad, macroeconomic level, unchecked spending by governments has left humankind holding an almost unfathomably large bubble of global debt. Meanwhile, the financialization of the global economy has resulted in unstable market dynamics prodded on by aggressive, self-interested traders. Their parasitic and insider gambling in derivatives, commodities, and housing markets, and their frontrunning and use of 'Black Edge' and 'Gray Edge' information, swing enormous amounts of wealth in their own directions while hardworking persons are left impoverished and facing exploitative usury as they try to make ends meet.

Educationally, there is an enormous disparity in access and concomitant opportunities. Logistically, globalization and interlocking and enmeshed supply chains result in massive food shortages in some areas and hyperinflation of food prices in others. Disruptions to the supply chain resulting from human reactions during the height of the Covid-19 pandemic demonstrate the extent to which the global logistics network is interlinked, with difficulties in one part of the world impacting the rest of the planet within days. With respect to health, an enormous gulf in access to care coincides with a lack of deep understanding regarding what fosters global public health. At the same time, human trafficking and sexual exploitation are virtually institutionalized at this point. Legally, the world has permitted multiple, opaque organizational forms such as multinational corporations (MNCs), translational corporations (TNCs), limited liability corporations (LLCs), non-governmental organizations (NGOs), and international non-government organizations (INGO) to function without adequate checks, all the while escaping any real accountability.

In addition to systemic problems, humankind also faces a variety of dynamic challenges. These involve rapid movement with pace, and their downstream consequences are all but impossible to predict. This list is quite long: climate change and extreme weather patterns; fires, floods, and droughts; mass extinction of species of animals and plants, which also threatens humanity (Dalton 2019); dwindling and outright loss of biodiversity; loss of cultural diversity; deforestation and devastation of the rain forests; desertification; swarms of marauding and rapidly reproducing desert locusts, which threaten millions of people (John and Feleke 2020); possible polar shifts; eruptions of volcanoes and supervolcanoes; potentially cataclysmic events such as largescale asteroid and comet impacts with Earth; potential solar flares, which, if on the scale of the Carrington Event of 1859, could devastate the electrical grid if it is not hardened first (Pompeo 2022); the rapid onset of dangerous terrestrial and undersea conflicts; pandemics; human dislocation, mass migrations, and a global refugee crisis; continued growth of the global population; demographic challenges including aging populations many members of which lack sufficient support; the rapid proliferation of irrational ideologies; unregulated development of synthetic biology, artificial general intelligence (AGI), and other forms of artificial intelligence (AI); unregulated and irresponsible genetic manipulation, genetic modification of organisms, and other forms of genetic and biological engineering, including genetic engineering of plants (UIA 2022); hyperinflation; worker dislocation; deteriorating logistics and information technology (IT) networks; crumbling highway and rail networks; increasing linkages between terrorist organizations and organized crime; increasing violent crime almost everywhere, with a tremendous surge in gun violence in the U.S.; the rise of kleptocratic, authoritarian, and totalitarian thinking and leadership—all of which have plutocratic (Ambrose 2015) and fascist elements—and corruption that reaches the highest levels of every government in the world.

One of the issues mentioned above—the loss of cultural diversity—is not discussed as often as many of the other topics. It nevertheless is a major issue to address. Many people in the so-called developed world often assume that 'progress' is inevitable, unassailable, and unavoidable—and that 'progress' means assimilation to the majority norm. However, when cultural diversity is lost, there is a corresponding loss of creativity within the human species. Historically, as the British pursued Empire, they possessed at least some self-awareness regarding their own impact on indigenous populations (Aborigines Protection Society 1837). But today, the export of Anglo culture into territories effectively controlled by the U.S.

and its NATO allies and the export of Han culture into areas effectively controlled by China occurs with an absence of almost any self-awareness from those who are exporting culture with respect to their impact on indigenous groups. At an intraspecies level, as cultures are assimilated into the majority norm, humankind experiences a devastating loss of creativity—indigenous knowledge is lost or, effectively, stolen, for example when pharmaceutical companies engage in biopiracy involving medicinal plants and restorative health regimes without providing benefits to the indigenous communities from which they harvest these plans and abscond information regarding their usage (Shiva 1999). This kind of 'assimilation' is quite different from the creative *acculturation* with ETI that is occurring now at an interspecies level—since acculturation, unlike assimilation, *increases* cultural diversity and creativity rather than limiting it (Andresen in preparation).

The seriousness of the overall global situation has not been lost on the political and business leaders who recently participated in the World Economic Forum (WEF) in Davos, Switzerland. These individuals use the term 'polycrisis' to refer to "the swirl of global emergencies that include economic slowdowns and rising inflation, the war in Ukraine and more" (Sorkin et al. 2023)—without of course acknowledging that the global elite have been unwitting and sometimes intentional architects of many aspects of polycrisis. The term polycrisis, which was coined in the 1990s, was used by Jean-Claude Juncker in 2016 when he was president of the European Commission. This term reflects the realization that systemic and dynamic challenges are not absolutes, and that different types of challenges often collide with and exacerbate one another (Sorkin et al. 2023).

Examples of systemic and dynamic challenges intersecting one another are when climate change leads to violent conflict, or when flooding causes dislocation and migration. Another example is when overreliance on coal exacerbates the negative effects of climate change (Bradsher and Krauss 2022). With respect to climate change, for example, the group in Davos voiced the need for a global framework to tackle climate-risk disclosures (Sorkin et al. 2023). The point is that any systematic or dynamic trajectory on its own, or any combination of these forces, has the potential to result in rapid and unforeseen consequences, leading to the potential extinction of humankind.

Often when systemic and dynamic challenges intersect, they amplify one another, which means that the potential *pace* of collapse of any one aspect of the system here on Earth—or the entire global system—is very difficult to predict. More than likely, we are underestimating the pace of potential collapse significantly. Of all the factors impacting the longevity of a species, it is this pace of change once things go off course—and human beings' tendency to underestimate it—that causes me the greatest concern. Since systems facing multiple pressures are vulnerable to rapid collapse, it is almost impossible to predict just how quickly a system *could* collapse. When change unfolds in volatile regions around the world, once a situation crosses an often seemingly invisible threshold, areas and even entire countries can collapse very quickly.

Within academia, existential risk is discussed at places such as the University of Cambridge's Centre for the Study of Existential Risk (CSER). Members of the existential risk reduction community study the risks from nuclear and biological weapons, climate change, and emerging technologies such as synthetic biology and AI—technologies that all pose incredibly high risks. Any one or more of them could lead to civilizational collapse or extinction, which makes sound regulation based on careful ethics discussion imperative. In addition, many powerful technologies are dual-use, meaning they can be helpful or harmful, so analyzing how much regulation is appropriate is critical. Some people even characterize nuclear technology as dual-use, though my own view is that there is no upside to nuclear weapons or to nuclear power, since both are deeply harmful to human society. Some technologists advocate for "differential technology development," which refers to speeding up technology that is seen to be safe or defensive in relation to technology that is seen to be harmful or offensive. On this, I agree with Belfield (2023) that speeding up technology in the absence of collective action, sensible regulation, patience, and wisdom will not keep humankind safe. My view is that we must regulate dangerous technologies, and that the most dangerous among them, such as technologies based on nuclear fission, should be abolished completely.

In addition to existential risk reduction, other approaches to understanding chaos and managing and mitigating risk are found in the field of complexity theory. In complex systems such as ours here on Earth, the impact of any sudden shock to the system can have unimaginable repercussions. Many different risk management and mitigation strategies have been developed that rely on studying the dynamics of complex systems, with systems theory approaches to global risk using one or more of several methodologies derived from complexity science, game theory, and mathematical optimization.

One may have heard of Black Swans or Grey Swans, for example, and of Dragon Kings. Black Swans are largescale shocks—i.e., unpredictable, massive events that have the potential to cause catastrophic consequences for a system. The 2008 collapse of the financial system often is used as an example of a Black Swan event. People have focused on preventative solutions to predict and mitigate Black Swans, even though according to the theory describing them, they are almost impossible to predict or to mitigate accurately. In contrast, Grey Swans are potentially very significant events, but the possibility of them occurring is thought to be predictable ahead of time and their probability of occurring is considered small.

Nassim Taleb (2012) argues that instead of trying to predict and mitigate Black Swans, we instead should focus on creating antifragile systems. An antifragile system is one for which shocks, randomness, disorder, change, etc., are good rather than bad—theoretically, anyway. The theory is that so-called antifragile systems become better and stronger than they were originally, and less susceptible to collapse, when faced with change and shocks such as Black Swan events. Again, in theory, the upside of shocks outweighs the downside in antifragile systems, so shocks are welcomed and even sought out. In contrast to antifragile systems, fragile systems are those for which shocks, randomness, disorder, change, etc., are detrimental, weakening the system and leading it towards collapse or ruin. In fragile systems, the downside of shocks and their negative effects outweigh the positive, upside effects of shocks, so fragile systems fall apart when Black Swans occur. Furthermore, fragile systems and the people who run them are averse to change, preferring instead to minimize disorder and to maintain the status quo.

But are any systems truly antifragile? That is the real question. To me, it appears that *any* system that is part of explicate order reality—i.e., the conventional reality that many people think constitutes the entirety of existence because they are consciously unaware of subtler implicate, etc., levels of order—inherently is fragile. My view is that no system can achieve *actual* antifragility once it has manifested in explicate order because everything in explicate order arises and then dissolves, like ripples on the ocean. The only question real questions are how long it will persist in an explicate order form, and the degree to which it will have a constructive rather than destructive impact on the ever-evolving whole.

The dynamic of arising and dissolution means that all explicate order manifestations inherently are fragile to a certain degree, with the salient question being to what extent. Therefore, regardless of the existence of a *theory* of antifragility, it is exceedingly unwise to seek out shocks, chaos, and disorder, since these will cause systems to collapse and dissolve more quickly. Only when implicate, etc., levels of order are understood as part of the system can we hope to achieve an antifragile outcome—and then that antifragile outcome will not be solely explicate order. It will be something much more profound, involving potentially an infinite number of subtle levels of order, as is true of everything that exists.

Along with Black and Grey Swans, the Dragon King is another important concept in conventional complexity analysis. The Dragon King concept involves a double metaphor—dragon, to signify something unique, and king, to signify something large in scale. All three of these concepts—Black Swan, Grey Swan, and Dragon King—are present in risk management literature (Glette-Iversen and Aven 2021). Of the three, Dragon King theory (Sornette and Ouillon 2012) is the most developed and useful.

Despite the variety of methodologies that exists within risk management literature, no version of conventional complexity analysis is close to being sufficient to understand, let alone to remediate, the systemic and dynamic challenges facing human society today. What we need is a correct ontology in accordance with which we can stabilize global society *as a whole* rather than one aspect (e.g., climate, the financial system, etc.) or one region (e.g., localities, nation-states, alliances of nation-states, etc.) at a time. In other words, none of the versions of complexity theory currently available *are broad enough* to address the myriads of explicate order issues present in the world today—and no versions of complexity theory available today even acknowledge the reality of implicate, etc., levels of order, which severely limits the usefulness of these theories and methodologies.

Any solution that focuses only on explicate order is partial, since it does not include implicate, etc., levels of order. Accordingly, partial, explicate order approaches only can stabilize a system—or an aspect of aspects of a system—in the short run. But although they can calm things down in the short-term, because the entire, deeply unstable system continues to push against the briefly stabilized elements, those briefly stabilized elements simply become unstable again later. For this reason, it makes no real sense to stabilize the entire global financial system, for example, if society is going to continue to permit the proliferation and 'modernization' of nuclear weapons to run off a cliff. We need *complete* solutions, not partial ones— and to be complete, any solution must consider as many elements of the broadest possible definition of a system at once, and it also must include subtle implicate, etc., levels of order. Perceiving and understanding these subtle levels of order is part of the process of developing an accurate ontology, and an accurate ontology is essential for the survival of our species. In many ways, then, the situation facing humankind is exceedingly simply. If we do not perceive and understand the nature of reality clearly and accurately, we will make missteps that lead to our own extinction. Reality is that direct.

Viewing existence from the perspective of an ontology based on the wholeness of reality—explicate, implicate, super-implicate, etc., levels of order together in holomovement—will lead us to optimize different aspects of existence as compared to what we focus on now. A utilitarian approach to ethics optimizes individualistic metrics such as power and profit, whereas a deontological approach based on wholeness optimizes social justice, fairness, goodness, resiliency, creativity, and love. Optimizing fairness means creating equal access to decision-making, resources, and security. Optimizing resiliency means recognizing that the most resilient system is the one with the most redundancy—i.e., a decentralized global system that emphasizes peer-to-peer interactions rather than those mediated by institutions. Too much centralization always is maladaptive, since centralization reflects fearful, fascist thinking and lends itself to plutocratic, kleptocratic, authoritarian, and totalitarian leadership structures, thereby stifling individual creativity and innovation. Optimizing creativity means moving beyond the explicate order status quo to imagine and realize viable, whole-based alternatives to fragmented consensus reality, which is obsessed with explicate order. Optimizing the good means creating safe and peaceful spaces in which individual creativity can flourish.

The survival of *Homo sapiens sapiens* as a species means very little if we do not have meaning in our lives—and the deepest source of meaning involves making a positive contribution to the whole. Contributing to the whole includes doing something make society safter and to remediate the inexcusable social injustice that currently characterizes human civilization. Social injustice is increasing rapidly, with the wealthy becoming unimaginably wealthy while the poor are becoming increasingly destitute. In fact, over the past two years, the richest 1% in the world have accumulated twice as much wealth as the rest of the world put together (Thériault and Ogola 2023). With dwindling land, resources, care, and overall economic and other forms of independence, vulnerable individuals find themselves driven to extreme actions such as mass migrations, which cause terrible suffering for hundreds of millions of people. Permitting this to continue by not doing anything to stop it is untenable, unethical, and immoral. By cultivating empathy, however, we become empowered to engage actively to reverse trends towards further societal injustice and potential collapse.

But *caring* is not enough. Even when people have good intentions and when they want to do the right thing, the actual reality is that actions taken based on an incorrect ontology almost inevitability end up doing more harm than good. No matter how much we care, our actions are ineffective in alleviating suffering and creating positive realities when we are not working from the correct perception and understanding of the nature of reality. This is because no matter how well-intentioned we are, the impact of our actions occurs in the context of a whole in which other people and members of other species are introjecting their own content into the implicate order. Since all of us literally are floating along in the same cosmic boat, shifting reality increasingly towards the good requires that as many beings as possible transform themselves in accordance with a correct ontology and that they act for the wellbeing of the whole. In this way, we will begin to mitigate the continued spillover of dysfunction and suffering into all aspects of society.

By correcting our ontology, we can maximize our chances for survival, avoid societal collapse, remediate social injustice, and participate in constructive, creative acculturation with ETI—all at the same time. It sounds good, but *how* do we do it? The main point is that *we must turn our ontology inside out* (or right side in, since our ontology is inside out as it is). Instead of thinking of the whole as a concatenation of parts and moving those parts around ruthlessly in the pursuit of so-called tactical and strategic advantage in one utilitarian scenario after another, we must learn to perceive and understand that the nature of reality is wholeness and to embody a deontological perspective befitting of the preciousness of existence.

Many people seem to have forgotten that existence is precious. They cheapen themselves and they view others as disposable. Lost almost entirely is a sense of reverence and gratitude for existence. We and other aspects of existence are not independent 'parts' adrift in a sea of nothingness. What we mistakenly perceive as independently existing 'selves' or independently existing 'things' really are exquisitely intertwined manifestations in a coherent whole in which everything participates in everything else. Remembering how precious existence really is naturally lifts the human spirit from despair to hopefulness—and hopefulness brings a strong sense of determination to do what necessary to reverse the societal collapse and extinction trajectories along which humankind currently is heading. To establish constructive paths in their place of maladaptive trajectories, human beings must learn that mind and matter are inseparable and that the nature of reality is wholeness. We also must orient ourselves around principles such as freedom and social justice rather than around the myopic pursuit of power, control, and wealth. Cultivating empathy, kindness, compassion, and love are central, and so is learning to perceive subtle levels of order beyond the explicate.

How do we turn our ontology the right way around? Here, let me describe the process in a conceptually intuitive manner. As described above, Bohm uses the metaphor of ripples on the surface of an ocean to help people understand the relationship between explicate and implicate, etc., levels of order. The ripples on the ocean metaphorically are analogous to manifestations in explicate order, while the ocean metaphorically is analogous to implicate order. Most human beings focus, often obsessively, on the ripples alone, while completely missing the ocean. In other words, our cognitive error is in mistaking explicate order as the entirety of existence. To turn our ontology the right way around, we must see the ocean as primary, not the ripples, and from there we must learn to perceive and understand the significance of even subtler levels of order as we increase our perception and understanding that the nature of reality is one of unbroken wholeness.

Turning one's ontology the right way around is similar to shifting one's focus from the foreground to the background in drawings in which different things appear when one emphasizes the so-called negative versus the positive space. Here, the explicate foreground "is the display" (Weber 1987), which is what most people see while they miss the background completely. But it is the background—implicate order—that is fundamental. Beyond this, too, exist even subtler levels of order, potentially to infinity. In actuality, foreground and background are inseparable, and together they comprise the holomovement—the continuous movement between explicate (foreground) and implicate, etc. (background) levels of reality. By learning to perceive the background and then moving it forward in our perception, we learn to emphasize the subtler implicate, etc., levels of order in our perception. This is turning our ontology the right way around.

Meanwhile, most people obsess over explicate order, and use tactical and strategic thinking to develop and implement a utilitarian approach to existence. Utilitarianism treats abstracted aspects of the foreground, including individual human beings, as chess pieces to be moved about the geopolitical and economic playing boards as one pursues control and dominance over others. Living in this cognitive box is a tragic waste of one's life. Nevertheless, as society becomes more routinized and people more obsessively focused on technology, people losing touch with the natural world and fail to realize that there is more to life than such maneuvering. Tragically, such individuals have lost their ability to perceive and understand implicate, etc., order subtleties. In other words, many people have lost their nuance as beings. To flourish more fully, they must regain it.

The educational process and other forms of socialization entrain people to focus only on the explicate order. In addition, mainstream media dictates to people what they should think, which leads to a reduction in independent thinking and the ability to perceive subtler aspects of existence. People emerge from this process of socialization not only thinking that explicate order is all that exists, but also thinking that reality is quite coarse because of the vulgarity of so much to which they are exposed. People miss the subtle almost entirely, and when they do have an unexpected experience of it, they often lose their composure because it seems so unfamiliar. Knee-jerk reactionism occurs when people project their psychodynamic shadows onto subtle realities, labeling them 'anomalous,' 'paranormal,' 'spooky,' and even 'demonic' or 'evil' simply because they seem so unusual. People do not realize that existence includes many exceedingly subtle aspects, and that in the whole, everything is enfolded in everything else. Subtle aspects of reality are precious and dear, and they are no more anomalous than you or I.

Missing the subtle aspects of existence—the implicate, super-implicate, etc., levels of order, and how they occur together with explicate order to comprise a whole—has caused humanity to go horribly and viciously wrong. People act with ruthless destructiveness towards one another and towards everything around them. As a species, we have taken this insanity to preposterous lengths as we engulf ourselves in suffering of our own creation. Now, as we inch forward towards an apocalyptic nuclear climax by which we could extinguish ourselves while taking other species with us and degrading Earth in the process, it is abundantly clear that we have lost our way. Now is the time to find it again.

All the viciousness, ruthlessness, aggressivity, and violence we see in human society today results in suffering—cycles and cycles of suffering. None of this is acceptable, nor should we for a moment succumb to the idea that conflict is 'part of our nature' as human beings, and that perpetuating and enduring the ongoing suffering we are experiencing as a species is an 'inevitable' part of the human experience. We can and must do much better than this.

The real cause of suffering is not a clash of civilizations or even a clash of fragmented ideologies. Suffering results from a failure to perceive and understand the nature of reality clearly and correctly. Even though civilizations and ideologies often do clash, and even though such clashes often are violent, what underlies this superficial, ideological cacophony is the deeper ontological error. People tangle themselves up in nets of multiple, fragmented, partial, and colliding worldviews which—because they are incomplete—inherently are incorrect. These fragmentary views of reality proliferate because they are unmoored in the whole. Then, the proliferation of fragmented thinking leads to many other types of proliferation, e.g., the proliferation of weapons, the proliferation of violence, conflict, and war, and the proliferation of anger, resentment, bitterness, and other negative emotions. Bohm puts it succinctly, "We cannot in the end do anything but destroy if we have a fragmentary approach" (Bohm 1986b, 40).

Everything that manifests in explicate order arises from an underlying reality—a ground beneath the surface, namely implicate, etc., levels of order. Becoming aware of these subtle levels of order facilitates conscious participation in a shared ontology of wholeness. This creates coherence at both intraspecies and interspecies levels and, also, at the level of the whole such that participation with others can occur in a fabric of universal coherence. This fabric of the "all-encompassing principle" of wholeness is present everywhere (Weber 1987)—we just need to see it, since perceiving and understanding the seamless interpenetration that is the whole and operating on the basis of this perception is the path to a moral, compassionate, and loving society. It also is the path to reversing trajectories of local and regional collapse occurring here on Earth.

ETI can help humankind transform to a new way of thinking. Even just thinking about ETI can help, because while humankind is mired in tactical and strategic thinking with respect to explicate order 'ripples,' ETI already thinks ontologically and understands the implicate order 'ocean.' In fact, the behavior of ETI's UAP is so sophisticated that ETI almost assuredly perceives and understands super-implicate, super-super-implicate, etc., levels of order, potentially to infinity. In other words, we can learn much from the ETI in our midst now, since this ETI already understand that a continuum of levels of order comprise the whole, how these levels of order work together, and how to operationalize them.

Another reason thinking about ETI can help humankind make the transition towards a correct ontology is because ETI's UAP displays—and other forms of communication from ETI such as genuine crop formations—are geometric in nature. This means that perceiving and thinking about them naturally extends our own understanding of advanced, sacred geometry (Andresen in preparation). Helping human beings extend their understanding of geometry is another reason UAP appear in swarms, which are biological and geometrical at the same time. One of the primary things we are being taught is morphogenesis, and one of the primary ways we are being taught involves geometry. We also are being shown information regarding the series of levels of order discussed by Bohm. Salient geometrical knowledge relating to morphogenesis and to wellbeing is encoded in UAP displays and, similarly, in ancient megalithic sites and monuments and in genuine crop formations. To help humankind transform, ETI is making is presence known to more and more people using such pedagogical displays, which involve novel, creative forms of expression. We should busy ourselves in understanding.

Since geometry is a central element of many UAP displays, and of genuine crop formations, ETI clearly is trying to expand humankind's awareness and understanding of sacred geometry. Sacred geometry describes geometrical knowledge that contains important information about the unfolded reality in which we live. Sacred geometry is a route towards understanding other physics principles more clearly, including in terms of equations based on geometric harmonics. Blanchette (2023), who emphasizes that all beings inhabit a matrix of wave forms, thinks that ETI/UAP are teaching human being a new way to think about quantum space. He observes that UAP are creating very specific shapes in the sky, and that at least one UAP display manifested a triangle based on the golden ratio, ϕ (1.6180339), as shown in a video taken by Linda A-Roraha in the Netherlands. Immediately upon watching the video, I noticed the similarity between the geometrical acrobatics of the lights and the type of geometry that often appears in genuine crop circles—with the circles in crop formations taking on an analogical placement to the lights in the sky in the triangular display. Even the feeling tone is the same between the video and genuine crop circle designs. Furthermore, the tetrahedron with the inscribed sphere described in the video is very similar to the 1991 Barbury Castle tetrahedron crop circle. In other words, ETI is providing human beings with two, similar ways to understand profound, geometrical knowledge.

So much that is truly stunning is occurring around us. Nobody need wait for the government, scientists, religious leaders, or anyone else to tell them what is real. Everyone can participate in creative acculturation with ETI themselves so they can learn to perceive and understand the coherent wholeness of existence directly.

Third Postulate: Transforming Ontology Unlocks ETI/UAP

Three: The path human beings must take to realize an ontological rather than tactical and strategic approach to existence is the same path that will help human beings: realize a society based on social justice; understand ETI/UAP more clearly and participate with

ETI/UAP more constructively; and explore the universe in a way that contributes positively to the whole. An accurate ontology will foster an effective, deontological approach to ethics that moves beyond the current, ineffective, status quo that is based primarily on utilitarian ethics. In contrast to a utilitarian approach, which optimizes criteria on behalf of certain individuals and/or groups, deontological ethics supports constructive rather than destructive outcomes for the whole.

ETI and humankind are *adjacent*. As two species that may share genetic and/or morphological elements, ETI and humankind *naturally* are coinciding now. One of the reasons for this is that human beings must learn what ETI already knows. ETI clearly understands that the nature of existence is wholeness, which humankind must learn if we want our species to survive.

The path we must take to transform our ontology so that *Homo sapiens sapiens* survives as a species and creates a socially just, moral, and loving society is the same path we need to follow to understand ETI/UAP more clearly and to explore the universe in a way that contributes positively to the whole. This is because the ontology that human beings need to act in a manner conducive to the wellbeing of the whole is the same one that ETI already is using.

Methodologically, I propose that there are at least two fruitful ways to study ETI/UAP. Both approaches study ETI and UAP together. The first approaches ETI/UAP from the perspective of the quantum theory, as described in this book. The second approaches the topic from the perspective of sacred geometry, as I describe elsewhere (Andresen in preparation). Combining both approaches would be very helpful, because by infusing the quantum theory with a greater understanding of geometry and extending it on this basis, one naturally moves towards a quantum theory that accounts for gravity and other macroscopic realities.

Throughout this book, I follow Bohm in referring to 'the quantum theory' rather than to 'quantum mechanics.' Bohm's elucidation of the quantum theory is decidedly not 'mechanical' or 'mechanistic' at all. As Bohm puts it, his theory is a quantum *nonmechanics* rather than a quantum mechanics (Bohm 1951, 167). Later in this book, I describe what is meant by quantum nonmechanics. Here, I simply wish to emphasize that in addition to studying ETI/UAP together, and in addition to taking the quantum theory as our starting point, we should not begin with just any interpretation of the quantum theory. Instead, we should use Bohm's quantum theory of wholeness and extend it as far as possible.

The reality of quantum entanglement and nonlocality underscores just how important it is to cultivate kindness, gentleness, compassion, and love, since everything exists together in a dynamic whole. Our beings are entangled with all beings. Even more fundamentally, everything is entangled with everything else. The degree of entanglement between humankind and ETI is so close and the intersubjectivity it engenders so seamless that both are almost unnoticeable. ETI may have been with *Homo sapiens sapiens* from the beginning of our species, and it may have operated on and around Earth and in Earth's waters long before that. In addition to feeling that ETI's presence is real, many people also have a deep, intuitive sense that ETI's presence is a longstanding one.

Rather than focusing too much on the technology behind UAP, we should shift our focus to how to develop peaceful and constructive forms of communication with one another, with ETI, and with other species. Pragmatically, we stand a better chance of advancing as individuals and surviving as a species if we help one another. This means we must strengthen our capacity for connection and cooperation rather than continuing to replay cycles of conflict, control, and domination.

We also should pay attention to what ETI is trying to teach us. ETI's pedagogical exercises demonstrate the wholeness of reality, since UAP displays break down rigid categorization, especially categorization in terms of dichotomies. ETI's pedagogical displays also help human beings perceive and view reality differently—as participation rather than interaction, interdependence rather than billiard ball type interactions, dynamic verbs rather than static nouns. All this teaching from ETI can help transform our species from one focused on explicate reality to one that understands implicate, super-implicate, etc., levels of order. This will help us step back from the precipice so we do not use weapons of our own creation to obliterate ourselves, other species, and large swaths of the natural world. It also will open our minds to the urgent need to redirect money and effort currently being dissipated on the design and manufacturing of weapons and on the violent conflicts and wars in which these weapons are used to solving the myriads of *global* challenges we are face right now.

Chapter 2:
Extraterrestrial Intelligence (ETI)

Size and Habitability of the Cosmos

Given the size and habitability of the observable universe, along with the virtual certainty that life is ubiquitous throughout, it is completely natural that an advanced ETI already is operating in our midst.

The term 'observable universe' refers to a spherical region of the universe that comprises all matter that can be observed. Assuming the universe is isotopic, the distance to the edge of the observable universe is roughly the same in every direction. This means that the observable universe has an approximately spherical volume centered on the observer. Every location in the universe has its own so-called observable universe, which may or may not overlap with the one centered on Earth. In a way, one can say that the universe has no absolute center, since the same cosmic expansion is seen from any vantage point. The realization that nothing is out of place in a universe with no actual 'center' and in which everything constantly is moving can help human beings accept that we are just one of many intelligent species in existence. Furthermore, as exoplanets continue to be discovered (Exoplanet.eu n.d.) and as scientists continue to discover new species here on Earth, the idea that life is ubiquitous in the universe is becoming more familiar to people.

Science today can only approximate the number of potentially habitable planets in the observable universe. Part of the reason why is because our ability to gauge the *actual* size of the universe is constrained by calculations that are based on the *observable* universe. Nevertheless, scientists have enough information to be able to say that the universe is enormous. One current estimate suggests that there may be as many as 20 trillion galaxies in the observable universe as seen from Earth (Siegel 2022). To approximate the number of stars such as our Sun estimated to exist in the observable universe, one can use the Milky Way galaxy as the template. Within our own Milky Way galaxy, there may be somewhere around 400 billion stars. Prior estimates were more around 100 billion, and numbers higher than 400 billion also have been proposed (Masetti 2015). One then can multiply 400 billion, i.e., the estimate for the approximate number of stars in the Milky Way, by 20 trillion, i.e., the number of galaxies in the universe. This yields an astounding, approximate 8 septillion (8 comma, followed by 24 zeros) possible stars such as our Sun in the observable universe. Theoretically, each one of these stars could host a planetary system such as our solar system, though to be conservative in one's calculation, one can take a much lower occurrence rate, such as one-quarter, or 25% (Tavares 2020; see also Bryson et al. 2020). Nevertheless, the number still is enormous, some 2 septillion, for the approximate number of potentially habitable planets in the observable universe that theoretically could support life. Letting these very large numbers sink in helps one understand how completely natural it is that ETI already is operating on and around Earth. In many ways, nothing could be more natural.

The idea that ETI is here already is buttressed by scientists' recognition of the general habitability of our universe and descriptions of the universe as a biophysical one thoroughly imbued with life and intelligence (de Duve 1995; Koerner and LeVay 2000). In considering the habitability of the universe, scientists extrapolate from its physical and chemical consistency to its presumed biological qualities. The fact that the universe displays the same physical and chemical conditions on average suggests that the universe also is a biological one. More recently, the discovery of numerous Earth-sized planets orbiting Sun-like stars also lend credence to the idea that we live in a biologically rich universe. Furthermore, the universe is characterized as having bio-friendly physical laws (Davies 2010).

Since life likely is ubiquitous in the universe as a whole, and since the universe has been shown to be physically consistent, it therefore stands to reason that life is ubiquitous in our own Milky Way galaxy. This realization was articulated by The International Academy of Astronautics (IAA) over two decades ago when it stated: "Subsequent advances in astronomy and the study of evolution have made it seem more probable that life, including intelligent life, may be widespread in the universe" (IAA 2000).

Not only do the numbers alone make it very probable that at least one advanced ETI already has made its way to Earth, but we also should not be the least bit surprised when other interesting things come our way, such as the interstellar object 'Oumuamua detected by the Pan-STARRS1 telescope of the University of Hawaii's Institute for Astronomy (IfA). The International Astronomical Union created the new category—I (for "interstellar")—to describe this category of object (IfA 2018; Meech 2018).

People increasingly are becoming comfortable with the extraterrestrial hypothesis (ETH) as an explanation for the origin of some of the UAP operating in our midst now. Like Giordano Bruno and others, many people who are open to the ETH also believe that humankind lives in a universe that is multiply inhabited by many different advanced ETIs. Such individuals have a much more open-minded view as compared to rare Earth adepts who hypothesize a virtually Ptolemaic, "geo-exceptional" world with a geo-surrounding galaxy that is devoid of intelligent life. Indeed, the ETI in our midst now may be part of a larger network of extraterrestrial civilizations (ETCs), such that one can propose a model of galactic civilizations of advanced ETIs. Here, the term "advanced intelligences" refers to active, techno-communicating (or otherwise-communicating) beings that form a local, advanced civilization. Such local, advanced civilizations could emerge from various solar systems across the galaxy. Some may have begun their evolution only a few million years before the evolution of *Homo sapiens* on Earth, whereas others may have begun their evolution tens or hundreds of millions of years ago and now may be extinct. On average, these advanced ETIs may have an order of magnitude lead time of one million years over *Homo sapiens*, which corresponds to less than 0.1% of the age of the Earth, the Sun, and most other stars and planets in the galaxy. A lead time of one million years is reasonable to posit since it is intermediate between the history of astrophysical objects and the history of *Homo sapiens* and of its predecessors and followers. Although this is a very small lead time on the scale of astrophysics and biological evolution, it is quite large when viewed from the scale of the evolution of human civilization (Gross 2022, 115, 121).

Regardless of how long ago an ETI arrived on Earth, by the time it did, it must have mastered change of location over vast, linear distances. There are at least three possibilities for how. The first possibility is that the ETI in our midst now has mastered general energy challenges that enable it to harness an enormous amount of energy for propulsion. The second possibility is that ETI understands the physics of traversable Einstein-Rosen bridges, colloquially referred to as 'wormholes' (Morris et al. 1988), or something similar, which could facilitate nonlinear movement through spacetime. A third even more esoteric idea that I am introducing here is that the ETI in our midst now understands morphogenesis, i.e., the process of creating and dissolving form. Of course, any combination of these three suggestions also is possible. For example, any ETI that understands morphogenesis almost assuredly also understands how to harness enormous amounts of energy and how to move in a nonlinear way.

Taking the first idea, namely that ETI has mastered general energy challenges permitting it to harness an enormous amount of energy for propulsion, it is helpful to note that the Kardeshev scale measures the technological level of a civilization in terms of how much usable energy it can harness. The scale originally was developed to describe Type I-III civilizations and later was extended later to Type IV-VI civilizations. Because human beings use dead plants and animals for energy, humankind is considered a Type 0 civilization and does not even register on the scale. According to the theoretical scale, a Type I civilization can harness all the energy available from a neighboring star; a Type II civilization can harness the energy of its own entire star; and a Type III civilization possesses knowledge of everything having to do with energy and can travel galactically (Creighton 2014).

Furthermore, even more speculative extension of the Kardeshev scale to Type IV-VI civilizations also has been proposed. Here, it has been suggested that a Type IV civilization can harness the power of its own supercluster of galaxies and eventually its universe of origin, effectively becoming immortal. Some people suggest that a civilization this advanced can tap into dark matter and manipulate the basic fabric of spacetime. A Type V civilization theoretically is advanced enough to escape its universe of origin and explore the multiverse. It also has mastered technology to a point at which it can simulate or build a custom universe. Finally, some people suggest that a Type VI civilization exists in the megaverse and can create and maintain the fundamental laws of universes. Such a civilization is posited to exist in an infinite number of simultaneously existing multiverses representing an infinite number of eventualities and all the laws of physics (Fandom n.d.-a; n.d.-b; n.d.-c). Although I do not subscribe to most of the ideas posited with respect to Type IV-VI civilizations, I think that extending the original Kardeshev scale beyond Type III civilizations does make sense. Doing so in a more grounded way, so to speak, requires future developments in our understanding of physics, however.

But mastering the physics of energy production and cosmic travel is the easy part. Much more difficult for *Homo sapiens sapiens* is mastering enormous energy usage and advanced physics without obliterating itself in the process. Despite clearly possessing very advanced knowledge, the ETI around us now has not obliterated itself and its surroundings, which makes it very impressive indeed. It also means that we can rest assured that this ETI in our midst embodies a kind, compassionate, and loving attitude towards the varied manifestations of the whole, including Earth, humankind, and the other species with which we share the planet.

Tremendous social cohesiveness, adherence to ethical principles, and spiritual maturity are required to master enormous energy usage and advanced physics without self-annihilating. These same qualities are necessary to undertake a coherent plan of action regarding how to utilize this energy, for example by using UAP to explore the universe. Here, it is imperative to realize that social cohesion is different from social control. For human beings, becoming socially cohesive without succumbing to plutocratic, kleptocratic, authoritarian, and totalitarian modes of social control is the real challenge in front of us, not the physics of energy production and cosmic travel alone.

With respect to social organization, ETI realizes that coercive, social control is anathema to a recognition that existence is a whole, and that the whole offers limitless opportunities for creativity. Furthermore, the simple fact that the ETI in our midst now tailors its encounter experiences to specific individuals and groups indicates that it has a decentralized rather than centralized approach to existence. This ETI realizes that coercive control and fascist thinking are antithetical a view of reality that recognizes wholeness for what it is—namely the opportunity for unlimited, creative expressions of meaning.

Origin of Life on Earth and Its Propagation Throughout the Universe

Learning more about ETI will help human beings understand the origin of life on Earth much more clearly than we presently do.

Philosophers and scientists have proposed two main theories—abiogenesis and panspermia—to account for the origin of life on Earth. Utilizing Aristotelian logic, abiogenesis posits that life emerged in situ on Earth and evolved independently of what occurred in the wider universe. In contrast, panspermia holds that life is widespread throughout the universe and that it arrives from outer space on planets such as Earth. On Earth and elsewhere, life then evolves by means of the transfer and interchange of microbiota, i.e., bacteria and viruses, in a vast, cosmic context (Wickramasinghe 2022b, 123). With respect to the two theories, panspermia is backed by the most compelling evidence, including a recent finding that the precursors to larger, biologically relevant molecules have been detected throughout interstellar space (Öberg et al. 2021).

Panspermia theory was very popular among academics, including scientists, during the first two decades of the twentieth century. The theory is consistent with the view that life is ubiquitous in the universe and that panspermia operates on a galactic scale. Theoretically, galactic panspermia would be supported if a survey of exoplanetary atmospheres across the Milky Way found that all, or almost all, habitable planets possessed biosignatures. Even though scientists supported panspermia previously, however, by the closing decades of the twentieth century, the theory of abiogenesis had gained in popularity. Sometimes discussions referred both to both theories.

In recent years, considerable evidence has been collected supporting panspermia as the correct theory explaining the origin of life on Earth. Two types of panspermia are possible—natural and directed. Natural panspermia refers to the idea that microbiota travel through the universe of their own accord, whereas directed panspermia refers to the idea that at least one intelligent species has seeded areas of the universe by directing microbiota to one location and/or another. It makes sense that an ETI with the technology to do so might try to use directed panspermia to settle a new planet, galaxy, or even a larger region of the universe. Such an ETI—which would have at least some morphological and genetic similarity to any species it seeded—might continue to monitor species it had a hand in seeding. It therefore may be incorrect that extraterrestrial morphologies, especially intelligence, are dissimilar to those found on Earth, as many people have proposed. In contrast, biologist Edward ("E") O. Wilson (n.d.) addresses the topic of ETI in a more nuanced manner, recognizing general commonalities between species of intelligent life. It also makes sense that if an ETI played a role in the development of hominins and perhaps other species on Earth, such an ETI could be expected to have established a permanent presence on and around Earth to monitor the goings-on of such species. It also would make sense that such an ETI would engage in pedagogical displays to help such species develop, which is consistent with the pedagogical content from ETI/UAP that human beings experience.

Some people suggest that the fact that all life on Earth shares left-handed chirality points towards directed panspermia over natural panspermia. Chirality, or handedness, refers to when an object or molecule cannot be superimposed on its mirror image by any translations or rotations. While all life on Earth does share left-handed chirality, theoretically this could result either from natural or from directed panspermia—or even from a different element common to life that has not yet been discovered. If biosignatures on a planet were to fall in an area that is not a habitable zone, this would indicate directed panspermia, but scientists have not detected this yet—though discussion of missions to search for biosignatures has occurred (Urton 2020).

During ancient history, the idea of panspermia was discussed quite openly, with such discussion extending back centuries. For example, ancient Egyptian papyri and engravings from the Old Kingdom in Egypt (ca. 2649-2130 BCE) refer to panspermia. The Ṛgveda (1500-1000 BCE) also contains references to panspermia (Temple 2007). Later, Vedic ideas with respect to the eternal nature of life made their way into Buddhist and Jain philosophies (Wickramasinghe 2022b, 121-22, inc. Fig. 1 on 124, 127).

In the West, the progenitor of panspermic theory was the pre-Socratic philosopher Anaxagoras of Clazomenae (500-428 BCE), whose ideas generally were consistent with freedom from theistic control. According to Theophrastus (b. 371 BCE), Anaxagoras held the view that the air contains the seeds of all living things. Similarly, Diogenes Lucretius (~3rd century CE) reported that Anaxagoras thought that the universe was comprised of particles, with the seeds of life carried across the universe and taking root wherever they fell on fertile soil (Theophrastus 1999, 163). Democritus (460-370 BCE) and Epicurus (341-270 BCE) also held rationalist explanations with respect to the origin of the universe and the origin of life. Both classical Greek thinkers eschewed the theological idea of a first cause. Instead, they supposed that all matter comprises invisible particles known as atoms and that all phenomena in the natural world, including life, result from such atoms moving, swerving, and interacting with each other in empty space in an infinite world. Around this time, the concept of an infinite and eternal universe often was discussed. Metrodorus of Chios (331-277 BCE), a student of Epicurus, held an almost modern panspermic view that spoke to the ubiquitous nature of life in the universe. Around 400 BCE, he wrote, "It is unnatural in a large field to have only one shaft of wheat and in the infinite universe only one living world" (Wickramasinghe 2022b, 121-22).

Soon after the theory of panspermia was discussed in the West by Anaxagoras, however, the theory began to encounter resistance. The powerful influence of philosopher Aristotle of Stagiera (385-323 BCE) played a major role in impeding the idea of panspermia from moving forward. Aristotle's early, conceptual proposal of abiogenesis was based on the idea that life arose spontaneously from non-living matter whenever and wherever the right conditions prevailed. Even though Aristotle did not invoke theistic intervention explicitly, his doctrine later was seen to be compatible with the idea of religious or theistic intervention in the origin of life. This helped propel Aristotle's own thinking forward—at the expense of panspermia. After Christianity was accepted in the Roman Empire under Constantine in the third century CE, Aristotelean philosophy was promoted and the ideas of Anaxagoras, Democritus, and Epicurus were rejected. Later, the Aristotelean worldview was expanded upon by Christian theologians and philosophers, especially Thomas Aquinas (1224-1274 CE), who advocated a geocentric model of the world that also included the concept of a physical universe firmly centered on Earth. The geocentric model persisted for several centuries until the Copernican revolution of the sixteenth century undid it, but because of continued theological support for the ideas of Earth-centered life and biology, other variations of geocentric ideas—such as abiogenesis—have persisted into modern times. Even though the historian Benoît de Maillet (1656-1738) wrote that the seeds of everything that live fill the universe—an idea reminiscent of the original ideas of Epicurus and Anaxagoras (Wainwright and Alshammari 2010)—reference to panspermia as a scientific proposition did not occur until the latter part of the nineteenth century (Wickramasinghe 2022b, 123).

The question of whether life originated on Earth or whether it was introduced from the wider universe is a very important one that merits full investigation involving scientific experiments and experimental verification. So far, all experiments that have attempted to demonstrate the spontaneous generation of life on Earth have produced null or at best ambiguous results. Putting it very directly, all attempts to demonstrate the validity of Earth-bound abiogenesis in the laboratories of the world consistently have failed (Deamer 2012). Furthermore, the spontaneous generation of life from non-living chemicals appears untenable given that it would require overcoming a super-astronomical information hurdle. In addition, new facts learned over the past five decades from astronomy, geology, space science, and molecular biology strongly challenge the validity of abiogenesis (Wickramasinghe 2022b, 125-27).

Although we have been discussing the origin of life, it makes sense to step back to ask what 'life' really is. This is a very complex question, as Erwin Schrödinger (1944) articulates. In order to function, a living system depends upon thousands of chemical reactions occurring within a membrane-bound cellular structure. Such reactions are organized in groups into metabolic pathways. Such reactions also can harness chemical energy from the surrounding medium in a series of very small steps, transporting small molecules into the cells, building biopolymers of various sorts, and, ultimately, producing copies of the cellular structure that possess the capacity to evolve. Batteries of enzymes comprising chains of amino acids play a crucial role as catalysts that precisely control the rates of chemical reactions. According to present biological theories, the information contained in the enzymes—which involves how amino acids are arranged into folded chains—is crucial for life. This information is transmitted via the coded ordering of nucleotides in DNA. Life *as we conventionally think of it* therefore depends upon enzymes, without which life as presently understood would not exist (Wickramasinghe 2022b, 124-25).

An RNA world may have existed that predated the DNA-protein world in which we live now. It is thought that RNA serves a dual role as both enzyme and genetic transmitter. If a few ribozymes are regarded as precursors of all life, one can estimate the probability of assembly of a simple ribozyme comprising 300 bases. This probability is 1 in 4^{300}, which is 1 in 10^{180}. *This is an exceedingly low probability*—almost unfathomably low. If one supposes that the universe is 13.7 billion years old, the probability of assembly of a simple ribozyme comprising 300 bases hardly can be supposed to have occurred even once in the history of the entire universe. It therefore is completely unsurprising that in almost half a century of experiments in laboratories around the world, no progress towards demonstrating the process of spontaneous generation of life has occurred (Wickramasinghe 2022b, 125; Wickramasinghe et al. 1996).

What the discussion above means is that abiogenesis as an explanation for the arising of life on Earth is all but mathematically impossible. The failure to witness any trend towards the emergence of a living system normally is attributed to the infinitesimal scale of the laboratory system when compared to the postulated terrestrial setting in which life is hypothesized to have arisen. However, even if one moves from the laboratory flask to all the oceans of Earth, one only gains a factor of ~10^{20} in volume. Even more, if one moves temporally from weeks in the laboratory to half a billion years, the gain still is only a factor of 10^{10}. This means that the probability calculation for a single ribozyme only increases by a factor of 10^{30}, which reduces the improbability factor stated earlier from 1 in 10^{180} to 1 in 10^{150}. On this basis, it becomes all but certain that the emergence of the first evolvable cellular life form was a unique event in the universe. Even if it did in fact occur on Earth for the first time, it must be viewed as essentially miraculous given how low the probability is of this occurring. The statistical probability calculations do not support the idea that the emergence of the first evolvable cellular life form can be repeated anywhere else, let alone in a laboratory simulation of the process. The only logical way to overcome improbabilities on this scale is to move to the largest system available, namely the universe itself (Wickramasinghe 2022b, 125).

Not only is the theory of abiogenesis improbable mathematically, but many tests of panspermia have yielded positive outcomes. In contrast to the failure to prove the theory of abiogenesis, the theory of panspermia firmly is rooted in data. From the 1970s on, scientists began to assemble a vast body of data and evidence from astronomy, geology, and biology that supports panspermia (Wickramasinghe et al. 1996; see also Wickramasinghe et al. 1999). Multiple, dynamical pathways available for interstellar and interplanetary transfers have been identified, and the survival properties of bacteria have been documented (Wickramasinghe et al. 2019). New evidence discussed in books and technical papers (J. Wickramasinghe et al. 2010; Wickramasinghe 2014; Wickramasinghe and Tokoro 2014a; 2014b; Hoyle and Wickramasinghe 2000) continues to verify prior predictions and to provide compelling evidence supporting panspermia "as the mode of origin and propagation of life throughout the universe" (Wickramasinghe 2022b, 125).

Evidence supporting panspermia has accumulated over decades. In addition to data described above, evidence supporting the theory includes:

the prediction and discovery in 1962 that carbon was the main component of cosmic dust; the identification in 1974 that organic polymers comprise the bulk of interstellar dust, which suggests they may result from the breakup of bacteria and viruses; and the prediction of the detailed mid-infrared absorption spectrum of interstellar dust based on prior laboratory experiments. This last item subsequently was verified by observations of the galactic infrared source GC-IRS7, which also established that panspermia was a process that satisfied a crucial Popperian test. Organic molecules are far from the simplest form of microbial life, and the improbability of their assembly into such microbes, which has been shown to be at a super-astronomical scale, points to an origin of life encompassing cosmological dimensions of space and time (Wickramasinghe 2022b, 126; Hoyle and Wickramasinghe 2000).

More recent evidence also strongly supports panspermia theory, including: the eruption of Comet Hale Bopp at large heliocentric distance of 6 AU (Wickramasinghe et al. 1996; see also Wickramasinghe et al. 1999); prediction of bacteria entering the stratosphere verified at a height of 41 km (Harris et al. 2002); Rosetta studies of Comet 67P/Churyumov-Gerasimenko that demonstrate consistency with the presence of bacteria (Wickramasinghe et al. 2015); the earliest evidence of life on Earth coming from the Hadean epoch during a time of comet impacts (Bell et al. 2015); and the fact that microorganisms have been found on the outside of the International Space Station (ISS) 400 km above the Earth (Grebennikova et al. 2018) without any idea regarding how such microorganisms could have been lofted from the surface of the Earth to reach the ISS (Wickramasinghe 2022b, 126).

The October 2014 discovery of a giant comet (C/2014 UN271) at 29 AU (astronomical unit) and the September 2021 discovery of a dramatic brightening episode provide opportunities to verify predictions of a biological comet. The eruptions of the comet at a heliocentric distance of 20 AU plausibly are explained as resulting from high pressure venting of the products of microbial metabolism in radioactively heated subsurface lakes. In contrast, the standard, non-biological model of comets does not account adequately for eruptions at such large distances from the Sun, where surface temperatures are as low as 60 K (Wickramasinghe 2022a).

In addition to explicit verifications of prior predictions regarding panspermia, after 2001, unmistakable "viral footprints" were discovered in human DNA and in the DNA of plants and animals that confirm the prediction from panspermia theory that cosmic viruses drive biological evolution on Earth (Hoyle and Wickramasinghe 1982; Wickramasinghe 2012). Two recent articles by Steele et al. (2018; 2019) summarize additional astronomical and biological data decisively supporting panspermia (Wickramasinghe 2022b, 126).

Another finding supporting panspermia theory is the simple fact that ultraviolet and infrared spectral signatures that could be regarded as having a connection with biology are present everywhere in the universe—from the solar system to the most distant galaxies, even to distances exceeding 8 billion light years. The total amount of such organic material in our galaxy alone amounts to nearly one-third of all the carbon in interstellar space. The possibility that all this organic material is the result of prebiotic chemical evolution is simply illogical, particularly in view of the combinatorial arguments mentioned above. Whenever similar spectroscopic features are found on Earth, they immediately are attributed to degradation products of biology, and it is known that well over 99.99% of all the organics on Earth indisputably are biogenic (Wickramasinghe 2022b, 125).

Yet another strong indication that the theory of panspermia is correct relates to Earth's oceans. The idea that life on Earth is connected to life in the entire universe is supported by comparing the incredibly large numbers one finds in astronomy to the total viral content of the oceans, which is estimated to be in excess of 10^{30}. Although the overwhelming majority of species identified in the oceans are informationally rich bacterial phages, a hitherto unknown component of other viruses also is included in this tally. While this number does not represent genetically distinct phages, it nevertheless is astoundingly super-astronomical, exceeding by more than a factor of a million the approximately 10^{24} total number of stars in the entire observable universe (Wickramasinghe 2022b, 126). In fact, it occurs to me that UAP interest in Earth's oceans may relate in part to the biological richness of these regions. ETI already may recognize this biological abundance, which human beings are only beginning to appreciate.

Scientists could attempt to demonstrate ongoing panspermia decisively using an experiment within the range of current technological capabilities. For example, space agencies could repeat an experiment done in 2001 by scientists in collaboration with the Indian Space Research Organisation (ISR) to collect microbiota at 41 km or higher to search for evidence of biological structures with a characteristic non-terrestrial isotopic signature (Wickramasinghe 2022b, 124-27).

Despite all the evidence in support of the theory of panspermia, and given that a repeat of the 2001 experiment described immediately above has not been undertaken, resistance to panspermia nevertheless persists. One reason may be because some people think that panspermia is at odds with Graeco-Roman and Judeo-Christian philosophical and theological traditions. Reviewing the panspermia/abiogenesis debate in relation to the actual data, one sees that considerable cultural filtering of evidence has occurred over multiple decades. This filtering has skewed public perception away from the virtual certainty that the theory of panspermia is correct. Such cultural filtering also occurs in other scientific disciplines, and it is something to be aware of so that cultural bias does not mislead people when they interpret scientific data (Wickramasinghe 2022b, 125-27).

In point of historical fact, control of science by the state and by other large organizations such as the Catholic Church has occurred for centuries— from classical Greece to the Ptolemaic epicycles in the Middle Ages. The Papacy even has exerted significant control over science in modern times (Merton 1973). One reason that the state and other large institutions have exerted some degree of control over science is because many largescale experiments are so expensive to conduct that they are beyond the reach of individual scientists. For example, space exploration of planets is conducted by the National Aeronautics and Space Administration (NASA) and other space agencies, the Hadron Collider is operated by the European Organization for Nuclear Research (CERN), and major genome sequencing projects in several countries are sponsored by large institutions and private biotech companies. The downside of such largescale and often corporate projects is that they push towards conformity, since some people incorrectly think agreement is required for social cohesion. But this view impedes genuine progress. A glaring example of this is NASA's stance with respect to extraterrestrial life. Even though NASA's declared mission is to search for extraterrestrial life, NASA has bowed to tremendous pressure over decades to support an undeclared premise that life originated in situ on Earth. This premise biases much of the research conducted in this area by NASA and by others (Wickramasinghe 2022b, 122).

How life originated on Earth and elsewhere in the universe relates to the cosmological question of whether the universe is finite or infinite. If the universe is infinite, then the information content of all life is an essential component of the universe, and it would make sense that life would be dispersed as viruses and bacteria available for assembly on every habitable planetary body that forms within the universe as a whole. The idea of an infinite universe is consistent with Vedic cosmology, which describes the universe as infinite in spatial extent and cyclic in time—a view that is quite close to modern models of an oscillating universe. It is important to note, however, that the Big Bang theory of the universe—which estimates the age of the universe to be 13.83 billion years—is by no means definitively proven. This current, consensus view in cosmology faces serious problems following the discovery of galaxy GN-z11 located at a distance of 13.4 billion light years—which would mean that this galaxy formed only 420 million years after the posited Big Bang origin of the universe (Jiang et al. 2020). Individuals such as Fred Hoyle, Geoffrey Burbidge, and Jayant Narlikar have discussed similar problems for the Big Bang cosmological model for three decades (e.g., Hoyle et al. 2000). In contrast to the consensus model, Sir Roger Penrose postulates a "conformal cyclic cosmology" according to which the universe undergoes an infinite number of cycles in which the Big Bang event 13.83 billion years ago (An et al. 2022) is the most recent cycle. In such a model, the origin of life and the origin of the universe are inextricably intertwined (Wickramasinghe 2022b, 127-28, inc. Fig. 2 on 128).

Search for Extraterrestrial Life (SETI)

Inspired by the size and scope of the universe and by the assumption that it is generally habitable, many scientists have undertaken the search for extraterrestrial life by means of various SETI (Search for Extraterrestrial Intelligence) initiatives.

An early version of SETI's mission statement demonstrates SETI's close connection to discussion of the ubiquitous nature of life in the cosmos:

> SETI, the Search for Extraterrestrial Intelligence, is an exploratory science that seeks evidence of life in the universe by looking for some signature of its technology. Our current understanding of life's origin on Earth suggests that given a suitable environment and sufficient time, life will develop on other planets. Whether evolution will give rise to intelligent, technological civilizations is open to speculation. However, such a civilization could be detected across interstellar distances, and may actually offer our best opportunity for discovering extraterrestrial life in the near future (SETI Institute 2004).

Presuming a universal cosmic biology, SETI initially undertook an empirical search for radio signals as the preferred method of detecting extraterrestrial life and later added optical search methods (Koerner and LeVay 2000, 241, citing to COSETI 1996, [see COSETI 2015<1996>]). Today, there are designs for telescope mirrors that could help scientists find biosignatures and technosignatures in the cosmos that are independent of radio signatures.

A major drawback of all SETI programs since the 1960s, however, is the fact that they are based on anthropocentric ideas with respect to how to define intelligence on a cosmic scale. The assumption has been that brain-based neuronal intelligence, augmented by AI, are the only forms of intelligence that can engage in SETI-type interactions. This does not account for how high levels of intelligence and cognition are inherent in ensembles of bacteria, nor the idea that such intelligence may be dominant across the cosmos and transferred by processes of panspermia. It also is possible that bacterial intelligence may be used by the planetary-scale bacterial system—i.e., the bacteriosphere—by means of processes of biological tropism to connect to extraterrestrial microbial forms independently of human interference (Slijepcevic and Wickramasinghe 2021).

Indeed, the so-called Fermi Paradox may contribute to some people assuming that SETI-type interactions only can be engaged by brain-based, neuronal intelligence (Slijepcevic and Wickramasinghe 2021). In 1950, Enrico Fermi is reported to have asked, "Where is everybody?" (Jones 1985). Given the size of the cosmos, Fermi wondered why our skies were not—on his view—filled with extraterrestrial emissaries. Even though Fermi personally did not witness ETI/UAP activity, even though many others have, many scientists have followed Fermi's lead in understanding the issue of UAP the way that Fermi did (Howell 2021). Probably because of Fermi's status in the scientific community, his quandary—which really was nothing more than a casual comment—attained the status of paradox, becoming known as "the Fermi Paradox." As Eric Korpela astutely notes, however, the Fermi Paradox "is neither a paradox nor a concept that originated with Fermi." It is only speculation (Korpela 2019, 738) (see also Andresen 2021, 197-98).

Although NASA contributed to SETI in the late 1960s and early 1970s, and, also, to the formal SETI initiative in 1992, Congress cut NASA's funding for SETI in 1993 (NASA History Office 2003; Stride 2005). Nevertheless, the non-profit SETI Institute has continued to search for intelligent, extraterrestrial life (SETI Institute n.d.). One method previously employed was SETI@home (n.d.), which offered a way for the public to participate in the search for ETI by harnessing computer power from millions of volunteers around the world to create a distributed method to analyze data from the SETI Arecibo radio telescope. Raw data was sent in packets to people's computers, which analyzed it using a special screensaver. Although the site no longer distributes tasks, SETI@home message boards continue to operate to conduct backend data analysis. A bibliography relating to SETI has been published as a NASA Reference Publication (Mallove et al. 1978). In addition, a continuously updated catalogue of SETI publications, which was commissioned by the International Academy of Astronautics (IAA) SETI Permanent Committee as part of the history of SETI working group, also exists (Dumas 2016, 5).

A recent SETI initiative—Solar System artifact SETI—focuses on finding artifacts of extraterrestrial civilizations (ETCs) in the Solar System. This initiative moves beyond the limiting assumption often found in astrobiology that one should search for microbial or, at best, unintelligent life. This assumption has held astrobiology back, especially considering that technological artifacts may be easier to find. To look for signs of extraterrestrial life 'out there,' astronomers can look for atmospheric signatures of biological activity, perhaps near a bright star, by using imaging techniques that factor in and account for the light from the star itself.

Even though we know that life has existed in the Solar System for eons on Earth, SETI searches for artifacts in the Solar System often presume that such artifacts would be of extrasolar origin. But if a technological and perhaps even spacefaring species were to have arisen in the Solar System previously, it could have produced artifacts or other technosignatures that have survived to the present day. It therefore makes sense to discuss the origins and possible locations of technosignatures of such a prior indigenous technological species that may have arisen on ancient Earth or on another body, such as a pre-greenhouse Venus or a wet Mars. In the former instance, such a civilization technically would not be 'extraterrestrial' in origin. In the latter instance, such a civilization would be extraterrestrial but not extrasolar. In the case of Earth, erosion and ultimately plate tectonics may have erased most if not all evidence of a truly ancient civilization if the species lived Gyr (a gigayear, or 1 billion years) ago. In the case of Venus, the arrival of its global greenhouse and potential resurfacing may have erased all evidence of the existence of a prior civilization on the Venusian surface. There also may be limitations relating to places where one may find remaining yet extremely old indigenous technosignatures, such as beneath the surfaces of Mars and the Moon, or in the outer Solar System (Wright 2017, 1).

Many factors go into whether one could find traces of a prior technological civilization on Earth. By way of analogy, current human civilization has impacted Earth sufficiently such that it has created a geological record of its technological activities, for example the Anthropocene (Zalasiewicz et al. 2011, as cited by Wright 2017, 4). But on a timescale of hundreds of Myr (million years) or Gyr, plate tectonics subduct almost all evidence for technology with the crust on which such evidence sits. This means that even if such evidence does exist, given enough time, it is erased entirely from the surface. On tectonic timescales, the parts of the surface that escape subduction also change substantially, so regions that are easily accessible today may have been practically inaccessible, for example under miles of ice, when a prior, technological species may have existed. Few or no signs of the technology of such a species therefore would exist. What this means is that the present-day detectability of technosignatures is a strong function of the age of the technosignature. Although historical records would reveal species less than a few thousand years old, archaeology would reveal technosignatures less than a few tens of thousands of years old. Furthermore, the geological record of the past few hundred million years might show a distinct layer if the technology had a widespread geological effect, as ours does. Past this, i.e., on Gyr timescales, any isotopic or chemical signatures of technology on the Earth's surface may be quite faint, so even if such technology existed at one time, it could be misinterpreted as natural (Wright 2017, 4).

If a prior technological species did exist on Earth, one can ask how long ago such a species could have been present here. Complex life has been common on Earth since the Cambrian explosion around 540 Myr ago. Prior to this, the fossil record only contains much simpler organisms, such as single-celled species and their colonies. One therefore might expect any prior intelligent species to be no older than the Cambrian explosion, though this conclusion is not foregone. Although one tends to associate intelligence with complex life that develops a nervous system using biological mechanisms that evolved during the Cambrian explosion, colonies of single-celled organisms may have been able to organize in complex ways prior to this and, also, to have achieved some manner of complex intelligence. It also is possible that there was a prior explosion of biological complexity in Earth's more distant past, before the fossil record becomes reliable. If this were the case, a form of complex life that left little or no fossil record nevertheless may have existed. A planet-wide cataclysm—either the same or different from what is presumed in this scenario extinguished any hypothesized species of intelligent life—could have destroyed all such prior, complex life. This in turn may have forced the biosphere to start over with the few single-celled species that survived, perhaps on a rock 'lifeboat' that may have ejected during the offending asteroid impact, as suggested by Wells et al. (2003). Accordingly, it would be difficult to find the first generation of complex life, since evidence for it may only exist in the most ancient rocks, if anywhere (Wright 2017, 5).

In addition to searching for prior examples of indigenous, intelligent life on places such as Earth, Mars, and Venus, it also is possible that life was or even still is present in the outer solar system. For example, scientists already claim to have found "[d]irect evidence of complex prebiotic chemistry from a water-rich world" in the outer solar system (Chan et al. 2018; see also Chan 2018).

Indeed, looking for extraterrestrial life in one region does not preclude looking for it in other areas. We can look for signs of life, including intelligent life, on and around Earth and in Earth's waters, in the inner solar system, in the outer solar system, elsewhere in the Milky Way galaxy, and/or elsewhere in the universe. To find ETI even closer to home, one simply can go outside and look up into the sky to see if one is fortunate enough to witness extraterrestrial UAP.

In the future, many explorations relating to ETI may come together such that people will be able to answer questions regarding the origin of life, the origin of the universe, and how and why life develops according to certain morphological patterns. Learning from ETI will help us better understand ourselves, and it also will provide us with insight into the universe we perceive around us and into existence as a whole.

Post-Detection Protocols

In 1989, the International Academy of Astronautics (IAA) adopted its "Declaration of Principles Concerning Activities Following the Detection of Extraterrestrial Intelligence" (Declaration 1989). The Declaration was endorsed by six major international space societies and, later, by many SETI investigators around the world (Billingham 2014, 16).

Item 3 of the Declaration states that those who find "credible evidence of extraterrestrial intelligence" first should inform the other parties to the Declaration, followed by the Central Bureau for Astronomical Telegrams of the International Astronomical Union, and, also, the Secretary General of the United Nations (UN) in accordance with Article XI of the Treaty on Principles Governing the Activities of States in the Exploration and Use of Outer Space, Including the Moon and Other Bodies. The document also instructs discoverers of credible evidence of ETI to inform the IAA, the International Telecommunication Union, Commission 51 of the International Astronomical Union, the Committee on Space Research of the International Council of Scientific Unions, the International Institute of Space Law, the International Astronautical Federation, and Commission J of the International Radio Science Union. Furthermore, the Declaration also calls for confirmation and monitoring regarding any data obtained of a detected ETI, along with the storage of this data to facilitate further analysis and interpretation (Declaration 1989, 1-2).

Although tremendous evidence supports the conclusion that an advanced ETI already is operating in our midst, to the best of my knowledge, none of the copious requirements laid out in the Declaration and described above ever have been followed. Nevertheless, the Declaration document further states that no one should respond to a signal from, or evidence of, ETI in the absence of "appropriate international consultations." It also comments, "a separate agreement, declaration or arrangement" will specify the procedures for such consultations (Declaration 1989, 2). Indeed, the Declaration's call for "appropriate international consultations" prior to any communication with ETI(s) is completely irrelevant given how widespread witness and experiencer reports are. Nevertheless, while erroneously assuming a central role for institutions, Item 9 of the Declaration articulates a process to specify the procedures relating to the detection of ETI:

> The SETI Committee of the International Academy of Astronautics, in coordination with Commission 51 of the International Astronomical Union, will conduct a continuing review of procedures for the detection of extraterrestrial intelligence and the subsequent handling of the data. Should credible evidence of extraterrestrial intelligence be discovered, an international committee of scientists and other experts should be established to serve as a focal point for continuing analysis of all observational evidence collected in the aftermath of the discovery, and also to provide advice on the release of information to the public. This committee should be constituted from representatives of each international institution listed above and such other members as the committee may deem necessary. To facilitate the convocation of such a committee at some unknown time in the future, the SETI Committee of the International Academy of Astronautics should initiate and maintain a current list of willing representatives from each of the international institutions listed above, as well as other individuals with relevant skills, and should make that list continuously available through the Secretariat of the International Academy of Astronautics. The International Academy of Astronautics will act as the Depository for this declaration and will annually provide a current list of parties to all the parties to this declaration (Declaration 1989, 2).

The Declaration stops short of suggesting how human society, en masse, 'should' in fact respond to a signal from an ETC—which is a good thing given that encounters already are occurring on an individualized and small group basis. In other words, any attempt to create a post-detection protocol now is too little too late given the number of credible witnesses who already have encountered ETI/UAP. Given that ETI/UAP clearly has chosen to communicate with individual human beings and small groups of people rather than with institutions—and given that this pattern is evident across hundreds of thousands if not millions of witness reports across the globe and dating back over a thousand years—the IAA's Declaration is nothing more than unnecessary and irrelevant bureaucratic production.

Regardless of the Declaration's superfluousness, people within formal institutions have tended to listen to academic exobiologists and astrobiologists when considering how to respond to evidence of ETI. Although how institutions relay to people that such evidence exists may have some impact on how people receive information, this is less true now than it may have been twenty or thirty years ago given that many people today rely upon alternative media as opposed to mainstream media and government statements to decide what they think about various topics. Nevertheless, over the past decades, institutions such as NASA and the IAA have discussed ideas for creating post-detection protocols.

In 1991-1992, NASA sponsored three workshops in the context of CASETI (Cultural Aspects of SETI) on the short- and long-term implications of contact with an extraterrestrial civilization (ETC). There, participants reflected upon the cultural, social, and political implications of contact (Doyle 1993; Billingham et al. 1994). Approximately twenty-five social and physical scientists considered contact with an ETC from four vantage points: history; behavioral science; national and international policy; and education. The historians suggest extrapolating from intraspecies to interspecies dynamics—which I think betrays exceedingly unimaginative and misleading logic, since intraspecies comparison are not the least bit instructive when thinking about humankind's contact with ETI. The behavioral scientists claim that social and religious groups will manifest diverse reactions to discovery of extraterrestrial life. The education specialists recommend that scientists make announcements concerning the detection of extraterrestrial life in the broadest possible way. Meanwhile, the policy specialists encourage researchers to develop comprehensive institutional policies beyond the IAA's post-detection protocol. Of note, during an annual meeting of the Congress of the International Astronautical Federation (IAF), which has hosted many SETI discussions, the IAA and the International Institute of Space Law (IISL) co-sponsored meetings on post-detection protocols and how to communicate important results from SETI. Nevertheless, most recommendations for formal, institutional, post-detection policies are irrelevant given that many individuals and small groups already report evidence of ETI. Indeed, institutional policies of any kind with respect to humankind's acculturation with ETI are inapposite. Although humankind's participation with ETI will end up impacting institutions and in many cases will lead to institutional disintermediation, humankind's participation with ETI is occurring with individuals and small groups, not at the level of institutions.

In 2000, the IAA made a special presentation to the UN Committee on the Peaceful Uses of Outer Space (COPUOS) regarding the IAA's position paper entitled, "A decision process for examining the possibility of sending communications to extraterrestrial civilizations" (IAA 2000; see also UN 2000, 2). COPUOS, which was established in 1959 by the UN General Assembly in accordance with UN Resolution 1472 (XIV), is housed within the Office for Outer Space Affairs within the United Nations Office at Vienna (UNOOSA 2023). The IAA's position paper provides historical context for more recent discussions of ETI/UAP. The paper chronicles how in the late 1980s, IAA's SETI Committee discussed how humankind should respond to the detection of a radio signal from an ETC. In 1990, *Acta Astronautica* devoted a special issue to this topic, entitled, *SETI Post-Detection Protocol* (Tarter and Michaud 1990). Among the papers in this collection, Finney (1990) considers the societal impact of a message from an ETC. Some twenty years earlier, Fasan (1970) considered legal and diplomatic issues associated with contact with an ETC.

In the foreword to IAA's 2000 position paper mentioned above, John Billingham, Chairman of the SETI Committee at IAA, writes the following:

> This open document is a proposal to begin serious international consultation on the question of future attempts to deliberately transmit electromagnetic signals from Earth to extraterrestrial civilizations [ETCs]. It was prepared over a number of years in the SETI Committee of the International Academy of Astronautics [IAA] by a special subcommittee under the leadership of Michael Michaud. It has been endorsed by the Board of Trustees of the Academy, which decided to make it a formal Academy Position Paper. It has also been

endorsed by the Board of Directors of the International Institute of Space Law. Both organizations consider that the questions raised in the document are of sufficient import to warrant sending it to many nations with a request that they consider bringing it to the attention of the Committee of the Peaceful Uses of Outer Space of the United Nations [COPUOS], for further study, and possible action, on behalf of all humankind. In September of 1996, the document was sent by the Academy to the sixty-three nations which make up this UN Committee (IAA 2000).

COPUOS issued a perfunctory response to the IAA position paper:

[T]he issue of the international process relating to possible communication with any eventually discovered extraterrestrial civilization, as had been discussed in the context of the presentation by the representative of IAA ... while not necessarily requiring immediate action, should be given serious consideration in connection with the future work of the Committee [i.e., COPUOS] and its Legal Subcommittee (UN 2000, 18-19).

To date, however, no such "serious consideration" has been given to the topic by COPUOS or by its Legal Subcommittee, thereby demonstrating the lack of effectiveness of institutions such as the UN with respect to participation with ETI/UAP.

We live in a much different reality than the one described in the Declaration and by the IAA, COPUOS, and most of the SETI community. In contrast with the general equivocation that often occurs in such environments— together with the tendency of individuals in such groups to display an arrogant, dismissive attitude towards historical, government documents and witness testimony—ETI itself has shown considerable wisdom by bypassing institutions and appearing to individuals and small groups instead. ETI's wisdom in taking a very egalitarian approach to acculturation with humankind means that no formal institution will lead the way. To the contrary, the fact that ETI/UAP tailors its interactions to individuals and to small groups demonstrates how accessible and just the phenomenon actually is.

Academic Discussion of Extraterrestrials

Intellectual discussion of extraterrestrial life goes back much farther than many people realize. Some very early discussion of UAP occurs over a thousand years ago in Vedic and other Sanskrit texts that refer to advanced, non-human beings travelling overhead in *vimāna*, a Sanskrit word that one can translate, for example as 'flying chariots' or 'flying palaces.' Many ancient texts of this genre describe non-human beings actively participating with human beings, towards whom they are very well disposed and to whom they teach topics such as mathematics, cosmology, healing, medicine, architecture, music, ritual, meditation, and yoga.

In Western academic discourse some centuries later, one finds Derham's (1714) 'astrotheology' discussion of what contact with an ETI would mean for human religions. Crowe's (1986) discussion of extraterrestrial life in the context of history of science, and Guthke's (1990) consideration of the topic from the point of view of intellectual and literary history. Tough (1991) is optimistic about the specter of human contact with ETI. His view is that given the size and habitability of the universe, such contact essentially is inevitable, and that because most stars in the Milky Way are older than our Sun—many of them millions of years older—human beings probably will find themselves relatively young as a species compared to other ETIs. Tough notes that discovering ETI will cause human beings to consider their place in the universe and to contemplate the future of human civilization, while exobiologist Heidmann (1995) focuses how detection of ETI will improve human understanding of biology. Beck (1971-1972) and Davies (1995) consider philosophical implications of the discovery of ETI, Regis (1985) looks at the impact of discovery of ETI on science, and Schenkel (1988) focuses on how ETI will challenge human society to change.

Much more recently, an international group of scholars contributed to *Extraterrestrial Intelligence: Academic and Societal Implications* (Andresen and Chon Torres 2022), an edited volume that extends discussion of ETI's impact on human society in multiple ways. In accordance with their academic specializations, the contributors to the volume address the topic of ETI/UAP from the perspective of multiple disciplines, including quantum theory, theoretical physics and chemistry, astronomy, astrophysics, space technology, philosophy, history, religion, theology, ethics, social science, and education. On the topic of academic disciplines per se, my own view is that ETI possesses a much more sophisticated understanding of how core disciplines such as physics, mathematics, chemistry, biology, history, etc., are interrelated (Andresen 2022a).

Mainstream discussion in the media today focuses almost exclusively on UAP (e.g., Creitz 2019; Picheta 2020) while failing to address the topic of ETI. Note, prior to recent usage of the term UAP, the two terms one found most often in literature on this topic were UFO (unidentified flying object) and USO (unidentified submerged object). All readers are familiar with the term UFO. The term USO refers to an object that can travel in the water, enter and/or exit the water, and, therefore, also often fly in the air (Sanderson 1970). Throughout this book, I sometimes refer to UFOs instead of UAP, especially if I am discussing a historical account and/or documentation in which the term UFO is used.

Many people also hedge their statements by suggesting that UAP may be nothing more than advanced, human craft. For example, even before its study began, NASA distanced its own study of UAP from discussion of ETI by stating, "There is no evidence UAPs [sic] are extra-terrestrial in origin" (NASA 2022). Evidently those individuals who are participating in NASA's study are unaware of the myriads of government documents that have been declassified on the topic of UAP indicating that many of these craft display technology beyond human reach. It is unclear how far NASA's study will progress given such a quotidian mindset—and, also, given that so little in funding has been allocated to it—i.e., not more than $100,000 out of NASA's 2022 annual budget of approximately $24 billion (NASA 2022).

In fact, elements within NASA have had knowledge of the UFO issue for decades. As far back as 1977, President Carter's Science Advisor recommended that NASA form a small panel to determine if there were any significant findings on UFOs since the Condon Committee's report had been released (Henry 1988, 94, 109). The Condon Committee is a shorthand reference to the University of Colorado UFO Project, which occurred from 1966 to 1968 under the direction of physicist Edward U. Condon. The Project, which was sponsored by the U.S. Air Force (USAF), delivered a highly circumscribed final report inasmuch as Condon's own conclusions, which comprise the first five pages of the report, do not follow logically from the case studies in the report itself (Page 1969, 1071). As Henry (1988) details, some months following the recommendation by Carter's Science Advisor, NASA responded on December 21, 1977 by proposing that "NASA take no steps to establish a research activity in this area or to convene a symposium on the subject" (Henry 1988, 95, 115-16).

In earlier years, NASA, like many other institutions, sponsored academic and think tank discussions regarding the implications for human society of contact with extraterrestrial life. These discussions included academics from many fields, such as astronomy, biology, astrobiology, philosophy, religion, etc. For example, in December 1960, the Brookings Institute published its own study on the implications of ETI for human society (Michael 1960).

NASA also sponsored the Brooking Institute's study, so that NASA would be compliant with a requirement in the National Aeronautics and Space Act requiring it to consider how the U.S. space program would impact American society. The Brookings Institute report states that at both individual and governmental levels, any information disseminated in tandem with religious, cultural, and social factors will shape people's reactions to radio contact with intelligent life. It also speculates that greater unity among humans could occur based on the perception of human uniqueness as compared to other life forms, or because human beings react in a unified way to something they perceive as 'alien.' According to the report, difficulties in deciphering a remote signal could lead people to respond somewhat passively, while a more sustained and explicit mode of contact could have more serious ramifications. The report states:

> Anthropological files contain many examples of societies, sure of their place in the universe, which have disintegrated when they had to associate with previously unfamiliar societies espousing different

ideas and different life ways; others that survived such an experience usually did so by paying the price of changes in values and attitudes and behavior (Michael 1960, 183).

My own position is that it is extremely misleading to take intraspecies examples from anthropology as one's template when considering how humankind and ETIs may participate together.

One objective of the Brookings study was to inform government officials what actions they should take in disclosing news about extraterrestrials to the public. The Brookings report offers many predictable and unimaginative recommendations, such as advocating that researchers consider previous historical and empirical studies regarding how humans react in the face of dramatic, unfamiliar events and social pressures in order to gauge how human beings might react to the discovery of extraterrestrial life. While the Brookings report often is blamed for why the U.S. Government (USG) has not disclosed information about the existence of ETI/UAP to the public, most people who make this claim do not quote the report directly. This may be because when one reads the report, it does not issue a strenuous warning that the public should not be told about UFOs. Indeed, the Brookings report contains few explicit remarks with respect to extraterrestrial life at all (Michael 1960, 42S, 44S, 182-184).

Given how little the Brookings report says about extraterrestrial life, it is remarkable the extent to which the mere existence of this report has influenced the direction of U.S. policy with respect to ETI/UAP. Interestingly, in its own discussion of the report found on its website, Brookings Institution itself seems somewhat, and perhaps legitimately, defensive about being held responsible by numerous media posts and elsewhere for the government's decision to withhold information on ETI/UAP from the public. While the language used on the website is politic (Dews 2014), it is not difficult to read between the lines.

How one human society reacts to another human society is completely irrelevant to the issue of how human civilization will change as a result of creative acculturation with ETI. Why? Because as a species, *Homo sapiens*

sapiens is notoriously violent and aggressive, with multiple examples of coercion between groups, particularly regarding people's beliefs. Forced religious conversion is one of the bloodiest aspects of human history. In contrast, an advanced ETI that already has passed over the threshold of violence and has managed to continue to exist despite knowing how to utilize immense founts of energy, or an ETI that has evolved in a way that did not involve aggression at all, will use pedagogy rather than violence and coercion to make its point. Any being that is truly intelligent will opt for pedagogy rather than violence and coercion, since if beings do not learn things for themselves, they do not sustain positive changes over the long run. Here on Earth, ETI clearly is using signaling and tailored pedagogy to help *Homo sapiens sapiens* become less self-destructive and more compassionate, altruistic, and loving. It is time we listen and get with the cosmic program. Humankind naturally exists in a universe that is replete with other intelligent species. We are returning to our roots when we acknowledge this, not encountering a vast unknown that will derail us as a civilization. If anything, creative acculturation with ETI will put human society back on track, since human civilization is derailed as it is.

Despite its shortcomings, the Brookings report does predict quite accurately that among human beings, scientists will have the most difficult time coping with advanced, extraterrestrial life, in part because an advanced ETI could possess knowledge that would overturn human scientific theories. While it is true that some scientists have been slow to enter the ETI/UAP conversation (Mosher 2017), that is not true across the board. Philippe Ailleris, a Project Controller at the European Space Agency, recently has called for serious scientific study of UAP (David 2020). Years before this, Professors James E. McDonald and Peter Sturrock also approached the topic of UAP from a serious, academic point of view. Based on considerable research and in Sturrock's case also on a personal UFO sighting, both scientists were certain that UFOs were real objects.

Sturrock, emeritus Professor of Applied Physics at Stanford University, has contributed significantly to discussion of UFOs (Sturrock 1974; 1987; 1999; 2001; 2004; Kaufmann and Sturrock 2004). Sturrock was involved in a workshop held at Stanford University on August 29 and 30, 1974 that brought together two groups of scientists interested in extraterrestrial civilization (ETCs). One group was comprised of scientists who study the UAP issue theoretically in terms of physical, astronomical, and biological knowledge and included scientists who search for extraterrestrial radio signals that may provide evidence for ETCs. The second group was comprised of scientists who pursue the UAP issue by analyzing eyewitness reports and photographs. The workshop included discussion of many topics, including: the existence of ETCs; interstellar travel; interstellar communication; human contact with an ETC; UFO evidence; evaluation of the UFO phenomenon; and assessment of human ignorance. The workshop participants generally agreed that the radio search for ETCs was a promising avenue of research and that study of UFOs was a justifiable pursuit (Carlson and Sturrock 1975a; 1975b).

Sturrock also had a leading role in efforts made by the Society for Scientific Exploration (SSE)—an organization he was instrumental in founding in 1982—to inquire into UFOs. With the support of Laurance Rockefeller, for example, Sturrock directed an effort that led to a meeting in New York in 1997. The proceedings of the meeting were published in the SSE journal (Sturrock et al. 1998). Prior to the 1997 conference, Sturrock conducted a study to determine the views of astronomers, astrophysicists, physical scientists, and mathematicians regarding UFOs (Sturrock 1994a; 1994b; 1994c).

Like Sturrock, McDonald also was very involved in studying UFOs. In addition, McDonald gave many public talks on the topic. On July 29, 1968, prior to the release of the Condon Committee report later that year, the House Committee on Science and Astronautics held an event in Washington, D.C., "Symposium On Unidentified Flying Objects." Statements were submitted by Dr. J. Allen Hynek, Prof. James E. McDonald, Dr. Carl Sagan, Dr. Robert L. Hall, Dr. James A. Harder, and Dr. Robert M.L. Baker, Jr. Prepared papers were submitted by Dr. Donald H. Menzel, Dr. R. Leo Sprinkle, Dr. Garry C. Henderson, Dr. Stanton T. Friedman, Dr. Roger N. Shepard, and Dr. Frank B. Salisbury (Committee 1968; for McDonald's presentation, see McDonald 1968). In 1969, McDonald made a detailed presentation at the General Symposium of the

American Association for the Advancement of Science entitled, "Science in Default: Twenty-Two Years of Inadequate UFO Investigations" (McDonald 1969).

During the 1968 event, "Symposium On Unidentified Flying Objects," Baker made this statement concerning the military tracking of space objects:

> There is only one surveillance system, known to me, that exhibits sufficient and continuous coverage to have even a slight opportunity of betraying the presence of anomalistic phenomena operating above the Earth's atmosphere. The system is partially classified and, hence, I cannot go into great detail in an unclassified meeting. I can, however, state that yesterday (July 28, 1968) I travelled to Colorado Springs (location of the [North American] Air Defense Command [NORAD] and confirmed that since this particular sensor system has been in operation, there have been a number of anomalistic alarms. Alarms that, as of this date, have not been explained on the basis of natural phenomena interference, equipment malfunction or inadequacy, or manmade space objects (Committee 1968, 131).

Given that Baker had classified access to one of—if not the most— sophisticated sensor systems operated by NORAD at the time, the USG clearly has collected data on unusual objects in space for decades. Of note, in 1981, NORAD changed its name from the North American 'Air' Defense Command to the North American 'Aerospace' Defense Command. Based on document declassification over the years, we also know now that many official documents generated by NORAD and by other elements within the U.S. Government (USG) mention UAP.

Also in the 1960s, the Space Science Board of the National Academy of Sciences (NAS) (1962) addressed biology in space as part of considerations involving the first Mars missions. Like the Brookings study, which also occurred in the 1960s, the NAS study was conducted at the request of NASA. It recognized that searching for extraterrestrial life could catalyze people to engage questions regarding their place in the natural world and the meaning and nature of life.

In September 1971, a joint U.S.-USSR meeting was held at the Byurakan Astrophysical Observatory in Soviet Armenia at which international participants from many fields including astronomy, physics, radio technology/radio physics, computer science and technology, chemistry, biology, linguistics, archaeology, anthropology, sociology, history, and cryptoanalysis discussed the scenario of gradual assimilation between humankind and ETI (Sagan 1971; 1973; Ambarzumian 1972). This was the first international conference on extraterrestrial civilizations (ETCs) and issues associated with contact. Participants included many internationally recognized scientists, including Carl Sagan, Freeman Dyson, Philip Morrison, Charles Townes, Frank Drake, John R. Platt, Marvin Minsky, and Bernard Oliver from the U.S.; Nikolai Kardashev, Iosif Shklovsky, Vitaly Ginzburg, Victor Ambartsumian (also spelled Ambarzumian), and Vsevolod Troitsky from Russia; and scientists from many other countries, such as Francis Crick from the U.K. The conference was organized jointly by the NAS, with assistance from the U.S. National Science Foundation (NSF) and the U.S.S.R. Academy of Sciences. Discussion occurred on the following topics: the plurality of planetary systems in the universe; the origin of life on Earth; the possibility of life on other cosmic bodies; the origin and development of technological civilizations; problems in searching for intelligent signals or for evidence of astroengineering activities; and the problems and possible consequences of establishing contact with ETCs. In the context of the event, Philip Morrison suggests that researchers would need to study a complex radio signal for a lengthy amount of time before they could hope to understand it. He states, "The data rate will for a long time exceed our ability to interpret it" (Morrison and McNeill 1973, 336-337). Inherently, this need not be the case.

NASA also co-sponsored a 1972 symposium, 'Life beyond Earth and the Mind of Man,' held at Boston University. Astronomy professor Richard Berendzen (1973) moderated the event, which included academics from diverse fields. These included Harvard biologist George Wald, then-dean of Harvard Divinity School Krister Stendahl, astronomers Philip Morrison and Carl Sagan, and anthropologist Ashley Montague. Then-Administrator of NASA, James Fletcher, also participated in the symposium. The theme of humankind's galactic heritage arises in the symposium discussion, with Berendzen articulating the optimistic view that discovery of ETI could bring humankind into its galactic heritage and could lead to improved social forms and institutions. He also mentions the possibility that discovering ETI could help humankind solve environmental crises and could improve technology, science, the arts, literature, and the humanities while also alleviating humankind's social and cultural isolation (Berendzen 1973, 49-50). Anthropologist Ashley Montagu emphasizes the importance of preparing for contact (Berendzen 1973, 25), which resonates with Tough's (1991) statement, "Seeking contact and preparing for successful interaction should be one of the top priorities on our civilization's current agenda." In fact, a positive approach to encountering ETI is both crucial and achievable. A recent Arizona State University study states that people are "upbeat" about the possibility of discovering life "outside of Earth" (Minton 2018; see also Davis 2018).

Consensus exists among many SETI scientists that contact with ETI could result in humankind learning a tremendous amount about a wide range of topics. Bernard Oliver points out that since galactic civilizations likely first formed four or five billion years ago, interstellar communication among ETCs may have been occurring for a long time. Extraterrestrials who participate in such interstellar communication will have accumulated an enormous body of knowledge handed down from the beginning of the communicative phase. The theme of humankind's galactic heritage arises in this discussion, too, with academics recognizing that by communicating with an ETC, humankind might find "the totality of the natural and social histories of countless planets and the species that evolved: a sort of cosmic archaeological record of our Galaxy" (Oliver and Billingham 1971, 31).

Similarly, SETI scientist Frank Drake (1976) also observes that participation with an ETI promises to teach humankind much about science, technology, art, and culture. Sagan and Drake (1975) claim that such a discovery would enrich humankind, while others note that a signal from an ETI has the potential to benefit humanity philosophically and practically (Morrison et al. 1977, 7- 8). SETI astronomers Thomas Kuiper and Mark Morris (Kuiper and Morris 1977) claim that the relative development of the two civilizations involved will influence the implications of a discovery of ETI. They argue that if contact occurs with a so-called superior civilization across essentially all of humanity and if this ETI were to make its store of knowledge available to humankind before humankind itself passes a certain threshold, culture shock would abort humankind's own development since human beings would absorb the knowledge instead of the knowledge enriching the galactic store. I disagree with this last point, especially given Bohm's statements regarding how information moves between explicate and implicate, etc., levels of order. In the holomovement, there is no substantive difference between gradual versus sudden Contact with an ETI—the only difference is in humankind's temporal perception of what is occurring. In other words, there are no epistemological or cognitive downsides of scenarios involving a sudden absorption of extraterrestrial knowledge when the absorbed technological knowledge is balanced with sufficient ethical and spiritual development such that technologies involving energy utilization—especially high energy utilizations—are not weaponized.

In 1993, a Bioastronomy Symposium was held in Santa Cruz, California. There, Almar (1995) posits that historical analogues of terrestrial contact are useful when considering how participation between humankind and ETI may unfold, which I again find to be an incorrect assumption. As I mention above, my own view is that analogues involving intraspecies dynamics between human groups are misleading templates for how creative acculturation with ETI actually is unfolding here on Earth. Previous research into human psychology and biobehavioral models do not help us predict and/or assess creative acculturation with ETI. Encounter and witness reports indicate that the ETI around us now is an immensely sophisticated species, which means that we must think about acculturation with this ETI in a more creative way rather than restricting ourselves to extrapolations from past and present instances of intraspecies dynamics.

Particularly in the case of an extraterrestrial species that is more intelligent and more emotionally and spiritually mature than human beings—which I think clearly is the case regarding the ETI in our midst now—creative acculturation between humankind and such an ETI will be enormously significant for human civilization. For one, it will catalyze human beings to engage in a self-reflexive reassessment of their place and role in the cosmos. Boylan (1996) recognizes this aspect of participation between humankind and ETI:

> The advanced development of the extraterrestrials can inspire us to reach beyond our own current horizons. The ETs can, and will, challenge us to develop ourselves more mentally and spiritually, and to stop worshipping material progress for itself. The ETs will challenge us to confront, honestly, effectively and unflinchingly, the ecological crisis our world is in, and to turn the damage around now, for it is very soon becoming too late. The ETs, with their selfless outreach across millions of miles of space to raise our consciousness, can set an example for us to extend our own renewed caring to the less fortunate members of our own human race, and to see our basic solidarity as one human people. And the ETs can make us aware that the precious gift of life is not confined to one planet, or one millennium, or one galaxy, and thus help us have an improved understanding of what the source of life is, and on what our deepest human destiny is grounded.

I resonate strongly with Boylan's positive tone and insightful comments. I also appreciate deeply his focus on empathy toward others. We can and must develop our capacity for empathy, which is the antidote to self-interested 'libertarian' philosophies that are cruel in their impact on less fortunate members of society and on those who are poor, vulnerable, and suffering. We need to help others, not try to 'control' them using insidious surveillance technologies.

In 2010, the topic of the UN's role in discussion of the ETI/UAP topic was discussed during a meeting of the Royal Society, "Towards a scientific and societal agenda on extra-terrestrial life." This meeting occurred during the Society's October 4-5 session, "Representing humanity? The possible role of the UN in the context of contact with extra-terrestrial life." Academics who participated included Prof. Frans G. von der Dunk, Dr. Mazlan Othman, Dr. Margaret Race, and Dr. Douglas Vakoch. At the time, Othman was Director of the UN Office for Outer Space Affairs (UNOOSA). The UN, however, did not want to engage the topic of contact with extraterrestrial life without consensus among its members. In any event, my view is that the UN is not the correct venue for the discussion of ETI/UAP. The phenomenon will continue to unfold in an egalitarian manner involving individuals and small groups, without people needing to defer to institutions such as the UN regarding how to participate with ETI/UAP.

While this chapter does not list every academic meeting or academic statement regarding ETI/UAP, it does provide a broad overview of the types of conversations that have occurred on this topic thus far while also demonstrating how far back such academic discussions extend. Although from the 1970s on, most academics have tended to downplay if not dismiss the extraterrestrial hypothesis (ETH) as a possible explanation for UAP, academia and mainstream science actively engaged the ETH in the 1960s in a National Academy of Science (NAS) review, in a symposium at the American Association for the Advancement of Science (AAS), and in publications in journals such as *Science*. Stigma surrounding UAP (then 'UFOs') gained momentum in the 1970s, however, causing many academics to back away from the topic even though many members of the general public continued to support the ETH as an imminently plausible explanation for many UAP. Now, however, many academics have begun to revisit the ETH as an explanation for certain UAP operating in our midst now. This is a positive development and it is likely to continue, especially since the presence of ETI/UAP is becoming more and more pervasive on and around Earth and in Earth's waters.

Chapter 3:
Unidentified Anomalous Phenomena (UAP)

ETI Visits D.C.

As if on cue, on Tuesday, November 26, 2019—two days before the Thanksgiving holiday in the United States—an unidentified object was sighted gliding down Pennsylvania Avenue in Washington, D.C. (Burris 2019). I say "as if on cue" because from my vantage point, I was researching the ETI/UAP topic just a few miles away precisely at the time this event occurred.

In response to the detection of the object, White House and Capitol buildings were put on lockdown and fighter jets were scrambled (Madrak 2019). Despite the rather robust response, mainstream media outlet CNN put out a flighty cover story that a flock of birds had caused the hullabaloo (Cohen et al. 2019). Suffice it to say, the United States of America does not scramble fighter jets for a flock of birds.

While rumors swirled that there was a government blackout of most journalistic reporting regarding the event, reporter Tom Rogan (2019a) systematically deconstructs CNN's absurd story, showing that it is completely implausible. Rogan makes many important points. First, the object was tracked entering the air defense area for the National Capital Region (NCR) at about 8:30 a.m. The radar contact indicated that the object at least entered the *thirty*-mile outer ring of the capital air defense area. This is one of the rings of the nation's most tightly monitored regions, namely the area immediately around the White House, the Capitol, where Congress has its sessions and offices, and the Pentagon. The contact was taken very seriously, and the Secret Service locked down the White House. The Capitol also was locked down. The fact that a lockdown even occurred further suggests that the contact either entered the capital air defense area's *fifteen*-mile inner ring, or at least was very close to it. The contact concerned commanders sufficiently that they launched multiple interceptor aircraft, and both fighter jets and a helicopter were scrambled to investigate. Of note, since first being flown in July 1972, many F-15 interceptors have been scrambled to intercept UAP, though none of the F-15s have succeeded in doing so given the sophistication and maneuverability of genuine, extraterrestrial UAP.

Rogan (2019a) carefully analyzes the capabilities of the Sentinel radar system that the U.S. Army uses to protect the White House, Capitol area, and Pentagon. The Sentinel radar system utilizes an extremely advanced X-band radar system that can discriminate individual birds, balloons, and weather from planes or drones. In fact, the Sentinel radar system is so advanced that it can differentiate a single bird from among many in a flock—let alone a flock as compared to some other type of object and/or phenomenon. Furthermore, the radar operators who utilize Sentinel and other platforms are very good at discriminating birds, since flocks of geese are a regular occurrence in the airspace over D.C. These radar operators also can identify balloons and other conventional objects, and they can separate radar contact merges. A 'merge' describes when more than one radar contact appears as one. Rogan states that although birds are common all day, every day, in the D.C. air defense area, the alert on the day the event occurred, November 26, 2019, "was very rare." Rogan could not "find a similar incident since someone decided to fly a gyrocopter up to the Capitol complex in 2015." NORAD ignored Rogan's questions regarding "the contact's origin, speed, track path, dissipation/loss-of-track point, and whether it hovered," referencing "operational security" as the reason it did not reply. Rogan concludes his article by pivoting to putative extraterrestrial UAP, stating that intelligent objects that do not belong to the U.S., China, or Russia are being tracked globally.

The November 26, 2019 event in D.C. is merely one among an estimated hundred thousand if not more UAP incidents that occur on and around Earth each year. From among these cases, I find this event in D.C. particularly interesting given its 'meta' overtones. For example, only a few days before the unidentified object was detected on November 26, on November 21, the *Washingtonian* announced that a special exhibit on UFOs would occur at the National Archives in Washington, D.C. In fact, the National Archives exhibit ran from December 6, 2019 through January 8, 2020 (Beaujon 2019).

The National Archives exhibit on UFOs displayed only two documents—a comic book rendering and a chart—both of which already had been shown online by the *Washingtonian*. Nevertheless, the exhibit was placed prominently on the main Rotunda level of the National Archives, near the Constitution, Bill of Rights, and Declaration of Independence. Other exhibits at the National Archives tend to be placed in a less prominent area, away from the Republic's foundational documents. Further highlighting the UFO exhibit, the two small documents were placed in an *enormous* showcase with a large wall hanging behind the showcase announcing the exhibit. This drew the public's attention to the UFO exhibit while not showing much in the way of actual documentation. It was as if the exhibit planners were broaching the topic of UFOs without providing too much information. However, a small placard stated that other documents on UFOs were available in the archives.

The decision of the exhibit planners at the National Archives to use a document from a comic book was interesting. It seemed to say two things at once—this topic is important, but perhaps it is nothing more than a joke. Nevertheless, the comic book describes facts that are quite substantive with respect to the famous July 19, 1952 sighting of UFOs over Washington, D.C. As the comic book states, according to "Harry G. Barnes, Senior Air Route Traffic Controller for the Civil Aeronautics Administration," who "was in charge of the National Airport, Washington, D.C., A.R.T. Control Center" on the night in question, and as validated by multiple other officials and also "the airport control tower radar operator" for the National Airport, Washington, D.C., "shortly after midnight on that date, seven pips [i.e., blips] appeared suddenly on the control center's scope." The document states that this occurred over restricted areas of Washington, D.C., including the White House and the Capitol (National Archives 2019, caps and italics omitted). The document also states that the event lasted for "14 minutes," during which at least one of the lights "zoomed out of our beam between sweeps! It accelerated from 130 miles per hour to almost 500 in less than 4 seconds…" These are very detailed and substantive statements.

Of note, on July 21, 2002, around the fifty-year anniversary of this event, the *Washington Post* published an article discussing the July 19, 1952 event together with other, similar UFOs events in the U.S. (Carlson 2002).

The second document in the 2019-2020 UFO exhibit at the National Archives was a chart depicting "the correlation between national media coverage of several highly publicized sightings and daily UFO sightings reported between June and September 1952." Again, the exhibit planners seemed to be hedging their bets by suggesting, albeit indirectly, that more media coverage of UAP causes people to think that they see UAP more often. Of course, the opposite interpretation also is possible, namely that the more people who see UAP, the more UAP reports are made, with more media coverage occurring as a result.

In fact, many UAP sightings occurred in 1952 (Eberhart 2022, 178-222). Many of these events occurred in Washington, D.C. and were witnessed by multiple people. This series of events often is referred to as "the Washington UFO flap of 1952" (UFOs 2019), or "the Washington, D.C. flap" (UAF 2015). In July 1952 in particular, hundreds of witnesses saw multiple UAP over multiple days. Radar scopes in the D.C. area tracked numerous UAP registering mysterious blips on July 19 and 20. Radar blips reappeared on July 27. The USAF scrambled interceptor aircraft to investigate but found nothing. Headlines screeched across the entire U.S., former President Harry S. Truman apparently was aware, and according to a CIA memo dated August 14, 1952 quoted below in chapter 4, the White House wanted to know what was occurring. Perhaps not knowing how to respond, the USAF and a later Civil Aeronautics Authority investigation passed the blips off to "temperature inversions" (G. Haines 1997 <1995>, 68, citing in fn. 13 on 80 to Klass 1983, 15 and to Good 1988 <1987>, 269-271; see also Good 1988 <1987>, 331, for discussion of the White House response). But this explanation is implausible given firsthand reports from hundreds of eyewitnesses who clearly indicate that the sightings were not caused by temperature inversions.

One of the things that interests me most about the November 2019 event is how it resonates at a meta level—spatially, temporally, and in many other ways—with the events of 1952. The 2019 incident occurred just a few weeks before the exhibit opened at the National Archives describing UFO events in 1952. Temporally, the comic book from 1952 describes essentially what happened again in 2019—a UAP flyover of Washington, D.C. Spatially, given the location of the National Archives on Pennsylvania Avenue NW, if the UAP in 2019 hovered over both the White House and the Capitol buildings, *it must have made its way past the National Archives building* (unless, of course, it blipped in and out)—which is precisely where the exhibit planners were setting up an exhibit describing more or less just that, but in 1952. Taken together, the events underscore how everything is interrelated in space and in time. Later in this book, and, also, elsewhere (Andresen in preparation), I discuss this theme in terms of Bohm's interpretation of the quantum theory. During a recent interview, I also mention how this same theme is discussed in ancient Sanskrit literature (Kalantarova 2021).

ETI's visits to D.C. also underscores Bohm's observation that similar forms resonate together at the level of implicate order (Bohm 1986a, 93), which by extrapolation means that similar forms also resonate together in explicate order. UAP events in D.C. are so entangled spatially and temporally that the very fact of them seems to be a message intended to help human beings understand how reality works. It is as if ETI is looking back on itself and encouraging human beings to do the same, while pointing out to human beings that existence itself is a complex, entangled stage for creative learning. Indeed, many of the pedagogical displays that ETI presents to humankind incorporate space/location, time/chronology, and the personal aspect of the ETI-humankind relationship at any specific point in space and time together in a mesh-like framework.

So, is it—*or is it not*—a stretch to suggest that in hovering over the White House and Capitol on November 26, 2019, ETI/UAP wanted a sneak peek at the National Archives exhibit describing its own very appearances in the neighborhood in 1952? It sounds fantastical, but at a deeper level, it is not. Maybe a 'sneak peek' per se was not the motivation, but ETI's 2019 visit could have been a nod of acknowledgement to the exhibit planners at the National Archives, with ETI indicating that it had 'noticed' the planning for the exhibit. This *pedagogy of presence* demonstrates how the facts of the matter present entanglement and nonlocality in a meta and self-reflexive way.

What is certain is that the ETI in our midst now has mastered the conjoined arts of synchronicity and the meta. ETI is communicating the unfolding, enfolding, and unfolding again of events, and how aspects of reality resonate with one another. Too, ETI's mode of communication involves an exceptional degree of self-referentiality, which further demonstrates the reflective nature of information and whole reality. Another example of such a demonstration is the famous 1561 UAP event in Nuremberg, then a Free Imperial City of the Holy Roman Empire and now part of Germany. There, witnesses on the ground saw what 'appeared' to be a battle of UFOs in the sky, which served as a dramatic reflection of the actual battle between Protestantism and Catholicism on the ground. One can wonder if ETI was reflecting humans' own ideological conflict back to them using UAP 'props' in the skies. In other words, interpreting the behavior of ETI/UAP requires considering the mutuality between the phenomenon and the situation in which it appears.

UAP activity often seems to spiral in a spatiotemporal way. In fact, synchronicity between the 1952 and 2019 events in Washington, D.C. reminds me of the discussion between Carl G. Jung and Wolfgang Pauli with respect to the nature of so-called coincidences—especially since there is yet another wrinkle to this story, no doubt one among many. As if in synchronous if not synchronicity-ous orbit—by which I mean the phenomenon itself and the human beings discussing it in the media—only ten days after the White House lockdown in reaction to detection of the unidentified object on November 26, 2019, the television channel History aired *UFOs: The White House Files* on December 6, 2019 (UFOs 2019).

Some people claim that Washington, D.C. has more documented UFO sightings and encounter reports per square mile than any other location on Earth (Stanley 2006; 2011). Regardless of whether the claim is true or not, Washington, D.C. certainly does have many reported sightings and radar detections of UAP. It is impossible to miss the fact that what occurred in the early 1950s in D.C. is similar to what happened in 2019, for example.

But the spatiotemporal spiral is wound a bit more tightly than that, since not only did multiple UFO displays occur in Washington, D.C. in July 1952, but *such displays also occurred in July 2002, on the 50ᵗʰ anniversary of the July 1952 occurrences*—ironically around the time the *Washington Post* published its article on July 21, 2002, mentioned above, which commemorated the fifty-year anniversary of the July 19, 1952 event (Carlson 2002). Here, ETI's actions mirror media coverage, just as ETI's actions mirrored the 2019-2020 UFO exhibit at the National Archives. Even more uncanny yet at the same time deeply reassuring is that ETI apparently is observing humans' tradition of honoring 50-year anniversaries, in July no less, which has obvious symbolic significance given the Fourth of July 'birthday' of the United States of America. What all of this discussion indicates is that ETI perceives and thinks in a very meta and self-reflexive manner, twisting and turning around and about itself and others with whom it is participating in cycles of deeply meaningful patterns—one presumes until the content of the ontological lesson has been assimilated, and all parties can move on to ontological lessons of even more creativity and depth.

A two-part *Nexus Magazine* article that I summarize below discusses many UAP sightings in Washington, D.C. As mentioned above, in July 1952, hundreds of eyewitnesses saw dozens of glowing, spherical UAP above the city, and newspapers around the world reported the sightings. Since the UAP had registered on radar over Washington, D.C., F-94 jets were scrambled to intercept them. The UAP easily outmaneuvered the jets and, also, hovered in the area for many hours throughout the night. One unidentified anomalous object is reported to have appeared and disappeared at will on at least two occasions. People were agog. In contrast, in July 2002, when glowing UAP returned—acknowledging the caveat that they may have returned before this—there was less media coverage, though a few media sources did report a brief but failed intercept of the UAP by F-16s (Stanley 2005, Part 1, 55-6).

Even though there was less media coverage of UAP activity in D.C. in 2002 as compared to 1952, a professional photographer has stated that he took clear, high-resolution, nighttime photographs of at least some of the UAP in Washington, D.C. in July 2002. He claims that on two separate nights, he captured color images on high-speed film. Humorously, his first sighting occurred on July 4, 2002—i.e., the Fourth of July holiday in the U.S. The first part of the *Nexus* article in which he is interviewed includes an excellent photo of an unidentified anomalous object over the Washington Monument on this date while the traditional display of July 4th fireworks is occurring. The photographer's second sighting occurred on July 16, 2002, when multiple UAP flew over airspace that is designated as restricted over the Capitol Building and the Washington Monument. On this date, the witness/photographer reports that UAP encircled and landed on the Capitol Building roof and the surrounding park area late that night. He also states that the large formation of UAP seemed to 'park' above the Capitol Building. The UAP emitted a strange energy, he says, which he claims to have captured on film. He states that a highly complex energy field signature resulted from the long exposure time of his photographs and the proximity of the UAP to the camera. Immediately prior to the UAP leaving the area, the photographer reports that they generated what he describes as a "wormhole" in space (Stanley 2005, Part 1, 55-6; see also 56 for the July 16, 2002 photograph of a blue spherical UFO that briefly lands on the Capitol Building roof, and, also, for a July 4, 2002 photograph of UFOs over the Washington Monument and Capitol Building; see 57 for a July 4, 2002 photograph of three UAP—or three lights that are part of a single unidentified object—over the Washington Monument).

The interviewer for the *Nexus* article and the photographer/witness discuss why so much media attention was given to UAP events in 1952, while relatively little media attention was given to similar events in 2002. They broach the idea that in 2002, U.S. media were focused on homeland security and the so-called war on terror. Meanwhile, multiple UAP incidents were occurring that year in the U.S. and elsewhere around the world. The interviewer and the photographer/witness also state that people in the media may be uncomfortable discussing anomalous events without being able to explain them, and that one's confidence level influences whether a person is willing to discuss UAP given the legacy of the USG's policy of denial, dissuasion, and deception regarding ETI/UAP. The interviewer also mentions that humankind would benefit from more education on the topic of UAP (Stanley 2005, Part 2, 49-51).

With respect to the event on July 4, 2002 over the Washington Monument, the photographer reports the presence of a blue object in the sky along with another unidentified anomalous object next to the Monument. The blue object, or one like it, was present again on July 26, 2002, in Waldorf, Maryland, approximately thirteen miles south of Andrews Air Force Base. Pursued by two F-16s, it left the planes "standing in the dust." Although the *Washington Post* carried a story, no photographs were shown of the object (Stanley 2005, Part 2, 51, 77).

Ten days after the events of July 16, 2002, UAP over D.C. finally made the national news on July 26. Shepard Smith, a Fox News correspondent at the time, now with CNBC, reported, "The nighttime skies over the nation's Capitol came alive with blue and orange lights streaking across the sky, so say a lot of panicked people who called to a radio station; no joke here." He went on to report that U.S. fighter jets pursued the UAP, and that NORAD confirmed to Fox News that two F-16s were scrambled but found nothing. The event was considered significant in part because the UAP flew over Andrews Air Force Base, which is used by U.S. presidents. Smith's national Fox News reporting was supplemented by a report from Brian Wilson, a correspondent who at the time was with Fox DC. Wilson stated:

> It's fair to say, Shepard, that there are a lot more questions than answers at this point, but something strange was going on in the Maryland night sky. At 1:00 a.m., the folks at NORAD saw something they couldn't identify in Maryland airspace, not far from the nation's Capitol. The track it was taking caused them some concern, so they scrambled two DC Air National Guard jets to check things out. Now, DC Air National Guard confirms that two F-16s from the 113th Wing were vectored to intercept whatever it was that NORAD was worried about. However, when the pilots got where they were supposed to be, they said they didn't see anything when they arrived on the scene (Stanley 2005, Part 1, 56).

There is more.

The next year, Fox News reported that on November 20, 2003, USAF fighter jets once again were scrambled and the White House was evacuated briefly (recall the similar reaction, also in November, on November 26, 2019) after "birds" (recall CNN's similarly disingenuous explanation for the November 26, 2019 event), or possibly "disturbances in the atmosphere," tripped radar that keeps watch on restricted airspace around the White House complex. Fox News quoted FAA spokesperson William Shumann, who claimed that it was a false radar target. But Shumann had no evidence for that view, which was based only on the fact that when the NORAD fighters arrived at the location, they found nothing. Certainly, NORAD fighters arriving at the location and finding nothing does not prove that the radar target was false—it simply suggests that UAP are faster than U.S. military jets. But what really stands out to me is the series of resonances between the events in November of 2003 and those in November 2019. In addition, the article discusses another UAP event in D.C., which occurred at 8:45 p.m. on May 2, 2005 when an eyewitness states that swarm of UAP passed undeterred through what was designated as restricted airspace over D.C. (Stanley 2005, Part 1, 57).

Part 2 of the *Nexus Magazine* article discusses the background of the photographer/witness interviewed by the article's author. Among many interesting topics, the two discuss why some people see UAP while others do not, and they note that at least some UAP events are very precisely tailored to certain individuals or groups of individuals. It is interesting to note that the photographer grew up on military bases, since his father was a senior master sergeant in the USAF. The photographer states that his father *was involved in nuclear warfare* and "knew something" (presumably about nuclear warfare, UAP, or both). Evidently some unusual UAP anomalies occurred at the Strategic Air Command (SAC) base at which the photographer's father was stationed. Also interesting, the photographer worked for six years at the White House. The photographer notes that the White House includes Air Force One, and he states about himself, "I do believe I have been tagged," noting that while ETI/UAP are observing humankind and have interacted with many people, they seem to participate in a relatively limited way with most people while they participate more intensively with specific people. The interviewer notes that that 'hot spots' exist for UAP activity, and that certain people with a history of close encounters seem to be drawn to these areas. Mentioning that the Capitol is a hot spot for him, the photographer adds, "But I feel that if I have pictures of them, and they are very advanced technically, then they must have *my* picture" (Stanley 2005, Part 2, 46-8, 50-1, 77).

The photographer also notes that news reports stated that two U.K. Royal Air Force (RAF) bases in England—RAF Lakenheath and RAF Sculthorpe—were involved in some controversial U.S. technologies, and, also, that UAP reports were made at both bases (Stanley 2005, Part 2, 51). In August 1956, for example, RAF Lakenheath was the site of a credible report involving both radar and visual confirmation of an unidentified anomalous object (Thayer 1971).

Returning to the D.C. incidents, the photographer, a technician skilled in capturing high-resolution photographs, earned a M.A. in imaging technology/engineering and worked for ABC News (Stanley 2005, Part 2, 49). He used various digital filtering processes to analyze his UAP images, though he notes, "There were just some things that my system could not analyse. For example, the object that I shot in the water at the Reflecting Pool at the Capitol: I could analyse some of the spectral data, but not all of it." He continues:

> It was too far out there for this particular technology to lock onto and perform variations on. I tried to do as many variations as possible for the sake of getting details from these objects, and some of the analyses of the variations were most intriguing and showed that there were other elements included in these things. For example, the green objects had a nucleus in them. And some of the nucleuses were completely different from those of the object that were on the ground. And there were two sets of objects on the ground (Stanley 2005, Part 2, 47).

The photographer describes two green objects sitting side by side on the ground, with the same shape and configuration as the objects in the sky. However, when he analyzed the objects, the nucleus of each one was red, whereas the nuclei of the objects in the sky were blue. The photographer also observes that UFOs do not necessarily have the shape of flying saucers but instead can have any shape and even can appear as orbs of energy. He also mentions that ball-shaped foo fighters—after the French *feu* for fire— have been reported for many decades (see also Chester 2007). With respect to the photographer's D.C. encounter near the Reflecting Pool in 2002, he states, "Now, the entity on the ground to the rear of the Reflecting Pool was the same entity in the very last shot that shows the objects warping out." His impression was that the UAP were on an "away mission"—UAP arrived, entities came out and went all around the Capitol, "did their survey and then they split." He adds, "whatever those objects in the water and air were, they had some intelligent entities inside them. It was obvious. At one point they were stationary, and the next point they were in motion" (Stanley 2005, Part 2, 47, 49-50).

The photographer notes that on the images taken at the Capitol in 2002, one could see that stationary UAP were discharging smaller UAP into the air. He also observed a small object above one woman's head that was spherical, the size of a quail egg, with the qualities of a pearl with opalescent light. After he photographed the object, he filtered the image, which he states showed that the object was modulating energy around itself—with every level of analysis showing a different level of energy. The interviewer likens this to plasma discharge technology, with fields of plasma existing within other fields of plasma. The photographer also notes that each object had its own signature, with variations on the surface that seem to be modulations of a complex energy field. The interviewer compares this to swirling such as on the surface of a whirlpool, and he expresses his view that plasma dynamics are an integral aspect of field propulsion. The interviewer also states that certain scientists and engineers told him that UAP must be 'warping space[time]' around the craft and could be channeling energy essentially 'freely' from the universe into a limited area (Stanley 2005, Part 2, 50).

About one of his photographs, the photographer states:

> One [image] shows four green objects in lateral and forward motion. It shows a green trace going up and a blue trace going lateral in its signature. According to the laws of physics as applied to light, if a

> green object moves in a given direction it is going to leave a green
> trace regardless of the direction it moves. Yet, this object left a blue
> trace as it moved laterally, against the known laws of physics. …
> While researching the history of green UFOs, I found cases
> describing objects essentially modulating colors. In one case, as one
> of the objects came closer to a military aircraft it turned completely
> green. … How does a solid object become elastic? The thing stretched
> like a balloon and then vanished and then reappeared then vanished
> (Stanley 2005, Part 2, 50).

The photographer also mentions that he took one image of an unidentified
aerial phenomenon that generated a ring of energy as it vanished. He states,
"It left a trail in the sky, like a vapor trail, but this energy trail affected the
air around itself in such a way that it left the air energized." Noting that the
Sun is a plasma of great magnitude, the interviewer likens the behavior of
the object to something being ionized, stating his view that ionization is
very common in plasma drives. The interviewer also states his opinion that
highly energetic fields that ionize the surrounding air apparently create
variations of colored light around UAP, which explains why they are so
incredibly bright (Stanley 2005, Part 2, 50).

The photographer also mentions that among the images he took, at least
one photograph shows a disc-shaped craft dissipating energy. He states
that energy clearly was discharged out of the object in the shape of a
balloon, and that the air around it was distorted. When the photographer
showed the image to friends, they were amazed, especially when they
learned that the photograph was taken at the U.S. Capitol in 2002. At that
time, the ban on commercial flights was still in effect for so-called restricted
airspace, and no flights were permitted to come anywhere near a federal
building. Yet, the objects were in this airspace and even on the ground.
"They landed that night," states the photographer. The interviewer states,
"It really messed with my mind when I first saw those images—especially
the one on the roof of the Capitol Building" (Stanley 2005, Part 2, 51).

Prevalence of UAP Sightings and Detection

ETI/UAP have been present across the globe for at least a thousand years, if not for significantly longer. Some of the oldest depictions of what appear to be extraterrestrial beings and UAP appear in petroglyphs and geoglyphs. Ancient reports from places including the Indian subcontinent and today's Middle East suggest that ETI has evolved in participation with hominin on Earth for thousands of years—perhaps tens or even hundreds of thousands of years, or perhaps even from the beginning of hominin evolution on Earth.

Many people erroneously gloss recent UAP history by stating that the current era of UAP activity began in the 1940s. As I mention elsewhere (Andresen 2022b, 308), this is incorrect. For example, UAP activity was prevalent in the 1930s. Even before that, numerous UAP reports were made in the nineteenth century. In 1880 in the U.S., for example, a UFO was witnessed by multiple people in Galisteo Junction (now Lamy), New Mexico (Tórrez 2004, 112-14). In fact, many UFOs were seen in the 1800s, particularly in 1896 and 1897 (Dunphy 2014; Arts 2003). Sightings continued into the twentieth century, e.g., a well-known UAP wave in 1909, and multiple sightings near Montgomery, New York extending back to the 1930s. Given that prior to the 1940s, no country possessed advanced propulsion of the type that UAP displayed during these events, the only logical conclusion is that at least some UAP are extraterrestrial in origin.

On February 25, 1942, a well-known UAP event referred to as the Battle of Los Angeles occurred in California, during which the U.S. military gunners—quite unnecessarily and, also, inhospitably—fired at an estimated fifteen UAP for many hours (Lucas n.d.; GPM n.d.). From the 1940s to the 1980s (with activity ongoing after that), many UAP reports were made in New Mexico. One important event occurred in March 1950 in Farmington when thousands of witnesses reported seeing hundreds of UAP. UAP also have been sighted around uranium mining and plutonium processing sites, including in New Mexico (Ridge 2005b).

Temporal clusters of UAP sightings also have occurred, e.g., between 1947 and 1952 in the U.S., and between 1953 and 1960 in France. One interesting feature of UAP events is the progression that one finds in how UAP present themselves. Included among many different shapes are 'spy balloons' reported in the Baltics (1890s), followed by 'airships' in the U.S. (1897), 'Zeppelins' in the U.K. (1910s), and 'ghostfliers' (1930s), 'foo fighters' (1944-1945), and 'ghost rockets' (1946) next—leading up to modern 'flying saucers' and 'Tic Tacs' (2004, and, also, in earlier years) (Svahn 2022-2023). I am particularly fascinated by accounts beginning as early as 1932 and continuing for years in the U.S., the U.K., and particularly in Scandinavia of what were referred to as "ghostfliers," "phantom fliers," and "mystery planes" (Keel 1970-1971, Part 1, 10-11; Chester 2007, 6-12). Thousands of observations of ghostfliers occurred in the 1930s in Sweden, for example.

During World War II, foo fighters were reported in both European and Far East theaters. Many U.S. pilots described them as bright lights trailing U.S. aircraft. Concerned that they might be Japanese or German secret weapons, the Office of Strategic Services (OSS)—which was established in June 13, 1942 and terminated on September 20, 1945, and which was the precursor to the Central Intelligence Group (CIG)—investigated but found no evidence indicating that UFOs were in fact adversarial weapons. An April 9, 1947 Intelligence Report from the CIG discusses ghost rocket reports it received from Sweden in 1946 (G. Haines 1997 <1995>, fn. 4 on 79). The CIG was established by presidential directive on January 22, 1946. In accordance with the Central Intelligence Agency Act, which supplemented the National Security Act of 1947, the CIG transformed into the Central Intelligence Agency (CIA) on September 18, 1947 (Pike 1996). The progression from ghostfliers to ghost rockets—particularly in Scandinavia (Gross 1988 <1972>; Aldrich 1998; Berliner 1976; Liljegren and Svahn 1987; Carpenter n.d.-a; n.d.-b; n.d.-c; Uda 2010; see also Wright 2019, 6)—mirrors to a certain extent humankind's own technological progression during that time period. Important, declassified documents regarding specific instances of UAP activity from the 1940s to the 1970s are collated by Uda (2010) and are essential reading for anyone interested in UFO history.

On September 23, 1947, only five days after the CIA had been established, Nathan F. Twining, Lieutenant General wrote a now-famous memo, the subject line of which was "AMC [Air Materiel Command] Opinion Concerning 'Flying Discs.'" Twining sent the memo to Brigadier General George Schulgen, who at the time was Chief of the Air Intelligence Requirements Division, Headquarters U.S. Army Air Forces, Washington, D.C. (Air Force n.d.). The memo states that many UAP seen in the skies by military personnel are not weather, astronomical, or other phenomenon but, rather, are real objects that merit further investigation. Twining famously writes, "The phenomenon reported is something real and not visionary or fictitious." He adds, "There are objects probably approximating the shape of a disc, of such appreciable size as to appear to be as large as man-made aircraft" (Twining 1947). Readers are encouraged to follow the link in the bibliography to read this document in its entirety.

On November 4, 1948, the United States Air Forces in Europe (USAFE) made a fascinating report in the form of a TOP SECRET document with respect to UAP activity in Europe. The document states that it is from "OI OB." It is possible that OI may mean "Operations Intelligence," and OB may refer to a suboffice within OI. The document reads:

> For some time we have been concerned by the recurring reports on flying saucers. They periodically continue to crop up; during the last week, one was observed hovering over Neubiberg Air Base [in Germany] for about thirty minutes. They have been reported by so many sources and from such a variety of places that we are convinced that they cannot be disregarded and must be explained on some basis which is perhaps slightly beyond the scope of our present intelligence thinking.

> When officers of this Directorate recently visited the Swedish Air Intelligence Service, this question was put to the Swedes. Their answer was that some reliable and fully technically qualified people have reached the conclusion that "these phenomena are obviously the result of a high technical skill which cannot be credited to any presently known culture on Earth." They are therefore assuming that these objects originate from some previously unknown or unidentified technology, possibly outside the Earth.

One of these objects was observed by a Swedish technical expert near his home on the edge of a lake. The object crashed or landed in the lake and he carefully noted its azimuth from his point of observation. Swedish intelligence was sufficiently confident in his observation that a naval salvage team was sent to the lake. Operations were underway during the visit of USAF officers. Divers had discovered a previously uncharted crater on the floor of the lake. No further information is available, but we have been promised knowledge of the results. In their opinion, the observation was reliable, and they believe that the depression on the floor of the lake, which did not appear on current hydrographic charts, was in fact caused by a flying saucer.

Although accepting this theory of the origin of these objects poses a whole new group of questions and puts much of our thinking in a changed light, we are inclined not to discredit entirely this somewhat spectacular theory, meantime keeping an open mind on the subject. What are your reactions?

TOP SECRET (USAFE 1948, typographical errors corrected).

Although this document is not widely known, it obviously is quite significant.

Progression in how UAP appear to human beings is a clear example of extraterrestrial pedagogy. ETI presents UAP in a manner that is similar to but often slightly more advanced than human technological capabilities at any moment in time. This is not an example of what Jung (1969) describes as the trickster element, though some people do become confused since ETI's pedagogy involves trying to help humankind realize how surreal reality actually is—especially given entanglement, nonlocality, and the subtle levels of order described by Bohm.

Given how many ETI/UAP events have occurred and how much information is available regarding such events, it is unnecessary to try to catalogue all of them here. However, noteworthy incidents that have occurred over the last fifty years include: _1986_ (Japan Airlines sighting over

Alaska) (Maccabee n.d.); _1987-1997_ (Gulf Breeze, Florida series of sightings); _1989_ (Belgian wave, which was investigated by the Belgian Air Force in consultation with the U.S. and other NATO countries); _1997_ (the 'Phoenix Lights' incident witnessed by multiple people in and around Phoenix, Arizona); _2000_ (independent police officers in southwestern Illinois witness rectangular or triangular UAP on January 5); _2004_ (Mexican Air Force detection of UAP); _2006_ (Chicago O'Hare Airport incident); _2007_ (two pilots witness exceptionally large, unidentified object off the coast of Alderney near the Channel Islands); _2012_ (El Bosque Air Force Base event in Santiago, Chile); and _2014_ (Chilean Department of the Navy records a video of UAP on November 11) (see The Black Vault Originals 2023a for declassified version of the video). These merely are a few of an exceedingly large number of ETI/UAP events one could mention.

Power and Its Precursors

Abstracting across a broad swath of historical references and a wide range of documented accounts of UAP, it becomes clear that UAP often are reported around sites associated with 'power' understood in concrete, metaphorical, and esoteric terms. The word 'power' has a range of meanings, from those that are literal—physical power that is either manufactured (e.g., nuclear power) or natural (e.g., tectonic force, or 'power')—to those that are metaphorical (e.g., political/geopolitical power) and esoteric (e.g., ritual power). With respect to the literal, physical meaning of 'power' as energy, precursors of power also exist, such as what is required to manufacture a nuclear weapon (e.g., the mining of uranium or the processing of plutonium).

Across the literature, at least eight types of sites are associated with UAP events: nuclear ICBM launch control facilities (LCFs) and weapons storage areas (WSAs); nuclear and other types of power plants; military installations and vessels, particularly those associated with nuclear weapons and nuclear power; uranium mining and plutonium processing sites; volcanoes; earthquake zones and other areas of high tectonic activity such as the Mariana Trench and the Atacama Trench; locations of geopolitical power, including the capital cities of the U.S. (Washington, D.C.) (Randle 2022), Russia (Moscow), and China (Beijing); and locations of ritual power, including ancient stone monuments such as Stonehenge, Avebury, Carnac, etc., and, also earthworks. In fact, Scotland—which is covered with ancient megalithic sites—also is the location of numerous UFO events (Swarbrick 2020). Some people even refer to Scotland as 'the Gold Coast of UFOs.' A common denominator of all these sites is 'power,' understood in various ways.

With respect to sightings of UAP near volcanoes, power plants, and military installations and vessels, ETI appears intent on mitigating and/or remediating risks faced by human beings, other species, and Earth itself from volcanic eruptions, accidents, or subterfuge involving the power grid, and, also, from intentional nuclear detonations, accidents, or deliberate sabotage involving nuclear sites and/or vessels (Andresen 2022b, 315).

In the case of stone monuments, research suggests that many if not all the major megalithic sites in southern England and in Carnac, France are in areas of higher permeability, which refers to concentrated flows within the groundwater (Johnson 2022). Accordingly, it most likely is a combination of the permeability of the groundwater is an important factor associated with UAP sightings at megalithic sites. Given that very intense, low-frequency magnetic fields have been associated with UAP activity (Meessen 2012), and given that ancient megalithic sites also are associated with magnetic anomalies, UAP activity near megalithic sites also must relate to energy utilization and enhancement, including that which is associated with higher permeability and, also, subtle forms and expressions of energy (Andresen in preparation).

Hessdalen, Norway, is a well-known UAP hotspot. A 2017 geological survey revealed the presence of strong magnetic anomalies in the Hessdalen Valley, which possesses several abandoned copper mines containing chalcopyrite and magnetite. The magnetite was not useful in copper production and was left in unused heaps around the valley (Gitle Hauge et al. 2016). In addition to old copper mines, the Hessdalen Valley also contains old zinc, sulfur, and iron mines. For example, one old copper and sulfur mine in the middle of the valley pushes acidic, sulfuric pollution into the Hesia River (Gitle Hauge et al. 2017).

In September 2007, university scientists and students associated with the Hessdalen Observatory, Østfold College, Norway, carried out a major sky survey in the Hessdalen area. The Hessdalen Observatory is an interactive UFO observatory. Over multiple years, its cameras and radar have documented many anomalous phenomena. At one point during the 2007 survey, a large, triangular-shaped light was photographed with a 30-second exposure. Optical grating in front of the lens created a continuous, optical spectrum that appeared to emanate from a solid object or a high-density plasma. The object was measured at 30,000 km/hour. The large triangular light also appeared to collect or otherwise absorb a smaller light that emerged underneath it (Visit Hessdalen n.d.). Reports also have been made of UAP landing in the Hessdalen area, including one landing that occurred in conjunction with what may have been soil sampling. A two-hour walk from the nearest dirt road, a two-ton piece of wet turf measuring 1.8 meters by 5 meters was cut with laser-like precision. It was lifted and placed a few meters away, with no sign of any machinery or people having passed through the surrounding grass and mud (TT 2010). This also is known to have occurred in Sweden and is referred to there as an 'Earth flow' phenomenon. Similar cases also have been reported in Ontario, Canada, where large triangles of soil approximately fifteen feet on a side and one foot deep have been cut and lifted as a single piece before being moved to another, nearby location.

Taken together, UAP activity and purported soil sampling in the Hessdalen region suggests that at least one aspect of ETI's activity may be a kind of scientific monitoring to remediate human environmental destruction on Earth—and, increasingly, in space (Andresen forthcoming). It makes sense that an advanced ETI would be interested in soil sampling in the Hessdalen region to determine the environmental impact of human mining activities there. It also makes sense that an advanced ETI would be concerned about how Earth is responding to environmental damage more generally. Sampling soil and undertaking other monitoring activities would be an obvious way to assess such environmental damage, which dovetails with reported instances in which UAP appear to map and/or survey areas beneath them. I also think that one reason some UAP may be surveying areas on and below the surface of Earth is to determine the location of underground water, which may relate to the observed correlation between UAP activity and ancient megalithic sites mentioned above.

UAP Witnesses

Hundreds of thousands if not millions of UAP sightings occur around the world each year.

Photo, video, radar, and other data attests to the reality of UAP. Witness accounts of UAP come from the general public, commercial and military pilots, individuals working in law enforcement, and military personnel more generally. For example, commercial and military pilots routinely witness UAP (Weinstein 2001). In addition, although the information is not often made public, data from the Federal Aviation Administration (FAA) in the U.S. frequently confirms UAP sightings. Furthermore, military sightings and detection of UAP have increased significantly in recent years. Independent radar data that often correlates with visual observations also exists in many cases. Common to almost all UAP reports is the sense that what is being observed or otherwise detected operates according to a technology that is currently beyond human capability. Military witnesses often state that the UAP they observe are so advanced that it appears the UAP are not acting in accordance with 'the normal laws of physics' (Phelan 2019). I discuss this topic later in the book in the context of Bohm's views on why the so-called laws of physics can and must change.

UAP are tremendously egalitarian inasmuch as they show themselves to people from all walks of life—everyone from schoolchildren to astronauts and cosmonauts have seen UAP. At least two former U.S. presidents have seen an unidentified anomalous object. In October 1969, former President Jimmy Carter, together with numerous other witnesses, saw an unidentified anomalous object from the ground when he was in campaigning for governor in Leary, Georgia. While Carter was preparing to give a speech at the Lions Club, he and a group of people were standing outside when someone in the crowd pointed to a UFO in the sky. Carter described the object as a light that was "very unique," saying, "I haven't seen it before or since." He adds, "There were about twenty of us who saw it." Years later, when he was president, Carter officially reported this sighting, which makes him the highest-ranking member of the USG to have made an official UFO report (UFOs 2019).

In the summer of 1974, when he was governor of California, former President Ronald Reagan together with two other individuals travelling in a plane with him and, also, the pilot, witnessed an unidentified white, zigzagging object moving towards the plane when they were in California airspace. In his statement regarding the event, the pilot, Bill Paynter, describes the object as several hundred yards away, appearing as a fairly steady light until it accelerated and elongated. Paynter states that the light then "took off. It went up at a 45-degree angle, very high rate of speed, and everyone on the plane was surprised. The UFO went from a normal cruise speed to a fantastic speed instantly." Some people speculate that this sighting may have contributed to Reagan's attempt to try to create a space weapons program, which he did when he announced the Strategic Defense Initiative (SDI) in March 1983. SDI was a computer-controlled missile defense program that was intended to use lasers and rockets to destroy 'enemy missiles' in outer space (UFOs 2019). Elsewhere (Andresen 2021), I discuss the disingenuous use of a 'threat' narrative regarding UAP as a pretextual rationale for the weaponization of space.

As I discuss elsewhere (Andresen 2021, 196, inc. fn. 4), numerous U.S. astronauts have observed UAP during their missions in space and, also, sometimes even when piloting conventional aircraft in Earth's atmosphere. In the 1960s, they include John Glenn, L. Gordon Cooper, James McDivitt, and Buzz Aldrin, and in the 1970s, Edgar Mitchell and Ron Evans. In addition, in 1973, astronauts Alan Bean, Owen Garriott, and Jack Lousma witnessed an unidentified, red object during the SKYLAB 3 mission (Maccabee and Sparks n.d.).

The first U.S. astronaut to orbit Earth, John Glenn, who later served as a U.S. senator, reportedly saw UAP in space. Later, he stated that the USG should study UAP seriously. Mercury astronaut Deke Slayton personally witnessed an unidentified anomalous object while he was flying a plane. Former astronaut L. Gordon Cooper discusses his experience in a letter quoted below. Cooper also is said to have spoken to former President Bill Clinton about UAP. Clinton reportedly mentioned the topic to the Secretary of Defense in conjunction with a meeting of the National Security Council (NSC) (for more on the NSC, see CRS 2022b).

Some U.S. astronauts have spoken about their UAP sightings publicly (Pasternack 2016), with both Cooper and Mitchell explicitly stating on camera that extraterrestrials are operating in and around visited Earth (Mazzola 2017). Mitchell strongly opposed the militarization and weaponization of Earth and space, and he suggested that UAP sightings were common near military bases because ETI was warning human beings about the dangers of militarization. "Let's hope that is exactly what the ETs, extraterrestrials, are trying to show us," he said, "We don't need to be this warlike civilization" (Pasternack 2016). Similarly, in a leaked email to John Podesta, White House Chief of Staff to former President Bill Clinton, Counselor to former President Barack Obama, and campaign chairman for Hillary Clinton, Mitchell wrote, "Remember, our nonviolent ETI from the contiguous universe are helping us bring zero point energy to Earth …. They will not tolerate any forms of military violence on Earth or in space" (Boyle 2016). The first U.K. astronaut to enter space, Helen Sharma, who visited the Soviet Mir space station in 1991, stated that ETI is real and may be present now on Earth (Picheta 2020).

In the late 1970s, as part of the efforts of the Mission of Grenada to move the UFO topic forward at the UN, Cooper, who at the time was a retired colonel in the USAF, wrote an extraordinary letter, dated November 9, 1978, to Ambassador Griffith, Mission of Grenada to the United Nations. The letter states:

> I wanted to convey to you my views on our extra-terrestrial visitors popularly referred to as "UFO's" …

I believe that these extra-terrestrial vehicles and their crews are visiting this planet from other planets, which obviously are a little more technically advanced than we are here on earth. I feel that we need to have a top level, coordinated program to scientifically collect and analyze data from all over earth concerning any type of encounter, and to determine how best to interface with these visitors in a friendly fashion. We my [sic, 'may'] first have to show them that we have learned to resolve our problems by peaceful means, rather than warfare, before we are accepted as fully qualified universal team members. …

I do feel that I am somewhat qualified to discuss them since I have been into the fringes of the vast areas in which they travel. Also, I did have occasion in 1951 to have two days of observation of many flights of them, of different sizes, flying in fighter formation, generally from east to west over Europe. They were at a higher altitude than we could reach with our jet fighters of that time.

I would also like to point out that most astronauts are very reluctant to even discuss UFO's due to the great numbers of people who have indiscriminately sold fake stories and forged documents abusing their names and reputations without hesitation. Those few astronauts who have continued to have a participation in the UFO field have had to do so very cautiously. There are several of us who do believe in UFO's and who have had occasion to see a UFO on the ground, or from an airplane. There was only one occasion [that I saw] from space which may have been a UFO. … (Cooper 1978).

Unapologetically and with no equivocation whatsoever, Cooper's letter supports the extraterrestrial hypothesis (ETH).

In South America, UAP activity is discussed very openly. As far back as people can remember, Chile, Argentina, and Perú have been the locations of significant UAP activity. For example, Marabamba, in Perú's central Huanuco region, is well known for UAP activity. People living in villages there frequently observe luminous balls of light in the sky over several days. Many UAP reports also come from Chilca, Perú, a beach resort 59 km south of Lima. There, considerable UAP activity occurs at Playa Punta Yaya, near Las Salinas, where the nighttime skies often are illuminated by fireballs and other types of UAP, some of which have been observed moving together as a group. People also describe UAP coming from a little mountain and diving into the sea, moving and changing speeds extremely quickly, and accelerating rapidly before disappearing.

Playa Punta Yaya in Perú is not far south of the equator. Looking across the South Pacific Ocean to the west, one sees Papua New Guinea, also located just south of the equator. There, in 1959, William B. Gill, an Australian Anglican missionary who ran a mission in Papua New Guinea, together with thirty-seven parishioners and staff, had a UAP encounter that lasted approximately four hours. Thirty-eight witnesses reported that they saw a UFO hovering over the mission station, and, also, that they saw humanoid beings on top of the UFO. Attesting to their sincerity, all these witnesses signed a document stating that their testimony was truthful. This event followed months of repeated sightings around the area by Stephen Moi, an assistant teacher, and by others (Gill 1978).

Recent Russian Discussion of UAP

On August 19, 2020, Russian cosmonaut Ivan Vagner stated that he detected a group of five UFOs when shooting a time lapse video while he was aboard the International Space Station (ISS) (TASS 2020). Note, TASS continues to use the traditional acronym UFO instead of using UAP. Vagner's video appears to show objects moving in low Earth orbit (LEO) (CTV News 2020). The event occurred during Vagner's first mission aboard the ISS, while it was passing over Antarctica and Australia. Vagner was recording video of the Aurora Australis, or the Southern Lights, when he captured what seemed to him to be an unusual phenomenon, i.e., four to five lights in a diagonal line. Russia's state space corporation Roscosmos, which is the main successor of the former Soviet space program, initially seemed to echo Vagner's assessment that the video showed something mysterious (Chronicle Herald 2020).

Vagner informed Russian state space corporation Roscosmos about his findings and he also sent the video to the Central Research Institute of Machine Building (TsNIIMash) in Koroylev, Russia for further analysis. Also on August 19, on Russia's Rossiya-24 channel, Roscosmos Spokesman Vladimir Ustimenko stated that the footage of the UFOs flying over the Southern hemisphere captured by Vagner had been sent to experts for analysis, though Ustimenko noted, "It is too early to make conclusions until our Roscosmos researchers and scientists at the Space Research Institute of the Russian Academy of Sciences [RAS] tell us what they think" (TASS 2020). A few days after the event, on August 22, 2020, U.S. reports suggested that the objects Vagner captured most likely were satellites (Howell 2020). A few weeks later, Vagner made statements indicating that he also accepted that conclusion.

With the exception of discussion of the event filmed by Vagner, which appears to have been an honest misidentification, official Russian media outlets have not addressed the UAP topic in significant depth in recent years. Although Russian media from time to time mention U.S. media coverage of the UAP topic, notably absent seems to be a serious counterintelligence push by Russia to muddy the waters regarding UAP. Nevertheless, there is no doubt that Russian leadership is aware of the ETI/UAP reality, and that Russian scientists are carrying out research on UAP. Almost two years after Vagner's honest misidentification of satellites for potentially extraterrestrial UAP, TASS announced on June 11, 2022 that RAS scientists are investigating the sightings of so-called UFOs in the history of humankind. The statement came from none other than General Dmitry Olegovich Rogozin, who at the time was Director General of Roscosmos. Rogozin added that he personally is familiar with the testimony of pilots who saw unusual phenomena during their test flights in the 1970s, and he noted that some of these pilots have drawn sketches of what they witnessed. In a televised interview with Rossiya-24 news channel, Rogozin stated:

> If we talk about specific facts of the so-called UFO [sightings], which might have taken place on Earth throughout the history of mankind,

which NASA speaks about, I would like to say that these studies have been conducted and are being conducted by our [Russian] Academy of Sciences [RAS] among others. The facts are collected and checked.

Rogozin also stated that while 99.9% of the sightings may be some atmospheric or other physical phenomena, he said, "However, we admit that such [extraterrestrial] phenomena could have taken place." He added, "Usually, what we are talking about appeared during a first test flight [of pilots]. I have talked to NASA, and there are proponents there too that we may be the object of external observation [by extraterrestrials]. I would like to believe it" (TASS 2022b; see also Browne 2022).

Of note, at the pinnacle of his career, Rogozin was Deputy Prime Minister of the Russian Federation in charge of the defense industry from 2011 to 2018. Prior to that, quite ironically, he served as Russia's ambassador to NATO from 2008 to 2011. However, on July 15, 2022, Putin dismissed Rogozin as Director General of Roscosmos because Rogozin was perceived as being too controversial (Tingley 2022b). Specifically, both NASA and the European Space Agency had complained about Rogozin's pro-war rhetoric (Foust 2022). Rogozin later came under attack by within Russia for his expensive 'wardrobe' selections (Berger 2022). Later, on December 21, 2022, he was injured in an attack on a hotel in Donetsk, Ukraine (Kirby 2022). Down but apparently not out, Rogozin reappeared again on February 2, 2023 announcing—predictably on his Telegram channel—that several Marker UGVs (uncrewed ground vehicles) had arrived in the Donbas to support his frontline soldier initiatives on behalf of Russia in eastern Ukraine (Bendett 2023).

With Rogozin otherwise occupied in Ukraine, it appears that Yuri Ivanovich Borisov, Rogozin's replacement as Director General of Roscosmos, is taking a more pragmatic approach to his position at the space corporation. On 12 April, 2023, Borisov stated that Russia's participation in the ISS has been extended until 2028 (AFP 2023). This is a welcome development inasmuch as it is preferable to have Russia continue to participate in the ISS rather than move toward China's initiatives in space. The real goal, however, is for all nations of the world to learn to cooperate in space rather than turning space into a domain in which current geopolitical conflicts are extended and exacerbated.

In addition to U.S. and U.K. astronauts, Russian cosmonaut Ivan Vagner has stated that he shot a time lapse video of a group of five UAP when he was aboard the International Space Station (ISS). This occurred when the ISS was flying over the southern hemisphere, specifically the Antarctic, at the peak of the Aurora Borealis (TASS 2020). Vagner states that the UAP video was submitted for review to experts at Roscosmos, which is Russia's state space corporation, or space agency. In turn, Roscosmos confirmed the video's authenticity (Chronicle Herald 2020; for the actual video, see CTV News 2020). It also has been reported that Vagner sent the video to the Central Research Institute of Machine Building (TsNIIMash) for further analysis. On Russia's Rossiya-24 television channel, Roscosmos spokesperson Vladimir Ustimenko confirmed the statement from TASS that the video was sent to experts for analysis (TASS 2020).

While Vagner and others in Russia appear to be taking an appropriately respectful stance towards the topic of UAP (TASS 2022b)–misidentification or not–an unnecessarily disrespectful comment was made on the topic of ETI/UAP by Russian state TV personality Vladimir Solovyov. Some people refer to Solovyov as "Putin's voice" because of his close ties to Russian President Vladimir Putin. In perhaps one of the oddest statements to date relating to the war in Ukraine, Solovyov warned the North Atlantic Treaty Organization (NATO), "If everything keeps progressing the way it is, only a couple of mutants in Lake Baikal will survive. The rest will be destroyed in a massive nuclear strike." Solovyov's mention of a nuclear strike obviously is irresponsible, reckless, and immoral. But another irresponsible aspect of his statement apparently was lost on the author of the *Newsweek* article in which the quote appeared. Despite correctly identifying Lake Baikal as "an enormous lake and one of the world's deepest, located in the mountainous Russian region of Siberia, north of the Mongolian border" (Van Brugen 2022), the author of the article apparently did not realize what Solovyov meant by "mutants."

The distasteful, slang word "mutants" sometimes is used by certain Russian individuals to describe extraterrestrial beings present here on Earth and in Earth's waters. Lake Baikal, which has the largest surface area of any freshwater lake in Asia, is the location of numerous sightings of both UAP and, also, of extraterrestrial beings. Some accounts date back to the nineteenth century, and one widely reported event occurred there in 1977. The Russian Navy, which is said to be on the forefront of UAP research, reportedly has detected gigantic underwater objects moving at amazing speeds. Discs are reported to have flown out of the water, gone into the sky, separated into different objects, and come back together. Deep underwater, the Russian Navy also is said to have detected UAP that emit various sounds. In one case, for example, people reported a sound like the croaking of frogs (PRPS 2022; MWL 2020; Siberian Times 2015; Quest TV 2020; TT 2010; see also Stonehill and Mantle 2020 <2016>; Wright 2019, 284-94 for discussion of other UAP cases in Russia). The intrigue surrounding Lake Baikal has not been lost on Vladimir Putin, who in 2009 climbed into a mini-submarine that dove almost a mile under the Earth's surface to the bottom of Lake Baikal (Parfitt 2009).

Here, my request to Vladimir Solovyov and to others in Russia is that they refrain from referring to extraterrestrials as "mutants." Name calling is a very poor way to begin a relationship. I do not even like the word "aliens," but the word "mutants" is truly beyond the pale. As supposedly intelligent beings ourselves, we can learn to be respectful towards others.

Despite it being unnecessarily rude in tone, Solovyov's comment nevertheless is instructive inasmuch as it indicates that knowledge of the ETI/UAP reality exists at the level of top leaders in Russia. The topic of ETI/UAP also must be somewhat pervasive in Russian society, otherwise Solovyov would not have made the comment, which presumes a certain base level of knowledge among listeners. Even further, Solovyov's comment suggests at least tacit acknowledgement that ETI is somehow more sophisticated than humans, since Solovyov suggests that ETI is 'the only species that would survive' if Russia unleashed a nuclear conflagration. Of note, historical reports indicate that Joseph Stalin, former Premier of the Soviet Union from 1924 to 1953, was interested in UFOs.

Based on the evidence to which he had access, Stalin concluded that UAP posed no danger to military forces (MWL 2020).

In a prior publication (Andresen 2022b, 291-92), I articulated the very legitimate USG concern that the release of existing classified information on UAP propulsion and/or materiel could create a route whereby this information would become available to peer adversaries of the U.S., i.e., China and Russia, and/or to other countries and groups. This would put the world at risk of a strategic reversal according to which the U.S. could be unseated in the global balance of power. This is a very real concern now, given that Russia and China have declared that they are operating as a bloc (President of Russia 2022). The concern is made even more imperative following an oblique comment made by Russian President Vladimir Putin with respect to Russia's initiatives to develop advanced weapons:

> We will continue to develop advanced weapon systems, including hypersonic and those *based on new physical principles*, and expand the use of advanced digital technologies and elements of artificial intelligence.
>
> Such complexes are truly the weapons of the future, which significantly increase the combat potential of our armed forces (Lock 2022, italics added).

Depending upon what may be meant by "new physical principles," one can understand the care shown by the USG in taking the strongest possible steps to safeguard classified information that may exist on UAP propulsion and/or materiel (Andresen 2022b, 291-92). In the long run, if human society succeeds in transitioning to a balanced, multipolar world in which human rights and democratic values are upheld across the globe, then it will be safer for existing classified information held by the USG on UAP to be released.

UAP Origins, Characteristics, and Capabilities

It is fascinating to consider the origins of extraterrestrial UAP. Where are they from? How old is the extraterrestrial civilization (ETC) that created the UAP we are seeing now? How long has the USG known that some UAP are extraterrestrial in origin? What agencies and/or branches of the USG hold such information? While society someday may receive definitive answers to questions such as these, in the meantime, we can describe UAP characteristics and capabilities as clearly as possible.

A partial list of UAP characteristics includes the following:

Acceleration/Deceleration: sudden acceleration from a stationary and/or near-stationary position; rapid climbs and almost instantaneous gains in altitude; virtual hover to rapid acceleration; seemingly instantaneous acceleration to very high and even high-Mach, speeds; speeds so high that the human eye cannot follow the objects; gains in altitude before moving to the horizon almost instantaneously; abrupt deceleration; etc.

Colors: reports of almost every conceivable color; many reports of silver, white, milky white, and black objects; plasma-like manifestations that may appear as a single color or with modulating colors; etc.

Density/Consistency: plasma-like; orbs witnessed dissolving in and out of one another; density similar to but less dense than dough, especially in instances involving UAP that do not appear to have occupants; reports of UAP dividing into sections and melding back together; UAP anywhere along the defined to amorphous continuum (Andresen 2022b, 301-4) separating into different objects and then rejoining into a single object; sections of UAP flying off in different directions; etc.

Effects: electromagnetic (EM) effects on nearby vehicles, watches, machines, etc., which stop upon the approach of an unidentified anomalous object and restart when the object departs (see NICAP n.d.-a; 1960)

Excretions: 'angel hair' excretions consisting of delicate, gossamer strands of material that appear to fall from some UAP, catching in trees or landing on buildings and usually dissolving when touched; plume of golden white plasma vented from an unidentified anomalous object; etc.

Formations: UAP observed flying 'in formation' (double entendre, since UAP *are* information); different formation shapes, such as V, reverse L, etc.; formations consisting of multiple objects, up to dozens if not more objects moving together; large objects sometimes circled by smaller objects; etc.

Gravity Control: apparent use of some form of gravity control; artificial gravity or force field that may guard occupants, if present, from high g-forces; etc.

Hovering: remaining stationary; hovering close to the ground (e.g., 150-500 feet about the ground); hovering over lakes or other body of water; etc.

Lift: no visible or apparent means of generating lift

Light Features: bright; luminous; luminescent; translucent; partially translucent; pulsating; solid lights similar to searchlights, which especially at night or half-light appear to scan the ground beneath them; beams of light, sometimes white, that emanate down, scan, and retract before UAP move away; lights of different intensity on the same object; strobing lights; lights moving in wave-like motion; light passing over witnesses; red and blue blinking lights (other colors also mentioned); multiple (e.g., five or six) bright yellow, triangular beams of light emitted from side of object, with light brighter at the apex and diminishing towards the ground; glowing lights emanating from inside UAP with windows; etc.

Maneuverability: fast and nimble movement; right-angle maneuvers; abrupt, right-angle and sharp, acute-angle turns at high speeds; 'tilt to go' attitude change before initiating movement in the direction of tilt; rapid climbs and returns within a few seconds; abrupt stops not far (e.g., 150 feet) off the ground; takeoffs at an angle; no visible control surfaces; triangular object that flipped and flew 'sideways' before 'righting' itself in an unusual way, with the object then conducting a series of turns without banking, descending slowly, and disappearing below tops of trees; disc shaped UAP rotating on their edges and moving on the vertical as opposed to the horizontal plane (Swords 2006, 12); object coming almost directly overhead and then pointing 'front' end down towards witnesses; etc.

Morphology: multiple forms, many of which are geometric (circles, spheres, ovals, discs, cubes, triangles, rectangles, cylinders, etc.); transmogrification (i.e., the ability to change shapes (Keel 1969)

Movement: rotating; spinning; changing direction abruptly ('as if on a dime'); moving straight up and/or straight down; smooth movement that is much less choppy as compared to movement of drones; rocking up and down like a tetter totter; revolving slowing around ninety degrees; darting; zigzagging; displaying unusual kinematic movement; etc.

Observability: low observability at multiple radar bands; some UAP do not appear on radar even though witnesses see them; sudden disappearance and reappearance; 'blinking' or 'winking' in and out; continuum from fuzzy to very distinct appearance (the 'defined' to 'amorphous' continuum I postulate, see Andresen 2022b, 301-4)

Propulsion: no visible signs of propulsion, yet indirect indicators of advanced propulsion; little or no visible signature

Sound (or lack thereof): noiseless; 'utterly silent;' extremely quiet; silent, but caused ice on lake to rumble and crack loudly; very slight humming sound; frog-like croaking, as an USO in Russia evocatively was described (TT 2010); etc.

Speed: very slow; very fast; high velocities; 'unusual' speed; hypersonic horizontal and vertical velocities; etc.

Transmedium Capability: covers features of both UFOs and USOs; ability to move between the liquid ocean, the gaseous atmosphere, and the vacuum of space (note, the vacuum of space exerts negative pressure and results in repulsive gravity, which could be a factor in UAP propulsion).

Gravity Control and Beyond

Many different ideas have been proposed to account for advanced UAP flight characteristics and propulsion capabilities, some more speculative than others. What UAP capabilities clearly demonstrate, however, is that ETI is not suffering from a broken physics. Recognizing that reality is a unified whole and that any absolute demarcation between macro and micro is artificial, however, ETI/UAP clearly has a physics view in which one, coherent theory works across scale.

In the current human understanding of physics, however, scale is not clearly understood. General relativity is understood as the mainstream theory relating to the very large, while the quantum theory is understood as the mainstream theory relating to the very small, and although many attempts have been made to unify these theories, the mainstream view is that none of these attempts have succeeded. Such attempts previously were referred to as the search for a Grand Unified Theory (GUT), or later, as the search for a Theory of Everything (TOE). Now, still realizing that a successor theory is necessary, physicists often use the phrase quantum gravity to discuss the attempt to understand how gravitational physics under general relativity and the quantum theory can come together. Despite much academic work in this area, especially since the early 1980s, many attempts to extend general relativity — for example string theory — have not been successful.

Outside of theoretical physics, other, engineering-oriented attempts have been made to overcome gravity. Sometimes these ideas are dismissed out of hand because they occur outside of mainstream, academic physics, though I do not think an engineering approach per se is incorrect methodologically, and it certainly can be pursued alongside more theoretical approaches. Some of the concepts one finds in the literature include antimatter reactors; electrogravitics; electric propulsion; gravitodynamics; gravity control/antigravity propulsion; gravity shielding; 'exotic' propulsion; and the possibility of different types of propulsion systems working together. Despite the creativity of many of these ideas, none appear to work—at least not yet.

Another speculative idea that has been proposed recently is that UAP create distortions in spacetime to create geodesics along which they move free of actual acceleration. Proponents suggest that such an explanation is consistent with quantum explanations that propose space is a network of quantum bits that all are entangled with one another. It has been proposed further that UAP propulsion may use a link between electromagnetism and the curvature of spacetime to influence the latter—i.e., that electromagnetic interactions influence the spatial entanglement network and introduce curvature, thus enabling geodesic motion (uaptheory 2022).

Many proposals for gravity control discuss the connection between gravity and electromagnetic fields. For example, it has been proposed that 'gravitational radiation theories' offer the key to gravity control. Articles along these lines mention concepts such as field control, artificial force fields, non-gravitational fields, and artificial gravitomagnetic fields. Attempts also have been made to explain how 'gravitational radiation' may impact devices similarly to how they are impacted by electromagnetic pulses (EMPs).

With respect to gravity control/antigravity technology at the macro level extending beyond small scale research on superconducting magnetic levitation, confusion exists as to whether humankind has achieved this capability in classified settings. Leaving this question to the side, and, also, leaving to the side the interesting topic of the relative status of USG research on gravity control as compared to research being conducted by peer adversaries such as China and Russia, even if knowledge of gravity control does exist in classified settings, this knowledge has not yet reached academia or the general public. Nevertheless, if human beings do learn to control gravity at some point, they could use this technology as a means of breakthrough propulsion to enable them to travel widely throughout the universe.

Many years ago, Robert Forward, a gravity expert at Hughes Aircraft Company—a company founded by Howard Hughes that ceased operations in 1997—predicted that at some point in the future, human beings would be able to create artificial gravity fields at will. In fact, many researchers have attempted to master gravity control. Literature on antigravity research mentions individuals such as Dr. Hermann J. Oberth, Burkhard Heim, Dr. Yevgeny (a.k.a. Eugene) Podkletnov, Prof. Martin Tajmar, Dr.-Ing. Clovis J. de Matos, Dr. Jochem Häuser, and Dipl.-Ing. Walter Dröscher—among many others.

Oberth, whose research was foundational to the fields of rocketry and astronautics, was as a mentor to Dr. Wernher von Braun (NASA n.d.). Oberth states that energy, inertia, and gravitational fields are aspects of the same thing and are impossible to separate from one another at a deep level. In 1954, Oberth gave a lecture, "Flying Saucers," in which he proposes many possible ways to think about what we now call UAP (Oberth 1954). There, Oberth broaches the idea that one could use fields of force that are not yet known to accelerate material objects in a manner similar to the acceleration caused by the force of gravity. Oberth also proposes that with an artificial gravity field, the field of force applies simultaneously to the passengers and to the spacecraft.

Heim worked at the Max-Planck-Institute (MPI) for Astrophysics in Göttingen, Germany. In 1955, during a period of intense gravity control propulsion research in the U.S. from 1955 to 1974, it was announced that Heim had made a contractual arrangement with Glenn L. Martin Company to help the company with its gravity control propulsion project (Talbert 1955a; 1955b). Heim's research on magnetogravitic propulsion led to his claim that his experiments provide a lead on antigravity involving an intermediate field that is neither electromagnetic nor gravitational in nature. Heim hypothesizes that if applied to space flight, his ideas would result in direct levitation, conversion of electricity into kinetic energy without any waste, and protection of the occupants and structures of antigravity vehicles against any effects resulting from acceleration of the vehicle, regardless of how great such effects may be (Deasy 2009; see also von Ludwiger n.d.).

Podkletnov claims to have designed and demonstrated gravity shielding devices consisting of rotating discs constructed from ceramic superconducting materials. Boeing's Phantom Works tried to verify Podkletnov's work, as did NASA (BBC 2002), though no official verification ever has been entered into the public domain. Some people say NASA simply failed to reproduce Podkletnov's claims. Given that Podkletnov is Russian, if his claims had been verifiable, one assumes that Russia's weaponry would tell the tale—which it certainly does not.

In March 2006, the European Space Agency (ESA) homepage reported on a study by Tajmar and de Matos (2006) stating that tangential acceleration measured by accelerometers around a rotating superconductor demonstrate artificial gravity. The apparatus constructed by Tajmar and de Matos was different from the one constructed by Podkletnov. Graham et al. (2007) report that they reproduced the results of Tajmar and de Matos. It also is possible that these results were reproduced in the results of the Gravity Probe B (GP-B) experiment (Häuser and Dröscher 2017, 494).

By mid-2009, Tajmar was working with Dröscher and Häuser, the two primary theorists working on Extended Heim Theory (EHT). EHT predicts extension of Tajmar's tangential force experiment (GME1) to a vertical force experiment (GME2). The idea is that by building a larger, non-superconducting, fermionic version of GME2, one can demonstrate the ability to lift a spacecraft. With respect to a new magnetic motor power source, the Irish company STEORN has an Orbo device that some people claim produces enough energy to power a Heim-Droscher-Tajmar drive (Deasy 2009).

My personal view is that the latest version of Extended Heim Theory (EHT)— which considers four subspaces that can be grouped in different ways to map onto the standard model of physics—is insufficiently advanced to explain what is happening with UAP. More promising may be an idea proposed by Häuser and Dröscher. Observing that no extreme gravitomagnetic field effects have yet to be produced, they propose a new set of thought (Gedanke) experiments based on the interference of matter waves, i.e., a so-called gravitomagnetic Aharonov-Bohm effect. Häuser and Dröscher claim that if the proposed gravitational Aharonov-Bohm effect is verified experimentally, it may decide the question of whether a gravity-like field exists (Häuser and Dröscher 2015; see also Bruhn n.d.). Of note, Dröscher has patented an interesting device based on his ideas (Deasy 2009; see Dröscher and Häuser 2007; 2015). Also of note, Dröscher and Häuser (2008) have published on gravity-like fields specifically in connection with space propulsion concepts.

NASA and other private companies conduct a wide range of research on alternative propulsion that does not involve propellants. Many ideas have been presented, such as Alcubierre warp drives (Alcubierre 1994); Lorentzian wormholes (Visser 1996); and EmDrive technology that uses the energy of trapped microwaves to create a force (BFD 2020c; see also Oberhaus 2019; 2020; Hambling 2020).

My view is that none of the aforementioned ideas fully account for the movement of extraterrestrial UAP. As interesting as gravity control technology is, since gravity manifests in explicate order, it still is an explicate order technology according to Bohm's framework. The ETI in our midst obviously can control gravity, but my view is that it is much more sophisticated than this. The flight characteristics and other interesting features of its UAP indicate that the ETI in our midst now understands the inseparability of consciousness and matter and implicate, etc., levels of order, and that it operates on the basis of an ontology of wholeness (Andresen 2022b). This is an entirely different level of discussion.

Essentially, I am proposing that UAP utilize at least two different forms of what can be called 'movement technology.' When they accelerate quickly and are seen and/or detected moving smoothly across spacetime, they are using some method of what we would consider to be breakthrough propulsion, which involves so-called gravity control. However, when they blip in and out, so to speak, something *beyond gravity control* is occurring—since gravity control, despite being novel from a human perspective, is not novel ontologically inasmuch as it is limited to explicate order.

In the case of the UAP operating in our midst now, being able to manifest and unmanifest at will is more than a cloaking maneuver. I also do not think this is a transition between putative 'dimensions.' Of note, Häuser and Dröscher (2017; 2019) observe that the concept of extra space dimensions conflicts with data from numerous experiments, particularly recent Large Hadron Collider data. Instead, my view is that ETI has found a way to operationalize the unfolding and folding back and forth between levels of order in such a way that it often appears to observers that UAP blip in and out seemingly instantaneously. In other words, ETI knows how to operationalize multiple, subtle levels of order—implicate, super-implicate, super-super-implicate, etc.—potentially to infinity. Here, UAP are *traversing the boundary between implicate and explicate orders, then moving back again to implicate order, etc., such that when they disappear in one location and reemerge in another, they are dissolving themselves at the explicate level and progressing through the implicate to reconstitute themselves in explicate order again, at another point in space, and with very little change in time.*

The upshot is that ETI/UAP demonstrate *both* breakthrough, gravity control technology, *and, also,* the ability to manifest, unmanifest, and remanifest in a very fluid way. Both movement technologies indicate a very sophisticated understanding of the reciprocal enfoldment of explicate and implicate, etc., levels of order and a remarkably profound understanding of the nature of reality. Manifesting, unmanifesting, and re-manifesting at will, ETI/UAP can change their morphologies essentially instantaneously, displaying a seamless integration with the whole.

Relatedly, I propose that *as energy in a system increases, one moves towards the deontological end of the continuum between utilitarianism on one end and a deontological approach to existence on the other.* This idea is one of the most important in the book, namely that as the overall energy in a system increases, everything becomes more and more finetuned. In the case of high energy systems, even the smallest movements can have very significant repercussions—precisely because one is using increased energy while engaging implicate, super-implicate, super-super-implicate, etc., levels of order. This translates to the ethical plane such that as energy utilization increases, one must move as far to the deontological side along the continuum between utilitarian and deontological perspectives as one is able to, perceptually and spiritually. The reason is that as the energy of a system becomes very high, one must act from more of a deontological perspective or otherwise risk blowback that, if intense enough, could obliterate one's context altogether. This realization applies at all levels from conventional geopolitics to ontological wholeness. Geopolitically, for example, as the energy associated with the weapons systems used in Ukraine increases, the entire conflict becomes more finetuned—with a small push in any one direction having the potential to cause enormous repercussions.

In the case of ETI/UAP, the insight above with respect to high energy systems becomes immensely significant. Any ETI that can manifest/unmanifest at will, thereby displaying that it understands morphogenesis in a very deep way, also can access enough energy to support almost unfathomable UAP acceleration and speed, since such knowledge is a precursor to the deeper knowledge involved with manifesting and unmanifesting. This also means that the ETI responsible for the UAP in our midst now *inherently* is not a threat to humankind. Knowing how to access almost if not unlimited energy without destroying one's entire civilization means that the ETI around us now understands the wholeness of reality. Such understanding *inherently* includes an ethical component, which is why ETI/UAP *inherently* are not a threat to humankind (Andresen 2022b, 304-305). Any ETI that arrives at Earth inherently will be kindly disposed to Earth and her inhabitants, simply because it will have mastered the balance between knowing how to utilize enormous amounts of energy while keeping its civilization going.

The ETI in our midst now is remarkably intelligence. With respect to the nature of intelligence, Bohm states, "this possibility of going beyond any specifiable level of subtlety is the essential feature on which the possibility of intelligence is based" (Bohm 1990, 282). Since it has progressed beyond a mere understanding of explicate reality, the ETI around us now operationalizes subtle levels of order. To reiterate my point from the preceding paragraph, this indicates that ETI is an *inherently nonthreatening intelligence* given its deep insight into the whole nature of existence. Experience and understanding of how everything participates in everything else makes aggressivity nonsensical, since being aggressive is tantamount to threatening oneself. Understanding the subtler levels of order, the inseparability of matter and consciousness, and the whole nature of reality therefore makes one a kinder and gentler being. The physics of the matter (pun intended) therefore deconstructs the narrative that any ETI that has reached Earth and is actively operating UAP here is—or even could be—a threat.

One also can conclude that any ETI that makes its ways to Earth inherently is nonthreatening by referring to Bohm's statements regarding the super-implicate level of order in relation to benevolence and compassion. With respect to the super-implicate order, Bohm himself states that it can be proposed that the super-implicate order is a kind of super-intelligence that is benevolent and compassionate, not neutral (Bohm 1986b, 40). Since this ETI in our midst now clearly understands subtle levels of order beyond explicate order—i.e., implicate, super-implicate, super-super-implicate levels of order—then it also is infused with these qualities of benevolence and compassion.

One can rest assured that the ETI in our midst now has a well-intentioned and magnanimous view towards humankind. Meanwhile, however, human beings continue to pose an enormous threat to one another. It is *intra*species conflict and violence that we must overcome if we want to create a society that is stable enough for advanced knowledge to be disseminated. This includes knowledge that ETI already possesses regarding how to utilize immense amounts of energy in a safe manner that benefits humankind, and, also, the whole.

UAP Press Conferences and Government Release of UAP Files

Over the years, many current and former employees from U.S. government agencies such as the FAA, experts on space, pilots, and other officials have provided personal testimony during press conferences regarding their direct knowledge of UAP. In 2001, a UAP disclosure press conference occurred at the National Press Club in Washington, D.C. (Greer 2017). A similar event occurred in 2007 when the Coalition for Freedom of

Information brought together a panel of high-level pilots and other witnesses during an international press conference also held at the National Press Club. The 2007 event included individuals from seven countries (the U.S., the U.K., Belgium, France, Iran, Chile, and Perú) and the Channel Islands (which is comprised of two bailiwicks that are self-governing Crown Dependencies of the U.K.) (Discogs 2007).

Although governments around the world have begun to release their UAP files. In the U.S., the CIA has released some of its UAP documents. For example, in the fall of 2016, documents that the CIA previously had released under the Freedom of Information Act (FOIA) but which had been held at the National Archives were sent to CIA's website (Wright 2019). The CIA released additional documents on UAP on January 7, 2021 (Greenewald 2021; Elliott 2021). In 2007, the U.K. released UAP files dating back at least to 1967. These files had been collected by DI55, which was a secret UFO branch of Defence Intelligence within the U.K. Ministry of Defence (Randerson 2007; see also The National Archives 2011; 2013; Norton-Taylor 2013). Furthermore, it has been reported that the U.K. has investigated UAP sightings since the 1950s (MOD 1951) and likely is in possession of tens of thousands of reports overall (TT 2010).

Beyond the U.S. and U.K., governmental, intelligence, and military organizations and committees in many other countries have collected and, in some cases released, files on UAP. A partial yet representative list includes: *Australia* (NAA 1957-1971); *Brazil* (Arquivo Nacional 2018; BUFOF 2015; BBC 2010a; Gevaerd 2005); *Canada* (Otis 2021; Li 2020; Library and Archives Canada 2007 <2005>); *Chile* (CEFAA n.d.); *Denmark* (Denmark 2016; Finnsson 2009); *France* (geipan n.d.; FUFOF 2015; New Scientist and Afp 2007; cnes 2006); *Mexico* (Yturria 2004); *New Zealand* (BBC 2010b; NZDF 1981); *Perú* (Montesdeoca 2019; Ramírez Muro 2014; Collyns 2013; Watson 2007); *Spain* (Watson 2007); *Spain and Portugal* (Ballester Olmos 1976); and *Uruguay* (Fuerza Aérea Uruguaya n.d.; Isgleas 2009) (see also TT 2010). Other countries that have released some UAP files include Argentina, Ireland, and Turkey (TT 2010).

In addition to general releases of UAP files, researchers have compiled databases available online that include details of witness reports. Such databases exist for Belgium (COBEPS 2021), Italy (CUN 2022) and Norway (Ballester Olmos and Brænne 2007), for example. In addition, a global database originally called 'Capella' that includes some 260,000 reports also has resurfaced (Basterfield 2018; 2022).

Furthermore, civilian organizations also compile tremendous amounts of UAP data. With respect to the U.S., important organizations include NICAP (National Investigations Committee On Aerial Phenomena) (NICAP n.d.-b), CUFOS (J. Allen Hynek Center for UFO Studies), MUFON (Mutual UFO Network), and NUFORC (National UFO Reporting Center). Another civilian organization in the U.S., the National Aviation Reporting Center On Anomalous Phenomena (NARCAP n.d.), has collected detailed descriptions of UAP from many different countries. Outside the U.S., similar civilian organizations that compile UAP reports are BUFORA (British UFO Research Association) in the U.K., GEP e.V. (Gesellschaft zur Erforschung des UFO-Phänomens, trans. Society for the Study of the UFO Phenomenon) in Germany, and APU (Asociación Peruana de Ufología, trans. Peruvian Association of Ufology) in Perú. Other such organizations exist around the world, including in China and Russia.

Over the years, various journals and periodicals have reported on UAP activity. These include *Journal of UFO Studies*, *Journal of Scientific Exploration, International UFO Reporter*, and *Flying Saucer Review* (FSR). FSR was published by Aerial Phenomena Research Organization (APRO). APRO (not to be confused with today's AARO, or All-domain Resolution Office, discussed below) was founded by Coral and Jim Lorenzen in January 1952 and remained active through 1988. I strongly recommend that people take time to consult reports from decades ago published in FSR, which are detailed and extremely interesting to read.

Policies among members of the Five Eyes (FVEY) intelligence alliance — which includes the U.S., U.K., Canada, Australia, and New Zealand — seem *somewhat* aligned in terms of document disclosure, as do policies among members of NATO. Interestingly, the Chief Science Advisor of Canada recently announced a "Sky Canada Project" to investigate UAP (The Peak

2023). While initially it was hoped that a scientific rather than national security lens would be instrumental in the Project (GCWUFO 2023a), a PowerPoint presentation (OCSA 2023) seems to suggest that Canada is being unduly influenced by the U.S.' national security agenda relating to UAP (GCWUFO 2023b; 2023c).

Although Japan is neither a member of FVEY or of NATO, Japan is a global partner of NATO, with which it has been in dialogue and acted in cooperation since the early 1990s. Although Japan has not released any UAP files to the public, the Japanese Defense Ministry has created procedures for reporting and collecting data on UAP (Siripala 2020). Significant UAP activity likely is occurring in Japan, since the USG suggested to Japan that the country collect data on UAP events and Japan quickly established a protocol for doing so (Johnson 2020).

Members of the U.S. military have stated that they have witnessed UAP in CENTCOM's (United States Central Command's) Area of Responsibility (AOR). This AOR includes twenty-one countries and spans more than four million square miles across Northeast Africa, the Middle East, and Central and South Asia. It is populated by more than 560 million people from twenty-five ethnic groups who speak twenty languages, with hundreds of dialects that transect national borders (USCENTCOM n.d.). Although to date, I am unaware of countries in this area officially releasing UAP files, many documented UAP reports do exist from these regions. In the case of Central Asia, it is likely that UAP files collected before the former Soviet Union collapsed in 1991 were routed to Moscow and likely remain there. From time to time, however, documents find their way to the surface, such as a September 1989 teletype describing a September 6, 1989 sighting of a UFO over Ürümqi, the capital of the Xinjiang Uygur Autonomous Region, which geographically is not far from the border of the former USSR. In addition, many unofficial statements suggest that Russia (Stonehill 1998; Stonehill and Mantle 2010; 2020 <2016>; 2017; MWL 2016; 2020; 2021) and China (Wright 2019, 267-68, 296) are studying the ETI/UAP phenomenon at a governmental level.

Ethical and Security Complexities Associated with UAP Information Sharing

Although society is becoming more open about the ETI/UAP topic, two constraints factor into how much information is the 'right' amount for the U.S. to share. These two constraints are the continuation of violent conflict around the world, and the attempts of a few people to attempt to capitalize on information about UAP for their own benefit. What can be shared now without jeopardizing social justice is that an extraterrestrial presence is operating in our midst. It is time that the 'mere existence' of this extraterrestrial presence should be acknowledged at an official level. Any of the major nation-states, such as the U.S., Canada, China, or Russia, could confirm this reality openly. This should occur as quickly as possible.

Meanwhile, however, the geopolitical climate remains too chaotic for the U.S. to share classified data on UAP propulsion and/or materiel. Doing so would create a potential scenario in which kleptocratic, authoritarian, and/or totalitarian regimes could attempt to weaponize such information. Meanwhile, people in the U.S. should not attempt to do so.

I also do not support attempts to 'globalize' UAP research under the auspices of the UN. We should not be naïve about the role that the UN plays in the world. Since the UN is an intergovernmental organization, one might assume that what occurs there facilitates cooperation between nation-states. Because of the existence of special interests operating within and alongside the UN, however, what occurs at the UN often facilitates supranational agendas that do not help any single nation-state, let alone nation-states collectively. Instead, many decisions made at the UN and at other organizations such as the World Bank, the International Monetary Fund (IMF), etc., simply favor elite groups to seek to concentrate power, wealth, and control over various aspects of society, such as the global money supply. Unfortunately, the quest for world domination has not yet gone out of style, which is why the UN and other powerful, international organizations have done very little to halt the proliferation of weapons.

Today's particularly unfortunate geopolitical reality involves two major blocs have formed across the world, with smaller countries picking sides.

NATO needs no introduction. But the group competing against NATO—the BRICS+20ish (Brazil, Russia, India, China, South Africa, potentially + Iran, Saudi Arabia, Mexico, Indonesia, Argentina, and scores of other countries that have expressed interest in membership)—may be more unfamiliar to people (Sguazzin 2023; MEM 2023; Devonshire-Ellis 2022).

Over the past year, China and Russia have strengthened political and military ties significantly. On March 20, 2023, as the war in Ukraine continued, Xi Jinping, President of the People's Republic of China, met with Russian President Vladimir Putin in Moscow (Magramo et al. 2023). The meeting underscores the extent of deepening economic and political ties between the two countries (Simmons and Ramzy 2023). On February 24, 2023, the Chinese Foreign Ministry released a 12-point position paper on a political settlement of the Ukraine war (CGTN 2023), which the U.S. does not support because it interprets the proposal as favoring Russia (Marlow 2023).

On March 31, 2023, Putin adopted a new foreign policy that identifies China and India as Russia's main allies. According to this new foreign policy, Russia, China, India, and other BRICS countries plan to create a new currency, in part to ameliorate the negative impact on Russia's economy of Western sanctions and other restrictions resulting from its role in the war in Ukraine. The new medium for payments will not defend either the U.S. dollar or the euro. Instead, the plan is to secure it using gold and other commodities such as rare-earth elements. Russia announced its new foreign policy less than two weeks after Xi visited Moscow to further cement the "no limits" partnership that Russia and China announced in 2022. Russia also announced that it will prioritize and enhance its role in groupings such as BRICS "to help adapt the world order to the realities of a multipolar world" (Mitra 2023).

But not everyone favors the idea of a multipolar world. Personally, I think a multipolar world is much more resilient than a centralized one—and resiliency is very important to the survival of *Homo sapiens sapiens*. Nevertheless, hawks in the U.S. repeatedly warn that either separately or in coordination with one another, China and Russia continue to seek permanent geostrategic advantage by overmatching the military forces of the U.S. and its allies. These same hawks emphasize the role of hypersonic weapons as key to the geopolitical ambitions of both China and Russia. It is accurate to note that hypersonic weapons can reach very fast speeds and that these weapons have a high degree of maneuverability within Earth's atmosphere. Hypersonic weapons include: missiles; Fractional Orbital Bombardment Systems (FOBS), which traverse part of an orbit; and hypersonic glide vehicles (HGVs), which may have persistent orbital capacities. Because they fly more than five times the speed of sound, hypersonic weapons are extremely difficult to counter. HGVs can carry kinetic or conventional warheads, and they also have the capacity to carry nuclear payloads. For example, Russian HGVs such as the Avangard can carry both nuclear and non-nuclear payloads and can be launched by massive RS-28 ICBMs and, also, by other missiles. If in fact HGVs were armed with nuclear weapons, they could violate the Outer Space Treaty of 1967, to which China and Russia, along with the U.S. and many other countries, are signatories. Hypersonic weapons also can be armed with electromagnetic pulse devices to disable a country's electric grid, which would result in massive casualties from loss of food, medical care, heating, and other basic services (Pompeo 2022).

Although neither Chinese nor Russian orbital HGVs threaten the survivability of America's nuclear triad, they nevertheless are destabilizing because they can maneuver at speeds that make interception by current anti-ballistic missile systems highly uncertain. In addition, HGVs can loiter in space, orbiting Earth while posing for months or even years as satellites before they may be summoned to attack without warning. The stealth of HGVs while in orbit and their swiftness once committed could delay the ability of the U.S. to determine the nature and the magnitude of an attack, though once an HGV is committed to attack, its heat signature becomes observable to a constellation of infrared satellites because of its immense speed. Accordingly, the U.S. is developing the Hypersonic and Ballistic Tracking Space Sensor (HBTSS) system, which is slated to be part of the Next-Generation Overhead Persistent Infrared Polar (NG-OPIR) program. The U.S. Space Force (USSF) has been awarded contracts for this multi-billion-dollar program, the first satellites of which may be operational by 2025 (Pompeo 2022).

Ironically, China has made great strides in the design of hypersonic weapons by using dual-use technology transfers from the U.S. with both civilian and military applications. In other words, stolen U.S. software, machinery, and data are at the core of China's advance. Even more, often without their consent, U.S. investors have funded HGV development in China and Chinese military programs by working with a complex web of Chinese front companies, subsidiaries, and exchange-traded funds (ETFs). In many instances, U.S. investment capital finances the research, development, and procurement activities of banned Chinese companies, which are linked directly to China's military (Pompeo 2022).

Like China, Russia also has considerable capability relating to HGVs. Putin recently sent a warship, Admiral of the Fleet of the Soviet Union Gorshkov (Russian Адмира́л фло́та Сове́тского Сою́за Горшко́в, henceforth Gorshkov), armed with hypersonic missiles, to the Atlantic and Indian oceans and the Mediterranean Sea—sailing right past the U.K. in the process. Meanwhile, Russia's prized Perm nuclear-powered submarines armed with Zircons (Tsirkons) are scheduled to enter service with the Russian Navy in 2026. In October 2022, test launches of the Zircon missiles from the Project 885 Sverodvinsk submarine conducted in the White Sea demonstrate that Zircon missiles can reach a speed of around Mach 9. Russia claims that the missiles have a striking range capability that can exceed 1,000 km and can evade current and future defense systems. Translated from Russian, Putin stated:

> I am sure that such powerful weapons will make it possible to reliably protect Russia from potential external threats and will help ensure the national interests of our country. I would like to stress that we will continue to develop the fighting potential of the Armed Forces and produce advanced models of weapons and equipment that will protect Russia's security for future decades (Crux 2023).

Also translated from Russian, Russian Defense Minister Sergei Shoigu stated:

> This ship [i.e., the Gorshkov], armed with the Zircon missiles, is capable of delivering pinpoint and powerful strikes against the enemy at sea and on land. At the same time, a feature of the Zircon hypersonic missiles is the ability to reliably overcome any modern or future air defense systems. In exercises, there will be training for the crew on deploying hypersonic weapons and long-range cruise missiles (Crux 2023).

The Gorshkov officially was commissioned in 2018 and was assigned to the Russian Navy's Northern Fleet. Gorshkov's commanding officer, Captain Igor Krokhmal, stated that the frigate also is loaded with variants of Kalibr cruise missiles. These include anti-ship and land-attack types of missiles, together with Zircons and a 130 mm main gun. Gorshkov class frigates can fire all these weapons and, also, Oniks supersonic anti-ship cruise missiles. Putin's deployment of the Zircon-armed Gorshkov has raised fears that Russia could fire Zircons at Ukraine from the Mediterranean Sea, which means that any missiles fired would fly over one or more NATO members during such an operation (Crux 2023).

One can extrapolate from current competition relating to hypersonic weapons to imagine how major world powers would handle knowledge of advanced UAP propulsion. Given brutal, ongoing competition between the U.S., China, and Russia over hypersonic weapons, it would be naïve to expect that anything different would happen with existing UAP knowledge and with future UAP research and development (R&D). China most likely is trying to steal any knowledge the U.S. may have on UAP propulsion and/or materiel, and Russia most likely is doing what it can to understand UAP activity that may have occurred in the former USSR or more recently in Russia.

Given the geopolitical climate, it would contravene U.S. national security to rush to the UN for UAP discussions. Instead, the U.S. should calmly and quietly continue any classified UAP research it may be doing without bowing to strangely loud demands for 'transparency.' Too much 'transparency' relating to UAP propulsion and/or materiel is dangerous right now. Besides, this clamoring after the USG to release classified data on UAP 'because the people have a right to know' is disingenuous. Such logic is not applied to other classified systems on which the USG is working. Clearly, the real agenda behind loud demands for transparency relating to UAP technology is a self-interested, elite move to consolidate power and control by moving the information in a domain where it is more accessible than it is now. Furthermore, there is no imperative to 'globalize' UAP research 'against an alien threat,' since no such 'alien threat' exists. Even if it did, it would not be enough to unite everyone. Most countries already realize that an extraterrestrial presence has been in our midst for decades, if not for considerably longer, and this knowledge has not united the world yet. More importantly, human beings need to learn to participate constructively with ETI rather than trying to muster a ragtag human assembly 'against' it.

In the immediate term, however, human beings need to learn to get along with one another, across the entire world, and without polarized conflict—especially conflict that is intentionally manufactured to be polarized. Having two large blocs of nations, NATO and the BRICS, is not conducive to the longevity of our species, since arraying two, powerful groups against one another invites competition and conflict. Instead, we must end the push towards globalization at the level of economics, banking, and trade; and end the push towards a hegemonic, unipolar world order at the level of geopolitics. Humankind needs to transition to a decentralized, multipolar world that embraces the beauty of a breadth of human cultures and traditions and provides the undergirding on which a multipolar, decentralized economy can emerge. This is the path to real social justice, it increases societal resilience, and it maximizes opportunities for creativity among human beings. In contrast, monoculturalism and a centralized world economy stifle human creativity and promote authoritarianism. Overly centralized systems also lack resilience in the face of shocks, which is evident when interlocking supply chains face logistical challenges, for example. Until the world is in a different place, without battles for hegemony, ongoing conflict, and socially unjust concentrations of wealth, power, and control, permitting information on UAP propulsion and/or materiel to be widely circulated is dangerous. We must avoid any scenario in which this information falls into the wrong hands—i.e., the hands of people who want to harm others. But there is no effective way to reveal certain information to the American people while keeping it out of the hands of foreign adversaries. For example, such a dystopian scenario could unfold if an elitist, supranational, and private organization with little if any genuine concern for the billions of other human beings on Earth were to gain access to this information and were to utilize it to dominate others.

If we are truly concerned about the longevity of *Homo sapiens sapiens*, human beings must mature ethically and spiritually so that data sharing on UAP becomes safe and so that autocratic regimes do not weaponize information on UAP propulsion and/or materiel and attempt to use it to replace liberal, democratic values with totalitarian ones. If Russia and China were to move away from totalitarian governance and further the social justice imperative within their own countries, this would increase freedom and creativity and would move us all towards a world in which information on UAP propulsion and/or materiel could be shared more freely without risking democratic values. Meanwhile, Western nations such as the U.S. and others also must assess themselves and must stop pushing a neoconservative, hegemonic, and unipolar agenda. Such action also detracts from creating the kind of cooperation humankind needs to permit information on UAP propulsion and/or materiel to be circulated more widely at some point in the future.

Chapter 4:
The U.S. Government (USG) Wades In

Early FBI and CIA Interest in UAP

The Central Intelligence Agency (CIA) and Federal Bureau of Investigation (FBI) in the U.S. have maintained longstanding interest in UAP, as I discuss elsewhere (Andresen 2022b, 282-86). I add details to that discussion below while noting that even both discussions together provide only a partial view of the full extent of the CIA's and FBI's decades-long interest in UAP.

During the years 1945-1950, a formalized intelligence establishment emerged in the United States, in keeping with the wary perception of foreign relations at the time. On January 22, 1946, President Harry S. Truman wrote a letter establishing an Intelligence Advisory Board consisting of the heads, or their representatives, of the USG's principal military and civilian intelligence agencies with functions relating to the national security, to advise the Director of Central Intelligence (DCI). Although no specific provision existed in the National Security Act of 1947 to continue this Board, or to form a successor, Section 303 (a) of the Act authorized the DCI, among others, to appoint such an advisory committee if it were deemed necessary to the functionality of the DCI and the CIA. On September 19, 1947, the DCI, then Rear Admiral Roscoe H. Hillenkoetter, then appointed an Intelligence Advisory Committee (IAC), which was for all intents and purposes a successor to the Board (DOS n.d.).

The FBI's early interest in UFOs occurred during the tenure of J. Edgar Hoover, who was Director of the FBI for almost fifty years from May 10, 1924 to May 2, 1972. Early evidence of recovered technology comes in the form of a July 1945 exchange between Clyde Tolson, Associate Director of the FBI from 1930 to 1972 and the long-time protégé and deputy of Hoover, and Hoover himself. On July 15, 1947, Tolson wrote to Hoover, "I think we should do this. 7-15." Although it is not altogether clear what 'this' refers to, Hoover replied, "I would do it but before agreeing to it we must insist upon full access to discs recovered. For instance in the Sw./Sov. case the Army grabbed it & would not let us have it for cursory information" (Fox 2004, timestamp 1:26:11). Although it also is unclear what 'Sw.' and 'Sov.' may refer to, recalling that the Roswell event occurred sometime between mid-June and early-July 1947, 'Sw.' may be 'Southwest.' It also is plausible that 'Sov.' refers to 'Soviet.' If these assumptions are correct, then Hoover's *assumption* may have been that a disc-shaped aircraft manufactured by the Soviets came down in the desert near Roswell, New Mexico.

Whatever did occur in the summer of 1947 in the desert near Roswell, any assumption that all UAP were manufactured by humans became less and less likely as the years went on. Starting at least as early as 1948, the USG was officially interested in UFO reports, which is demonstrated by the USAF's initiation of three separate projects to study UFOs—Project Sign, Project Grudge, and Project Blue Book. Project Sign was established in 1948 and initially was called Project Saucer, perhaps suggesting that at least one saucer had been seen, detected, and/or recovered. Project Grudge was established in February 1949. The most well-known of the three initiatives, Project Blue Book, was established in March 1952 (see also Wright 2019, 113-14). As I discuss elsewhere (Andresen 2022b, 289-91, 311), in its investigation of UAP, the USAF often worked in conjunction with the USAF Office of Special Investigations (OSI), and, also, with the CIA. By the time Project Blue Book was terminated on December 17, 1969, Robert C. Seamans, Jr., Secretary of the USAF, announced that no U.S. military agency would continue reporting or receiving reports on UFOs. He also stated, "No UFO reported, investigated and evaluated by the Air Force was ever an indication of threat to our national security" (Marker 2019).

Interestingly, earliest known example of use of the term "Unidentified Aerial Phenomena" of which I am aware comes from a January 31, 1949 Office Memorandum of the United States Government from a Special Agent in Charge (SAC) at the FBI to the Director of the FBI, who as mentioned above was J. Edgar Hoover (Uda 2010). Returning to the topic of 'discovered' and/or 'recovered' craft, a famous USG document supports the interpretation that the U.S. does in fact have at least three extraterrestrial UAP in its possession. On March 22, 1950, Guy Hottel, from

SAC [Strategic Air Command] in Washington, D.C., wrote a memo to Hoover, who was Director of the FBI at the time. The memo states:

> An investigator for the Air Forces stated that three so-called flying saucers had been recovered in New Mexico. They were described as being circular in shape with raised centers, approximately 50 feet in diameter. … [T]he saucers were found in New Mexico due to the fact that the Government has a very high-powered radar set-up in that area and it is believed the radar interferes with the controlling mechanism of the saucers (FBI 1950).

Hottel had been named acting head of the FBI's Washington Field Office in 1936. At the time he wrote the memo, he was special agent in charge (FBI 2013).

During World War II, the USG relied heavily on technical intelligence. In July 1945, shortly before the war ended in September of that year, the Air Materiel Command T-2 Intelligence Department was founded at what then was called Wright Field. Later, in 1948, Wright Field and Patterson Field merged to become Wright-Patterson Air Force Base (WPAFB), located east of Dayton, Ohio. One of T-2's responsibilities was to identify foreign aircraft and equipment. On May 21, 1951, the Air Technical Intelligence Center (ATIC) was founded at WPAFB "as a field activity of the Assistant Chief of Staff for Intelligence." During the 1950s, ATIC analysts conducted computer analysis of aircraft in what was cutting-edge work for the time. Of note, ATIC later was renamed the Aerospace Technical Intelligence Center on September 21, 1959. In July 1961, this group morphed into the Foreign Technology Division (FTD) under the new Air Force Systems Command (NASIC n.d.). Of note, one should not confuse either Air Technical Intelligence Center or Aerospace Technical Intelligence Center with the Advanced Technical Intelligence Center (also ATIC), a nonprofit corporation established in 2006 currently operating at a facility located near Wright Patterson AFB.

After the events described above in chapter 3 during which UFOs were witnessed in July 1952 by hundreds of people in Washington, D.C., the CIA became more involved in monitoring UAP than it had been previously. To investigate the topic of UFOs, the CIA formed a special study group within the Office of Scientific Intelligence (OSI) and the Office of Current Intelligence (OCI), both which were part of the Directorate of Intelligence (DI). Established on December 31, 1948, OSI served as CIA's focal point for the analysis of foreign scientific and technological developments. Established on January 15, 1951, OCI's primary function was to provide current, all-source intelligence to the President and the National Security Council (NSC). Later, in 1980, OSI was merged into the Office of Science and Weapons Research. A July 29, 1952 memorandum documents the formation of the special study group on UFOs. The memo was sent from Ralph L. Clark, Acting Assistant Director, OSI, to Deputy Director for Intelligence (DDI) Robert Amory, Jr. On January 2, 1952, DCI Walter Bedell Smith created the Directorate for Intelligence (DDI) composed of six overt CIA organizations—Office of Scientific Intelligence (OSI), Office of Current Intelligence (OCI), Office of Collection and Dissemination (OCD), Office of National Estimates (ONE), Office of Research and Reports, and Office of Intelligence Coordination. The DI's function is to produce intelligence analysis for U.S. policymakers (G. Haines 1997 <1995>, 68, inc. fns. 14 and 16 on 80). Also of note, OSI may have conducted some cover-up and counterintelligence efforts relating to UAP, which is plausible given the ethos of that period.

Several documents contain proof that the CIA has taken an active role in monitoring the UFO phenomenon. For example, the July 29, 1952 memorandum mentioned above—which was "classified Eyes Only"—states the following:

> In the past several weeks a number of radar and visual sightings of unidentified aerial objects have been reported. Although this office has maintained a continuing review of such reported sightings *during the past three years [i.e., since 1949]*, a special study group has been formed to review this subject to date. O/CI [Office of Central Intelligence] will participate in this study with O/SI [Office of Scientific Intelligence] and a report should be ready about 15 August (Good 1988 <1987>, 330, italics added).

A few days later, on August 1, 1952, Edward Tauss, then acting Chief of OSI's Weapons and Equipment Division, wrote an informal memo to Philip Strong, Deputy Assistant Director, SI (Special Intelligence). The memo recommends that the CIA continue to monitor the UFO situation in coordination with ATIC (Air Technical Intelligence Center), which is discussed in more detail later in this chapter. In this memo, Tauss reported on behalf of the group at the CIA reviewing information on UFOs that most sightings could be explained easily (G. Haines 1997 <1995>, 68, inc. fns. 15 and 16 on 80). However, Tauss also writes:

> [S]o long as a series of reports remains "unexplainable" (interplanetary aspects and alien origin not being thoroughly excluded from consideration) caution requires that intelligence continue coverage of the subject. ... It is recommended that CIA surveillance of subject matter, in coordination with proper authorities of primarily operational concern at ATIC [Air Technical Intelligence Center], be continued. *It is strongly urged, however, that no indication of CIA interest or concern reach the press or public,* in view of their probable alarmist tendencies to accept such interest as "confirmatory" of the soundness of "unpublished facts" in the hands of the U.S. government (Good 1988 <1987>, 331, italics added by Good).

In the memo, Tauss clearly recommends that the CIA continue to monitor the UFO situation in coordination with ATIC. Upon receiving the report from the special study group that had been created within the OSI and the OCI to review the situation reported by Tauss, Robert Amory, Jr., Deputy Director for Intelligence (DDI), assigned responsibility for UFO investigations to OSI's Physics and Electronics Division, with A. Ray Gordon as the officer in charge (G. Haines 1997 <1995>, 68, inc. fns. 15 and 16 on 80).

According to the Minutes of the Branch Chief's Meeting from August 11, 1952, the intention was that each branch in the Physics and Electronics Division contribute to the investigation, and that Gordon coordinate closely with ATIC. Amory relayed DCI Walter Bedell Smith's directive that the group focus on the national security implications of UFOs. As an aside, some people contend that on August 1, 1950, Smith replaced James Forrestal as a member of MJ-12 [Majestic 12]. Forrestal, who died under exceedingly suspicious circumstances, was the last Cabinet-level U.S. Secretary of the Navy and the first U.S. Secretary of Defense. It also has been stated that the CIA was, and perhaps still is, a member of an Incident Response Team "to investigate UFO landings, if one should occur." In 1997, it was claimed that this team had never met, though it also was admitted that apparent lack of documentation at the CIA on its UFO-related activities in the 1980s leaves the situation "somewhat murky" (G. Haines 1997 <1995>, 68, inc. fns. 17 on 80, 90 on 83, and 93 on 84).

According to an unsigned CIA memorandum dated August 11, 1952 and titled, "Flying Saucers," the CIA study group met with Air Force officials at WPAFB and reviewed their data and findings. Claiming that ninety percent of the reported sightings were easily accounted for, the USAF characterized the other ten percent as "a number of incredible reports from credible observers." The USAF rejected theories that the sightings involved U.S. or Soviet secret weapons development or that they involved "men from Mars," claiming that no evidence supported these ideas. Instead, USAF briefers sought to explain UFO reports as the misinterpretation of known objects or as little understood natural phenomena (G. Haines 1997 <1995>, 68, 71, inc. fn. 19 on 80).

In August 1952, the CIA established a special study group to study UFOs. According to an August 14, 1952 unsigned memo with the same title—"Flying Saucers"—as the one mentioned above, USAF and CIA officials agreed that since outside knowledge of the CIA's interest in UFOs would cause people to take the issue more seriously, the topic should be damped down. This concealment of CIA's interest contributed to later charges of a CIA conspiracy and cover-up regarding UFOs (G. Haines 1997 <1995>, 71, inc. fn. 20 on 80). Classified secret at the time, the memo describes a briefing regarding its own internal survey regarding the USAF's study on UFOs. The August 14, 1952 memo reads:

> During the past weeks, with the phenomenal increase in the number of Flying Saucer reports there has been a tremendous stimulation of

both public and official interest in the subject. Requests for information have poured in on the Air Force, including an official query from the White House …

At this point, OSI felt that it would be timely to make an evaluation of the Air Force study, its methodology and coverage, the relation of its conclusions to various theories which have been propounded, and to try to reach some conclusion as to the intelligence implications of the problem—if any. In view of the wide interest within the [Central Intelligence] Agency, this briefing has been arranged so that we could report on the survey. It must be mentioned that *outside knowledge of Agency interest in Flying Saucers carries the risk of making the problem even more serious in the public mind than it already is, which we and the Air Force agree must be avoided* (Good 1988 <1987>, 331, emphasis added by Good).

The memo continues, "we have reviewed our own intelligence, going back to the Swedish sightings of 1946" (a reference to the ghost rockets mentioned above in chapter 3), and it lists the various types of UFOs reported to the USAF:

Grouped broadly as visual, radar, and combined visual and radar, ATIC [Air Technical Intelligence Center] has two major visual classes—first, spherical or elliptical objects, usually of bright metallic luster, some small (2 or 3 feet across), most estimated at 100 foot diameter and a few 1000 feet wide. There are variants in this group, such as torpedos, triangulars, pencils, even mattress-shapes. These are all daylight reportings.

The second visual group, all night reporting, consists of lights and various luminosities, such as green, flaming-red or blue-white fire balls, moving points of light, and luminous streamers.

Both categories are reported as single objects, in non-symmetrical groups and in formations of various numbers.

Reported characteristics include three general levels of speed: hovering; moderate, as with a conventional aircraft; and stupendous, up to 18,000 miles per hour in the White Sands Incident. Violent maneuvering was reported in somewhat less than 10%. Accelerations have been given as high as 20 g's [*sic*]. With few exceptions, there has been a complete absence of sound or vapor trail. Evasion upon approach is common (Good 1988 <1987>, 331-32).

One clearly can conclude from this document that even as early as 1952, the USAF studied UFOs in a detailed manner.

According to yet another unsigned memo, also titled "Flying Saucers" but dated August 19, 1952, the CIA study group also considered "the USSR's possible use of UFOs as a psychological warfare tool." The group searched the Soviet press for UFO reports. Finding none, the group concluded that the absence of reports must have been the result of deliberate, Soviet government policy (G. Haines 1997 <1995>, 71, inc. fn. 21 on 80).

Early in its history investigating UFOs, the USG recognized the psychological warfare 'value' of the topic (Huyghe 1979). DCI Smith "wanted to know what use could be made of the UFO phenomenon in connection with US psychological warfare efforts." Smith expressed his opinions at a meeting in the DCI Conference Room attended by his top officers, as documented in an August 20, 1952 memorandum, "Flying Saucers," Deputy Chief, Requirements Staff, Foreign Intelligence (FI), memorandum for Deputy Director, Plans, Directorate of Operations Records (Information Management Staff, Job 86-00538R, Box 1). Smith wanted to know whether the USAF investigation of flying saucers was sufficiently objective and how much more money and work effort were necessary to determine the cause of what he thought was a small percentage of unexplained UAP. Smith believed "there was only one chance in 10,000 that the phenomenon posed a threat to the security of the country, but even that chance could not be taken." Smith also thought it was CIA's responsibility by statute to coordinate the intelligence effort required to 'solve' the UFO issue (G. Haines 1997 <1995>, 68, inc. fn. 18 on 80).

The CIA study group was concerned about the national security implications of the UFO topic with respect to the former Soviet Union, particularly given the USSR's increased capabilities during the Cold War. According to two memoranda from H. Marshall Chadwell, Assistant Director of OSI, for Smith, dated September 17 and September 24, 1952, both titled "Flying Saucers," and, also, Chadwell's memorandum for Smith dated October 2, 1952, Chadwell considered the issue of UFOs so important "that it should be brought to the attention of the National Security Council, in order that a communitywide [sic] coordinated effort towards it [sic] solution may be initiated." The CIA study group was concerned that the Soviets could use UFO reports to touch off mass hysteria and panic in the U.S. The study group also was concerned that if the Soviets deliberately overloaded the U.S. air warning system such that it could not distinguish 'real' targets from 'phantom' UFOs, the Soviets might gain a surprise advantage in a potential nuclear attack (G. Haines 1997 <1995>, 71, inc. fn. 22 on 80 citing to two Chadwell memorandum for Smith, as per above; to Chadwell, memorandum for DCI Smith, October 2, 1952; and to Klass 1983, 23-26).

In November 1952, Chadwell also urged Smith to establish an external research project of top-level scientists to study UFOs, as documented in Chadwell's memorandum of November 25, 1952 and, also, in Chadwell's memorandum, n.d., "Approval in Principle - External Research Project Concerned with Unidentified Flying Objects" (see also Philip G. Strong, OSI, memorandum for the record, "Meeting with Dr. Julius A. Stratton, Executive Vice President and Provost, MIT and Dr. Max Millikan, Director of CENIS" [Center for International Studies]). Strong believed that in order to undertake such a review, an external research project would require the full backing and support of DCI Smith. After this briefing, Smith directed DDI Amory to prepare a NSC Intelligence Directive (NSCID) for submission to the NSC on the need to continue the investigation of UFOs and to coordinate such investigations with the USAF, as documented by Chadwell's December 2, 1952, memorandum for DCI, "Unidentified Flying Objects" (see also Chadwell memorandum "Approval in Principle," mentioned above) (G. Haines 1997 <1995>, 71, inc. fns. 24 and 25 on 80).

Note, the MIT Center for International Studies—then CENIS and now CIS—was founded in 1951 in response to Cold War conflict between the U.S. and the former USSR (MIT CIS n.d.).

Chadwell's December 2, 1952 memorandum for DCI, with attachments, documents some of the substantive aspects of Chadwell's December 1952 briefing to DCI Smith on UFOs. Chadwell urged action because he was convinced that "something was going on that must have immediate attention" and that "sightings of unexplained objects at great altitudes and traveling at high speeds in the vicinity of major US defense installations are of such nature that they are not attributable to natural phenomena or known types of aerial vehicles." Chadwell drafted a memorandum from the DCI to the NSC and a proposed NSC Directive to establish the investigation of UFOs as a priority project throughout the intelligence and the defense research and development communities (G. Haines 1997 <1995>, 71, inc. fn. 23 on 80 citing to Chadwell, memorandum for DCI with attachments, December 2, 1952; to Klass 1983, 26-27; and to Chadwell, memorandum of November 25, 1952).

On December 4, 1952, DCI Walter Bedell Smith recommended to the IAC that it investigate UFOs. H. Marshall Chadwell, Assistant Director of the USAF Office of Special Investigations (OSI), had reviewed the UFO topic and ATIC's active program relating to UFOs. The IAC agreed that the DCI should instruct selected scientists to review and appraise the available evidence on UFOs in the context of relevant scientific theories. Major General John A. Samford, Director of Air Force Intelligence, offered full cooperation. This catalyzed the formation of the Scientific Advisory Panel on UFOs. The group informally was known as the Robertson Panel because it was headed by mathematician and physicist Howard P. Robertson (G. Haines 1997 <1995>, 71-72, inc. fns. 26-30 on 80).

The Robertson Panel held meetings and on January 17, 1953 issued a brief, two-page report entitled, REPORT OF THE SCIENTIFIC PANEL ON UNIDENTIFIED FLYING OBJECTS (The Black Vault 2001, 28-29). This document often is referred to as "the Robertson Panel Report." A second, more detailed document often referred to as "the Durant Report," was released on February 16, 1953. Its formal title is *Report of Meetings of*

Scientific Advisory Panel on Unidentified Flying Objects Convened by Office of Scientific Intelligence, CIA, January 14-18, 1953 (The Black Vault 2001). The final two pages of this document are the Robertson Panel Report. The Durant Report is more detailed than the Robertson Panel Report inasmuch as the former also includes information on the Panel's history and proceedings. The Durant Report was written by Frederick C. Durant III, an expert in rocketry and spaceflight and a past President of the American Rocket Society. Durant, who was a CIA officer working with OSI, served as Secretary to the Robertson Panel, attending its meetings and writing the summary of the proceedings mentioned above (G. Haines 1997 <1995>, 72).

The Robertson Panel submitted its report to the IAC, the Secretary of Defense, the Director of the Federal Civil Defense Administration, and the Chairman of the National Security Resources Board. Unfortunately, the Panel recommended that a public education campaign should be undertaken to reduce public interest in UFOs. Panel members were concerned that potential enemies contemplating an attack on the U.S. might exploit the UFO phenomena by issuing false UFO reports to disrupt U.S. air defenses, and that the U.S. air defense system could be overwhelmed by such false UFO reports. Following the Panel's meetings, CIA officials stated publicly that no further consideration of the subject appeared warranted, although they continued to monitor sightings in the interest of national security. Philip Strong and Fred Durant from OSI also briefed the Office of National Estimates on the findings. Following the Robertson Panel findings, the CIA stopped its effort to draft a NSCID on UFOs (Haines 1997 <1995>, 72, inc. fns. 34-35 on 80-81; see also The Black Vault 2001, 29), which Smith had directed Amory to do, as mentioned above (G. Haines 1997 <1995>, 71, inc. fn. 25 on 80).

USG consideration of the broader geopolitical considerations of the UFO topic continued at a high level throughout the 1960s, especially as regards the former Soviet Union. Mr. Allen Dulles, who was Director of Central Intelligence (DCI) from February 26, 1953 to November 29, 1961 (CNN 2021), is reported to have told former President John F. Kennedy, who served from January 20, 1961 to November 22, 1963, that the CIA had been sending decoy devices into Soviet airspace to test Soviet radar capabilities. During these missions, the CIA became aware that the Soviets could not distinguish aircraft and ballistic missiles from UFOs. Dulles was concerned that the Soviets might mistake an unidentified flying object for a U.S. missile, and, on the bais of such an error, that the former Soviet Union potentially could initiate a nuclear strike against the U.S. (UFOs 2019).

On November 11, 1963, the former Soviet Union launched an unmanned spacecraft as part of its Kosmos 21 mission, but the spacecraft did not make it out of Earth's orbit. On November 12, 1963, the day after this unsuccessful rocket launch, Kennedy took two significant actions. He issued a National Security Action Memorandum instructing NASA to work with the Soviet space program, and he also issued a memo to the CIA instructing it to share secret documents and other information on UFOs with the Soviets. Ten days later, on November 22, 1963, Kennedy was assassinated, and his instruction to the CIA to share UFO information with the Soviets was not carried out (UFOs 2019). To be clear, I am not insinuating a theory for the assassination of former President John F. Kennedy—and ten days seems like a very short amount of time to plan such an act. The reality is that at the time he was assassinated, Kennedy had many powerful enemies both inside and outside the U.S. That being said, the abovementioned dates are rather remarkable.

Brennan and Woolsey on UAP

Today, many people within the USG have begun to discuss the ETI/UAP in a relatively open manner, as I discuss elsewhere (Andresen 2022b, 284-85).

Both Ambassador R. James Woolsey, the Director of Central Intelligence (DCI) from February 5, 1993 to January 10, 1995, and Mr. John O. Brennan, the Director of the CIA from March 8, 2013 to January 20, 2017 (CNN 2021), have indicated that an ETI may be responsible for some UAP operating on and around Earth. Woolsey's and Brennan's views on UAP are quite pertinent, especially given the long history of their knowledge of the topic. In late 1993, while he was DCI, Woolsey ordered a review of all files on UFOs, and Brennan participated in this review. On September 30, 1993,

Brennan wrote a memorandum entitled "Requested Information on UFOs" (Brennan 1993), which he gave to Richard Warshaw, Executive Assistant to Woolsey at the time. This information comes from interviews with an OSWR (Office of Scientific Weapons Research) analyst on June 14, 1994 and an OSI (Office of Scientific Intelligence) analyst on July 21, 1994 (G. Haines 1997 <1995>, fn. 90 on 83).

When he wrote the September 30, 1993 memo, "Requested Information on UFOs," Brennan either was an analyst or an overseas case officer (Jackson 2009)—presumably the former. Some years later, Warshaw himself became a member of the External Referral Working Group (ERWG) established by the intelligence community (IC) to confer and discuss former President Clinton's April 17, 1995 Executive Order 12958, Section 3.4 of which required that all permanent records twenty-five years or older be declassified by April 17, 2000 (FAS n.d.). Also of note, in his article, "CIA's Role in the Study of UFOs, 1947-90," Dr. Gerald K. Haines states, "Except where noted, all citations to CIA records in this article are to the records collected for the 1994 Agency-wide search that are held by the Executive Assistant to the DCI" (G. Haines 1997 <1995>, fn. 3 on 79). In other words, Brennan's effort to collate material on UFOs for the CIA was significant—perhaps even more so given that Haines reports, accurately or not, that Haines himself found almost no documentation regarding the CIA's involvement in the 1980s relating to UFOs (Haines 1997 <1995>, fn. 90 on 83). Also of note, when wrote his article on the CIA's role in studying UFOs, Haines was Chief Historian of the National Reconnaissance Office (NRO), an agency within the U.S. Department of Defense (DOD), a post he held from 1994 to 1998. Prior to that, Haines was a historian at the National Security Agency (NSA) from 1986 to 1992. At a time subsequent to writing his article on UFOs, which was first published in a classified format in 1995, Haines became Chief Historian of the CIA, a post he held from 1998 to 2002.

Brennan discusses UAP in two separate interviews (Brennan 2020; 2021). During both interviews, Brennan makes it clear that he is open what is referred to as the extraterrestrial hypothesis (ETH) as a way of explaining the origin of certain UAP (Brennan 2021). In his December 16, 2020 interview, Brennan makes a forthright comment regarding information on UAP that is in the possession of the USG. He states:

> I think some of the phenomenon we are seeing continues to be unexplained and might in fact be some type of phenomenon that is the result of something that we don't yet understand and that could involve some type of activity that some might say constitutes a different form of life (Brennan 2020; see also Lewis-Kraus 2021).

Brennan's statement provides a clear indication that he is open to the ETH as an explanation for some UAP.

Brennan's second interview during which he discusses UAP (Brennan 2021) occurred on June 24, 2021, the day before the Office of the Director of National Intelligence (ODNI) released its *Preliminary Assessment: Unidentified Aerial Phenomena* on June 25, 2021 (ODNI 2021). Retired U.S. Army (USA) General Wesley K. Clark, who served as Supreme Allied Commander Europe of the North Atlantic Treaty Organization (NATO) from 1997 to 2000, conducted this interview. When Clark asked Brennan about UAP, Brennan responded in a manner suggesting openness to the ETH and, also, indicating that his views on the matter are based on photographic, video, and radar evidence to which he has had access.

Woolsey's comments on UAP also seems to indicate that he also has seen photographic, video, and radar evidence of extraterrestrial UAP. Woolsey's opinion is informed further by the personal testimony of a close, trusted associate whose airplane experienced unusual effects in conjunction with an unidentified anomalous object. Woolsey's comments are so noteworthy that I quote them at length here:

> There have been, over the years now, events of one kind or another, usually involving some kind of aircraft-like airframe [i.e., UAP], that has put a number of people into a situation where they'd say—like me—I never thought there was anything to all this, it always seemed pretty far out to me—don't know what the reality is—but—and fill in the 'but.'

There was one case in which a friend of mine was able to have his aircraft stop at 40,000 feet or so and not continue operating as a normal aircraft. What was going on? I don't know. Does anybody know? We'll have to look into it. There have just been enough things *like* that that have occurred that I think there will be a lot of examination of what's going on over the course of the several months, or, maybe years, but maybe months …

I have a friend who had been in an aircraft—40,000 [feet] is not crazy, I mean, it's just—other pilots looking at this know that this is not a wild and crazy idea—but I talked to someone whom I respect who says that there was some event at which an aircraft was paused, and, basically, that's all I know about that. …

These people have reported very curious behavior by aircraft, and it may be something real that is an extraordinary change for some unheralded reason, or it may be a complex set of different views of what's going on in the world of cyber and so forth. I just don't know. But I'm not as skeptical as I was a few years ago, to put it mildly. Something is going on that's surprising to a series of intelligent, aircraft-experienced pilots, and we'll just have to see what it is. …

I have been—not in the presence of—but I've been in conversations with several individuals who have been close to an aircraft performing in an extraordinary fashion, that performs in a fashion that has not yet come to be something that people are comfortable with or expecting to see.

That's about as far as I'd like to go, which is openness to new things, willingness to examine them, hope that we can be friendly and able to deal with a wide range of behaviors in terms of dealing with our fellow human beings or with other creatures if they exist.

And I think we ought to be open to new possibilities—some new possibilities are frauds—I don't think this one is. But over history there have been a lot of frauds, and we could have another one—I don't think so. ...

[I]f we get close to it, we ought to get up to it with a sense of both friendliness and skepticism, and we shouldn't neglect either one. We should not get into a mode in which we say only idiots could come up with a crazy damned idea like this, but we also shouldn't get into the mode of saying, 'Hey this is cool, it's really exciting—yeah, let's go.' No. I mean, 'No' on both of those counts.

Let's let the scientists and engineers examine, assess, play it straight, and see if there's some—anything—out there that we all would want to know about. ... I think we ought to be cautious but open, in a limited fashion, open to understanding what's taking place (Woolsey 2021, paragraph breaks added for readability).

When I transcribed Ambassador Woolsey's remarks, it became evident the extent to which he was reaching to describe something that was all but indescribable from within the context of his prior worldview. Here, I sincerely want to commend Mr. Woolsey for his genuineness and willingness to engage the phenomenon. I also appreciate his aspiration "that we can be friendly and able to deal with a wide range of behaviors in terms of dealing with our fellow human beings or with other creatures if they exist" (Woolsey 2021). Kindness, and openness are important traits to embody in all contexts, including in the context of ETI/UAP.

Recent DOD and Congressional Activities Relating to UAP

DOD recently acknowledged an official program that investigated UAP, namely the Advanced Aerospace Weapon System Applications Program, also referred to as the AAWSA Program, or AAWSAP (NYP 2022).

Attachment 1 of AAWSAP documentation from July 18, 2008 articulates how the Defense Intelligence Agency (DIA), which is part of DOD, was involved in AAWSAP:

The Acquisition Support Division (DWO-3) of the Defense Intelligence Agency (DIA) has the responsibility to provide guidance and oversight to the Department of Defense (DoD) acquisition process along with leveraging the DoD Intelligence Community to coordinate, produce and maintain projections of the future threat environment in which U.S. air, naval, ground, space, missile defense and information systems operate. In order to accurately assess the foreign threat to U.S. weapon systems, a complete as possible understanding of potential breakthrough technology applications employed in future aerospace weapon systems must be obtained (NYP 2022).

Attachment 1 also states the Program's objectives:

One aspect of the future threat environment involves advanced aerospace weapon system applications. The objective of this program is to understand the physics and engineering of these applications as they apply to the foreign threat out to the far term, i.e., from now through the year 2050. Primary focus is on breakthrough technologies and applications that create discontinuities in currently evolving technology trends. The focus is not on extrapolations of current aerospace technology. The proposal shall describe a technical approach which discusses how the breakthrough technologies and applications listed below would be studied and [shall] include proposed key personnel that have experience in those areas (NYP 2022).

Another program at DOD, the Advanced Aerospace Threat Identification Program (AATIP), also collated information on UAP. Although the topic has been discussed in multiple venues, the precise relationship between AAWSAP and AATIP remains unclear. Many media reports state that AATIP received $22 million to fund its work, though this money formally may have been allocated to AAWSAP. This $22 million in funding apparently came from a black budget hidden in plain sight in an unclassified Department of the Navy document (BFD 2020c).

In addition to formal DOD programs, USG research on UAP often occurs within private companies linked to the military and IC. On contract to the USG, such companies are found in the U.S. and abroad. For example, AAWSAP awarded its prime contract to Bigelow Aerospace, located in North Las Vegas, Nevada, which in turn subcontracted some of the work to other individuals and entities. In past decades, and in some cases extending to the present, other companies also may have been involved in UAP research. Historically, these companies included Braddock Dunn & McDonald (BDM) and Thompson Ramo Wooldridge (TRW). Today, companies such as Northrop Grumman, Lockheed Martin, Boeing, and others may be involved in UAP research, with the intent that the USG acquire breakthrough technologies ahead of other countries.

Although AAWSAP and AATIP are no longer operating as formal programs, USG investigations of UAP continue today. UAP often are seen and detected by radar and other instruments operated by the U.S. military in conjunction with its operations across the globe. In accordance with numerous actions by Congress over the past few years, members of the U.S. military and IC are being called upon to report what they know about UAP.

Some relevant USG activities relating to UAP are described below, along with other noteworthy events that have occurred during the four-year time frame from April 2019 to April 2023.

April 24, 2019, Spokesperson for DOD's Office of the Deputy Chief of Naval Operations for Information Warfare confirms that U.S. Navy is drafting UAP reporting procedures. Joseph Gradisher, Captain, U.S. Navy (ret.), who at the time was spokesperson for DOD's Office of the Deputy Chief of Naval Operations for Information Warfare, stated that UAP enter airspace designated for the military as often as multiple times per month. He also stated that the recent increase of sightings "prompted the Navy to draft formal procedures for pilots to document encounters" (Paul 2019).

June 17, 2020, Senate Select Committee on Intelligence submits the Intelligence Authorization Act (IAA) for Fiscal Year 2021 (FY21). The Act, which reveals that the Office of Naval Intelligence (ONI) monitors UAP activity, includes a provision for the preparation of a public report to provide a "detailed analysis of unidentified aerial phenomena data and intelligence reporting collected…by" the Office of Naval Intelligence, geospatial intelligence, signals intelligence, human intelligence, measurement and signals intelligence, and the FBI (116th United States Congress 2020; Duncan 2020).

The final version of the Act was folded into the National Defense Authorization Act for FY 2021 (NDAA FY21), a $2.3 trillion spending bill that essentially was guaranteed to pass. When it did, DOD had a 180-day countdown by which it and designated entities were required to produce a public report on what they knew about UAP (Cowen 2020). The NDAA FY21 required that the report be unclassified, though it did permit DoD and the IC to include a classified annex when they presented the report to the Senate intelligence and armed services committees (Duncan 2020).

August 4, 2020, DoD establishes the Unidentified Aerial Phenomena Task Force (UAPTF). A DoD press release from August 14, 2020 states:

> On August 4, 2020, Deputy Secretary of Defense David L. Norquist approved the establishment of an Unidentified Aerial Phenomena (UAP) Task Force (UAPTF). The Department of the Navy, under the cognizance of the Office of the Under Secretary of Defense for Intelligence and Security (OUSD(I&S)), will lead the UAPTF (U.S. DOD 2020a).

May 6, 2021, the National Reconnaissance Office (NRO) releases a PowerPoint presentation, "Recent Sentient Highlights." The document discusses a highly classified NRO system called 'Sentient' that detected a possible Tic Tac shaped UAP (The Black Vault Originals 2023b; for more on Sentient, see Scoles 2019).

June 25, 2021, Office of the Director of National Intelligence (ODNI) issues its *Preliminary Assessment: Unidentified Aerial Phenomena*. ODNI issued this document in response to s requirement articulated in Senate Report 116-233, which accompanied the IAA for FY 2021. The requirement was that the Director of National Intelligence (DNI), in consultation with the Secretary of Defense (SecDef), submit an intelligence assessment regarding UAP and progress made by the UAPTF in understanding UAP (ODNI 2021, 2).

Data in the *Preliminary Assessment* is limited to USG reporting of incidents that occurred from November 2004 to March 2021. The document was prepared for the Congressional Intelligence and Armed Services Committees and was drafted by the UAPTF and the ODNI National Intelligence Manager (NIM) for Aviation (NIM-Aviation), with input from:

- Under Secretary of Defense for Intelligence and Security (USD(I&S));
- Defense Intelligence Agency (DIA);
- Federal Bureau of Investigation (FBI);
- National Reconnaissance Office (NRO);
- National Geospatial-Intelligence Agency (NGA);
- National Security Agency (NSA);
- Air Force, Army, Navy, Navy/Office of Naval Intelligence (ONI);
- Defense Advanced Research Projects Agency (DARPA);
- Federal Aviation Administration (FAA);
- National Oceanographic and Atmospheric Administration (NOAA);
- National Geospatial-Intelligence Agency (NGA);
- ODNI/NIM-Emerging and Disruptive Technology;
- ODNI/National Counterintelligence and Security Center; and
- ODNI/National Intelligence Council (ODNI 2021, 2).

Of the 144 UAP reports discussed in the *Preliminary Assessment* that originated from USG sources, the document only identifies one reported UAP with high confidence and leaves 143 unexplained (ODNI 2021, 4). Interestingly, 18 incidents described in 21 reports involved observations of unusual UAP movement patterns or flight characteristics. The *Preliminary Assessment* states, "Some UAP appeared to remain stationary in winds aloft, move against the wind, maneuver abruptly, or move at considerable speed, without discernable means of propulsion. In a small number of cases, military aircraft systems processed radio frequency (RF) energy associated with UAP sightings" (ODNI 2021, 5).

After ODNI released its *Preliminary Assessment*, the U.S. Senate Select Committee on Intelligence (SSCI) requested representatives from NORAD, U.S. Northern Command (USNORTHCOM), and U.S. intelligence community (IC) entities to discuss UAP in a classified setting.

May 18, 2021, former U.S. President Barack H. Obama II makes televised a remark regarding UAP. Obama stated, "there is footage and records of objects in the skies that we don't know exactly what they are. We can't explain how they moved, their trajectory. They did not have an easily explaininable pattern" (Obama 2021).

October 19, 2021, former U.S. senator and the current Administrator of NASA, Clarence William ("Bill") Nelson II makes a remark regarding UAP during video interview. Nelson stated that U.S. Navy (USN) pilots have reported more than three hundred UAP sightings since 2004 (Nelson 2021).

November 11, 2021, the current DNI, Ms. Avril D. Haines, states that UAP may have "come extraterrestrially" to Earth (A. Haines 2021).

November 23, 2021, DoD establishes the Airborne Object Identification and Management Synchronization Group (AOIMSG). To provide oversight of AOIMSG, USD(I&S) was directed to lead an Airborne Object Identification and Management Executive Council (AOIMEXEC). The UAPTF immediately was transitioned to AOIMSG as its successor, with the UAPTF ceasing to operate (Hicks 2021; U.S. DOD 2021).

December 27, 2021, the NDAA (National Defense Authorization Act) for Fiscal Year 2022 (FY22) is signed into law. The NDAA FY22 is codified as Public Law 117-81, 135 Stat. of the 117th Congress (NDAA FY22). Discussion of UAP specifically occurs in Title XVI, SPACE ACTIVITIES, STRATEGIC PROGRAMS, AND INTELLIGENCE MATTERS, Subtitle E—Other Matters, Section 1683, ESTABLISHMENT OF OFFICE, ORGANIZATIONAL STRUCTURE, AND AUTHORITIES TO ADDRESS UNIDENTIFIED AERIAL PHENOMENA (NDAA FY22, 531-33, 578-83).

Section 1683(b)(1) designates the duties of the Office established under (a). A partial list of these includes: (1) Developing procedures to synchronize and standardize the collection, reporting, and analysis of UAP incidents across DOD and the IC; and (6) Coordinating with other departments and agencies of the Federal Government as appropriate, to include the Federal Aviation Administration (FAA), the National Aeronautics and Space Administration (NASA), the Department of Homeland Security (DHS), the National Oceanic and Atmospheric Administration (NOAA), and the Department of Energy (DOE).

Section 1683(c) discusses response to and field investigations of UAP. Here, (1) states that the Secretary of Defense (SecDef), in coordination with the Director of National Intelligence (DNI), shall designate one or more line organizations within DOD and the IC with appropriate expertise, authorities, accesses, data, systems, platforms, and capabilities to respond rapidly to and conduct field investigations of incidents involving UAP.

Section 1683(d) discusses scientific, technological, and operational analyses of data on UAP, stating that the SecDef, in coordination with the DNI, shall designate one or more line organizations with primary responsibility for scientific, technical, and operational analysis of data gathered by field investigations conducted under (c) and data from other sources, including with respect to the testing of materials, medical studies, and development of theoretical models to better understand and explain UAP.

Section 1683(e)(2) discusses an intelligence collection and analysis plan, stating that the office established under (a), which we now know as the All-domain Anomaly Resolution Office (AARO) (but which originally was AOIMSG), acting on behalf of the SecDef and the DNI, shall supervise the development and execution of an intelligence collection and analysis plan to gain as much knowledge as possible regarding the technical and operational characteristics, origins, and intentions of UAP, including with respect to the development, acquisition, deployment, and operation of technical collection capabilities necessary to detect, identify, and scientifically characterize UAP.

Section 1683(f) outlines a science plan, saying that the head of the Office, which we now know as AARO, on behalf of the SecDef and the DNI, shall supervise the development and execution of a science plan to develop and test, as practicable, scientific theories to:

(f)(1) account for characteristics and performance of UAP that exceed the known state of the art in science or technology, including in the areas of propulsion, aerodynamic control, signatures, structures, materials, sensors, countermeasures, weapons, electronics, and power generation; and

(f)(2) provide the foundation for potential future investments to replicate any such advanced characteristics and performance.

Section 1683(h)(1) requires that not later than October 31, 2022, and annually thereafter until October 31, 2026, the DNI, in consultation with the SecDef, shall submit a report on UAP to the appropriate congressional committees These committees are defined later in Section 1683(l)(1)(A)-(D) as the Committees on Armed Services of the House and Senate; the Committees on Appropriations of the House and Senate; the Committee on Foreign Affairs of the House and the Committee on Foreign Relations of the Senate; and the Permanent Select Committee on Intelligence of the House and the Select Committee on Intelligence of the Senate.

Section 1683(h)(2) lists the elements required in the report, which include, per (A)-(C), numbers of UAP-related events, and per (D), an analysis of data relating to UAP collected by means of: (i) geospatial intelligence (GEOINT); (ii) signals intelligence (SIGINT) [traditionally the domain of NSA]; (iii) human intelligence (HUMINT) [traditionally the domain of CIA]; and (iv) measurement and signature intelligence (MASINT) [traditionally the domain of DIA].

As a note, GEOINT is the analysis and visual representation of security related activities on Earth. GEOINT is produced by integrating imagery, imagery intelligence, and geospatial information. The NRO designs, builds, and operates imagery satellites, while the NGA processes and uses the imagery (NWC 2022).

<u>Section 1683(h)(2)(M)-(O)</u> specifically discusses reporting requirements relating to nuclear installations, nuclear weapons, and nuclear-powered vessels. I discuss this in detail in chapter 7.

<u>Section 1683(h)(3)</u> specifies that each report submitted under (h)(1) shall be submitted in unclassified form but may include a classified annex.

<u>Section 1683(i)(1)</u> calls for regular hearings on UAP through 2026, stating that not later than 90 days after the enactment of this Act and not less frequently than semiannually thereafter until December 31, 2026, the head of the Office, which we now know as AARO (but which was AOIMSG when this legislation was passed), shall provide classified briefings on UAP to congressional committees specified in <u>(l)(1)(A), (B), and (D)</u>. These are: (A) the Committees on Armed Services of the House of Representatives and the Senate; (B) the Committees on Appropriations of the House of Representatives and the Senate; and (D) the Permanent Select Committee on Intelligence of the House of Representatives and the Select Committee on Intelligence of the Senate.

May 17, 2022, Congress holds a public hearing on UAP. Although the formal requirement in Section 1683(i)(1) of the NDAA FY22 is for classified briefings, Congress also held a public hearing on UAP, which occurred on May 17, 2022 (HPSCI 2022a; 2022b).

June 9, 2022, NASA announces that it is undertaking an independent study on UAP (NASA 2022).

July 20, 2022, U.S. Senate Select Committee on Intelligence (SSCI) submits in draft form Senate Report 117-132, entitled "REPORT together with ADDITIONAL VIEWS," on the Intelligence Authorization Act (IAA) for Fiscal Year 2023, i.e., S. 4503. The SSCI submitted this report after having considered the original IAA for FY23, which is the bill to authorize FY23 appropriations on behalf of intelligence and intelligence-related activities (U.S. SSCI 2022, 1). After the language of the IAA was revised, the Act was folded into the NDAA FY23.

A section of the July 20, 2022 report is entitled "Modification of Requirement for Office to Address Unidentified Aerospace-Undersea Phenomena." Even though this terminology was not accepted into the final NDAA FY23, the language of this section of the report is illuminating. The first sentence reads:

> At a time when cross-domain transmedium threats to United States national security are expanding exponentially, the Committee is disappointed with the slow pace of DoD-led efforts to establish the office to address those threats and to replace the former Unidentified Aerial Phenomena Task Force as required in Section 1683 of the National Defense Authorization Act for Fiscal Year 2022 (U.S. SSCI 2022, 12).

Let me deconstruct and unpack this sentence. First, the use of the word "threats" is unnecessary and distasteful. There is no evidence that UAP are a threat, and it is manipulative to continue to state otherwise. It is irresponsible to tell people that something extraterrestrial is a 'threat' because it happens to appear in a certain region of spacetime that one's country has designated as 'restricted airspace' because we live under the burden of a militaristic paradigm (see also GCWUFO 2023c). As I argue throughout all my publications on this topic, considerable, tangible evidence exists that UAP *are not a threat*. In this book, I even discuss, including in physics terms, why ETIs *inherently* are not a threat. The militaristic presumption that anything unknown—including an extraterrestrial unknown—is a 'threat' is completely counterproductive when discussing ETI/UAP. It also is out of alignment with viewing ETI/UAP in a much broader context than one determined by countries' legislative and national security apparatuses.

Second, in the citation above, the word "transmedium" refers to phenomena that can operate in the air and/or in the water, as I discuss above in chapter 3.

Third, in the citation above, the word "cross-domain" refers to activity that occurs across what the U.S. military defines to be its 'warfighting domains.'

The sentence above therefore means that the presence of UAP is "expanding exponentially" across the six domains in which the U.S. military conducts operations—subsurface naval (subsurface), surface naval (sea), ground (land), air, space, and cyberspace.

The report from which the citation above is taken refers to extraterrestrial UAP, not merely to UAP more generally speaking. That is clear when the report describes what is meant by a new term it introduces, "unidentified aerospace-undersea phenomena." Of note, although this new term was not accepted into the NDAA FY23 (which redefines UAP as Unidentified Anomalous Phenomena, as explained below), the language of Senate Report 117-132 nevertheless is telling:

> The formal DoD and Intelligence Community definition of the terms used by the Office shall be updated to include space and undersea, and the scope of the Office shall be inclusive of those additional domains with focus on addressing technology surprise and "unknown unknowns." Temporary nonattributed objects, or those that are positively identified as man-made after analysis, will be passed to appropriate offices and should not be considered under the definition as unidentified aerospace-undersea phenomena (U.S. SSCI 2022, 12-13).

In the passage above, instead of simply using the word "extraterrestrial," the USG is using an unnecessarily redundant euphemism, "unknown unknowns." But the fact that the sentence invoking "unknown unknowns" is juxtaposed against the following sentence in which "man-made" objects clearly are subtracted from the equation makes the intention of the document writers clear—i.e., they are pointing to an extraterrestrial origin for certain UAP. Using euphemisms for the word "extraterrestrial" is unhelpful. People should simply start discussing extraterrestrial UAP openly and honestly, without recourse to such ambiguous language.

July 20, 2022, DoD renames and expands the scope of AOIMSG to the All-domain Anomaly Resolution Office (AARO). DoD established AARO because the NDAA FY22 includes a provision to establish an office with broader responsibilities as compared to those originally assigned under

AOIMSG. AARO exists within the Office of the Under Secretary of Defense for Intelligence & Security (OUSD(I&S)). Dr. Seán M. Kirkpatrick, former Chief Scientist at the Defense Intelligence Agency's Missile and Space Intelligence Center, is the Director of AARO (U.S. DOD 2022c).

On January 11, 2023, Kirkpatrick made a presentation, available now online, entitled, *The Defense Department's UAP Mission & Civil Aviation*, to the National Academies of Sciences, Engineering, and Medicine's Transportation Research Board (Kirkpatrick 2023). The word 'domain' in the AARO name refers to the six domains mentioned above into which the U.S. military divides its operations.

December 17, 2022, DoD renames the acronym UAP from Unidentified Aerial Phenomena to Unidentified Anomalous Phenomena. AARO considers UAP across all domains in which the U.S. military operates. AARO also works with the military departments and the Joint Chiefs to expand UAP reporting beyond aviators to mariners, submariners, and so-called space Guardians (Lopez 2022).

December 23, 2022, the NDAA (National Defense Authorization Act) for Fiscal Year 2023 (FY23) is signed into law. Discussion of reporting procedures for UAP occurs in Section 1673, Unidentified Anomalous Phenomena Reporting Procedures (NDAA FY23, 565-68).

Section 1673 (a)(1) calls for the establishment of a secure mechanism for authorized reporting by the SecDef, acting through the head of AARO and in consultation with the DNI, of: (A) any event relating to UAP; and (B) any activity or program by a Federal Government department or agency, or a contractor thereof, relating to UAP, "including with respect to material retrieval, material analysis, reverse engineering, research and development, developmental or operational testing, and security protections and enforcement."

Section 1673 (a)(2) states that "the mechanism for authorized reporting establishing under paragraph (1) prevents the unauthorized public reporting or compromise of classified military and intelligence systems, programs, and related activity, including all categories and levels of special access and compartmented access programs."

Section 1673 (c)(1) amends Section 1683 of the NDAA FY22 by replacing the word "aerial" with "anomalous." Henceforth, UAP therefore means "Unidentified Anomalous Phenomena" rather than "Unidentified Aerial Phenomena."

January 12, 2023, ODNI (Office of the Director of National Intelligence) releases the unclassified version of its first official UAP report, entitled *2022 Annual Report on Unidentified Aerial Phenomena* **(ODNI 2022).** This is the first actual 'report' that responds to the Congressional requirement stated in the NDAA FY22, since ODNI's first public document on the topic of UAP was its *Preliminary Assessment*, which it released prior to the NDAA FY22 requirement for 'reports' per se. The release of the 2022 report was delayed from its official October 31, 2022 deadline, as stipulated in the NDAA FY22, Section 1683 (h)(1) (see above).

The 2022 *Annual Report on Unidentified Aerial Phenomena* makes the following points:

The Numbers: Since ODNI's *Preliminary Assessment* of June 25, 2021 mentioned 144 UAP reports over seventeen years of reporting from 2004 to 2021, with 143 being of unknown attribution at the time that ODNI wrote that document, the *2022 Annual Report* states that 247 new UAP reports and an additional 119 UAP reports discovered or reported for the prior period had been made. This totals 510 UAP reports as of August 30, 2022 (ODNI 2022, 2). Of the 366 newly-identified reports (247+119=366), 195 could be attributed, which leaves 171 "uncharacterized and unattributed UAP reports" (366-195=171). In other words, at the time ODNI's *2022 Annual Report* was written, ODNI and its partner agencies were unable to characterize or to attribute 171 UAP (ODNI 2022, 5).

The Authority: As a result of AARO's broad scope of authorities and responsibilities, it is the primary USG entity tasked with attributing UAP when possible. The purpose of this requirement is to ensure that UAP detection and identification efforts span DoD and its interagency partners (ODNI 2022, 2).

Coordination of Effort: The *2022 Annual Report* emphasizes coordination of effort across the entire USG, including but not limited to DoD and the intelligence community (IC). According to the *Report* (ODNI 2022, 3), AARO works with:

- National Intelligence Manger for Aviation (NIM-Aviation);
- Under Secretary of Defense for Intelligence and Security (USD(I&S));
- Defense Intelligence Agency (DIA);
- Federal Bureau of Investigation (FBI);
- National Reconnaissance Office (NRO);
- National Geospatial-Intelligence Agency (NGA);
- National Security Agency (NSA);
- U.S. Army (USA), U.S. Navy (USN), U.S. Marine Corps (USMC), and U.S. Air Force (USAF);
- Federal Aviation Administration (FAA);
- National Aeronautics and Space Administration (NASA);
- National Oceanographic and Atmospheric Administration (NOAA);
- Department of Energy (DoE);
- ODNI/NIM-Emerging and Disruptive Technology (NIM-EDG);
- ODNI/National Counterintelligence and Security Center (NCSC); and
- ODNI/National Intelligence Council (ODNI/NIC).

This coordination of effort reflects considerable, though not total, organizational alignment within the USG for the purpose of studying UAP.

Future Data Collection: ODNI's *2022 Annual Report* states that since the June 2021 publication of its *Preliminary Assessment*, ODNI has developed strategic guidance to enhance further collection on UAP so that working together, AARO and ODNI can collect and report on UAP in a comprehensive manner for the IC. Of significant interest, the *2022 Annual*

Report states, "NIM-Aviation will remain the IC's focal point for UAP issues, while AARO is the DoD focal point for these issues and activities" (ODNI 2022, 4).

Attributed UAP: Among the 195 attributed UAP were Unmanned Aircraft System (UAS) or UAS-like entities, balloon or balloon-like entities, and so-called clutter, which includes birds, weather events, and airborne debris such as plastic bags (ODNI 2022, 5, inc. fn. 2).

Unattributed UAP: ODNI's *2022 Annual Report* states that 171 UAP reports are unexplained (ODNI 2022, 5), even after what was considerable effort by ODNI, DoD, AARO, and all the partner agencies and entities to attribute as many UAP events as possible prior to the publication of the document.

No Adverse Events or Health Effects: Most reports of unattributed UAP originate from USN and USAF aviators and operators who witnessed UAP during their operational duties. ODNI states, "To date, there have been no reported collisions between U.S. aircraft and UAP." ODNI also states, "Regarding health concerns, there have also been no encounters with UAP confirmed to contribute directly to adverse health-related effects to the observer(s)" (ODNI 2022, 6).

Factors Impacting UAP Detection: The report indicates that multiple factors affect the observation and detection of UAP, including weather, illumination, atmospheric effects, and accurate interpretation of sensor data. The report also straightforwardly states that "a select number of UAP incidents may be attributable to sensor irregularities or variances" (ODNI 2022, 3), which is an indirect way of implying that these factors do not account for all 171 unattributed UAP reports.

February 4-12, 2023, the U.S. shoots down four objects (Trevithick 2023b).
- February 4, 2023, Secretary of Defense Lloyd J. Austin III issues statement that a USAF fighter shot down a Chinese high-altitude surveillance balloon off the coast of South Carolina (U.S. DOD 2023).
- February 10, 2023, DOD detects and shoots down a second, high-altitude object over Alaska (Garamone 2023).

- February 11, 2023, the U.S. shoots down a third, high-altitude object over Canada's Yukon Territory (Bowman and Doubek 2023).
- February 12, 2023, the U.S. shoots down a fourth, high-altitude object over Lake Huron in Michigan (Vergun 2023).

The appearance of balloons on the UAP scene was foreshadowed in ODNI's *2022 Annual Report*. The document was released on January 12, 2023, less than a month before a USAF fighter shot down China's high-altitude surveillance balloon. As discussed above, ODNI's *2022 Annual Report* describes 163 attributed UAP "characterized as balloon or balloon-like entities" (ODNI 2022, 5). Given that there is no confusion about the human-made nature of "balloon or balloon-like entities," the numerous, media stories calling these objects "UFOs" (see for example Cobb 2023; Parks et al. 2023) clearly are making a disingenuous attempt to conflate human-made objects with genuine, extraterrestrial UAP.

February 13, 2023, the USG announces that it has created an Interagency Task Force to study UAP. According to Rear Admiral John F. Kirby (Ret.), current Coordinator for Strategic Communications at the National Security Council in the White House, "the President through his National Security Adviser [Jake Sullivan] has today directed an interagency team to study the broader policy implications for detection, analysis, and disposition of unidentified aerial objects that pose either safety or security risks" (MSNBC 2023). The group also includes Secretary of State Antony Blinken, Secretary of Defense Lloyd Austin, and Director of National Intelligence Avril Haines. The task force will engage "their relevant counterparts to share information" and "to try to gain their perspectives" at a global level, while the administration will brief members of Congress and local officials (Judd 2023).

Following the February 13, 2023 announcement, some people expressed confusion with respect to perceived overlap between the net interagency task force and AARO (Trevithick 2023a). Marco Rubio (2023) tweeted, "Why is the White House creating a new 'interagency team' to monitor,investigate & report on unidentified aerial objects when we already have @DoD_AARO which we helped create over two years ago?"

Some people wonder if AARO will be relegated to reviewing historical documents while the new, interagency group run out of the White House will take a more dominant role in decision-making relating to UAP.

February 14, 2023, the *Wall Street Journal* reports that AARO was underfunded. The article states, "The Pentagon office set up to detect and identify mysterious objects, such as the three shot down by the U.S. jet fighters over the past week, was mistakenly underfunded" (Youssef and Wise 2023).

February 16, 2023, many U.S. senators co-sign a letter requesting robust funding for AARO. The letter is sent to Deputy Defense Secretary Kathleen Hicks and Principal Deputy Director of National Intelligence Stacey Dixon. The senators express a view in the letter that AARO is underfunded, and they request that additional funds be allocated to the Office (USS 2023).

March 8-12, 2023, Three banks related to cryptocurrency sector collapse (Sigalos 2023; see also Chapman 2023). Silvergate Capital Corporation, Silicon Valley Bank, and Signature Bank all collapse.

March 16-26, 2023, First Republic Bank is subject to a private sector rescue. On March 16, a group of U.S. lenders comprised of Bank of America, Citigroup, JPMorgan Chase, Wells Fargo, Goldman Sachs, Morgan Stanley, BNY Mellon, PNC Bank, State Street, Truist and U.S. Bank commit to deposit $30 billion of cash into First Republic Bank (Business Wire 2023; Egan et al). By March 26, the amount of the capital infusion had more than doubled. First Republic Bank apparently has received at least $70 billion from other banks and from the Federal Reserve to boost its liquidity (Revell 2023). Weeks of failed rescue attempts followed, with JPMorgan eventually winning a bid to buy First Republic on May 1, 2023 (Anand et al. 2023).

March 19, 2023, UBS announces it is buying Credit Suisse (Thompson 2023).

Elsewhere (Andresen in preparation), I discuss what appear to be parallel agendas to centralize control over the global money supply and, also, to centralize control over information and decision-making relating to UAP.

What I find particularly interesting about UBS' takeover of Credit Suisse—since the latter certainly is not the first Swiss bank to help ultrarich Americans dodge U.S. taxes (Mark 2023)—is that it occurred only two months after employees of Credit Suisse made statements appearing to signal the institution's interest in participating as banker to a multipolar world. On January 17, 2023, Axel P. Lehmann, Chairman of the Board of Directors at Credit Suisse, made a presentation to the 53rd annual meeting of the World Economic Forum (WEF) in Davos, Switzerland. During his talk, Lehmann stated, "I think the world is probably fundamentally changing, and we are entering into a, I will call it a 'multipolar world'" (Bloomberg 2023). Only a few days later, on January 20, 2023, Zoltan Pozsar (2023) wrote a succinct yet powerful article in *Financial Times*. At the time, Pozsar was Global Head of Short-Term Interest Rate Strategy at Credit Suisse. His 2022 Credit Suisse website biography describes his position then as Managing Director, Investment Strategy and Research, and, also, founding member of the Shadow Banking Colloquium of the Institute for New Economic Thinking (Credit Suisse 2022). One important theme in Pozsar's article is encapsulated in its title, "Great power conflict puts the dollar's exorbitant privilege under threat." Two months later, Credit Suisse was the subject of a takeover by the more 'compliant' UBS. In other words, as soon as Credit Suisse began to venture out far enough into the brave new multipolar world, it ceased to exist as a relatively independent major, Swiss banking establishment.

Interestingly, even as far back as 2017, Credit Suisse Research Institute also famously authored the Getting Over Globalization" report (CSRI 2017). Prior to that, in 2014, Nannette Hechler-Fayd'herbe, Head of Investment Strategy at Credit Suisse, did an interview in which she frames discussion of geopolitics in terms of the emergence of a multipolar world, understood as several different poles of economic interests backed by military assertiveness. She also contextualizes this discussion in terms of the financial market perspectives of Ukraine, the Middle East, and the China Seas (Credit Suisse 2014). In other words, Credit Suisse actively has been positioning itself as banker to a multipolar vision, not a unipolar one. It therefore is no real surprise that it was the most prominent European bank to be taken over during the March 2023 banking crisis.

March 28, 2023, members of Senate Armed Services Committee call on greater funding for AARO. Following the release of the Biden administration's budget request, lawmakers attending a Senate Armed Services Committee hearing called for more funding for AARO. Secretary of Defense Lloyd Austin stated that DOD had requested $11 million to fund AARO's research for fiscal year 2024 (FY24), and he pledged to fully fund AARO in the future. The actual FY24 AARO budget figures are confusing, however, since on March 14, 2023, Pentagon spokesperson Susan Gough told *Military Times* the figures were classified. In addition, in the FY23 budget, it has been suggested that the Biden administration only funded AARO's basic operating expenses (Perez 2023).

Michael J. McCord, Comptroller of the Department of Defense, stated he was unaware the office needed more funding and was under the impression there was "adequate funding" for the newly formed organization. It also appears that what DOD requested in funding in the FY 24 budget is different from what AARO itself requested internally within DOD for its own funding (Pollina 2023).

April 3, 2023, The White House website publishes a document from the Planetary Defense Interagency Working Group of the National Science & Technology Council entitled, "National Preparedness Strategy & Action Plan for Near-Earth Object Hazards and Planetary Defense" (PDIWG 2023; WH 2023). The Planetary Defense Interagency Working Group is the current name of what previously was the Detecting and Mitigating the Impact of Earth-bound Near-Earth Objects (DAMIEN) Interagency Working Group (IWG). It falls under the jurisdiction of the Committee on Homeland and National Security, National Science and Technology Council (NSTC). Originally, in January 2016, the DAMIEN IWG was convened "to define, coordinate, and oversee goals and programmatic priorities of Federal science and technology activities related to potentially hazardous near-Earth objects (NEOs)" (PDIWG 2023, ii). I analyze the April 3, 2023 document elsewhere (Andresen forthcoming).

April 19, 2023, the U.S. Senate Committee on Armed Services (SASC), Subcommittee on Emerging Threats and Capabilities, held a meeting to receive testimony on the mission, activities, oversight, and budget of the All-domain Anomaly Resolution Office (SASC 2023).

Dr. Seán M. Kirkpatrick, Director of AARO, was the sole witness in the hearing. Kirkpatrick explained that AARO has more than three dozen experts working in four areas: operations; scientific research; integrated analysis; and strategic communications. He stated that a small percentage of UAP cases have anomalous signatures, and he showed a slide entitled, "Collecting UAP Reporting Trends from 1996-2023" (SASC 2023).

As of the week of the hearing, AARO had 650 cases. Each case is subjected to a multistep analytical process that includes prioritization, packaging data, and sending the data for analysis to two groups—an IC (intelligence community) group of analysts, and a scientific/technical group composed of scientists and engineers. One element of data packaging involves determining what data has been gathered and what data one needs—e.g., radar, electro-optical (EO), thermal, overhead, etc.—to analyze the case. Feedback from the two groups is adjudicated, case recommendations are made, AARO writes up the case, and AARO then sends the case to a senior technical advisory group for peer review of the case recommendation. This is sent back to Kirkpatrick, who makes a determination on the case. Approximately 20-30 cases have made it about halfway through the analytical process, with a handful having made it through to peer review with signed case closure reports. In addition, many cases are uploaded to a classified web portal so that individuals can collaborate on resolving them (SASC 2023).

Kirkpatrick outlined his view of the range of hypotheses for what UAP are, from adversary breakthrough technology at one end (the area of IC competency), to known objects such as hypersonics in the middle (the realm of measurement), all the way to extraterrestrial UAP at the other end (where he thinks the academic community may be helpful). If a case were to show evidence of extraterrestrial origin, Kirkpatrick stated that AARO would work with interagency partners at NASA to notify USG leadership of its finding (SASC 2023).

Kirkpatrick also stated that with respect to mission and goal, his vision is that at some point in the future, an AARO should not be needed. His view is that DOD should be able to normalize into its existing processes and work with other organizations what AARO is tasked with doing now. Kirkpatrick also commented briefly on NASA's current UAP study, which focuses on unclassified data sources that may augment the classified data sources used by AARO. He stated that the NASA study may help one understand if by specifying a signature, for example climate science satellites looking at Earth would be helpful in seeing UAP. But he added that such platforms often do not have the required resolution of approximately 1-4 meters, since for example climate civil satellites have a larger resolution than this. Numerous other topics of a technical nature also were discussed at the hearing, which is available to watch online (SASC 2023).

Stovepipes and Schisms

Elsewhere (Andresen 2022b, 291-93), I describe the current schism between branches of the USG on the topic of UAP, including the long-standing alignment between the USAF and CIA. Although some USG data on UAP that derives primarily from the USN is being shared in both classified and public settings, much of what the USAF and the CIA know about UAP is likely to remain carefully protected. Providing too much information to Congress creates a situation in which highly classified information could be leaked from members of Congress to foreign persons or entities. Indeed, the susceptibility of members of Congress to supernational pressures is one of the reasons Congress marches to the divergent beats of a gleaming batterie of oh-so-many discordant drums.

Tension between Congress—certain members of which want access to information—and those elements within the USG that do not want to disclose it is captured in the language of Section 1683(i)(4) of the NDAA FY22:

> INSTANCES IN WHICH DATA WAS NOT SHARED.—For each briefing period, the head of the Office established under subsection (a) shall jointly provide to the chairman and the ranking minority member or vice chairman of the congressional committees specified in subparagraphs (A) and (D) of subsection (k)(1) an enumeration of any instances in which data relating to unidentified aerial phenomena was not provided to the Office because of classification restrictions on that data or for any other reason (NDAA FY22, 582).

One can intepret this passage to indicate that Congress is pointing specifically at the USAF/CIA—mostly at the USAF—for not releasing information on UAP. This interpretation is buttressed by language one finds in ODNI's *Preliminary Assessment* of June 25, 2021. Although that document states that it was prepared with input from many elements within the USG, including the USAF (ODNI 2021, 2), another statement occurs a few pages later in the document: "The UAPTF [UAP Task Force] is currently working to acquire additional reporting, including from the U.S. Air Force (USAF)" (ODNI 2021, 7). This singling out of the USAF is quite blunt for those familiar with how disagreements are aired in Washington, D.C.

In truth, however, the USAF may not be holding as much information about UAP as some people suspect. Both DoD and the IC operate Special Access Programs (SAPs) with compartmented elements. These programs, which are subject to review by a Special Access Program Oversight Committee (SAPOC), receive in excess of $100 billion per year in funding. DOD's Defense Advanced Research Projects Agency (DARPA) and other entities within DOD also operate other classified programs. DOD has some oversight of these programs, for example those operated by the National Security Agency (NSA) (CDSE 2021).

In addition to regular SAPs, so-called waived SAPs also exist. Waived SAPs are not subject to routine oversight. In fact, the legislation governing them specifically protects waived SAPs from broad disclosure (Andresen 2022b, 292-95). The Senior Review Group (SRG) is responsible for ensuring that SAPs are not duplicated, and the Senior SAP Working Group (SSWG) coordinates, deconflicts, and integrates SAPs, addresses policy, oversight, and management relating to them, and provides recommendations to the SRG (CDSE n.d., 3).

Work on behalf of waived SAPs sometimes is conducted by private groups within the USG, often in so-called black programs operated by the military and/or the IC. Anyone working in a waived SAP must hold an extremely high security clearance. Furthermore, while some work on waived SAPs may occur at USG installations, work on these programs also occurs on behalf of the USG at installations owned and/or run by contractors, for example private aerospace companies. For example, the National Intelligence Program (NIP) and the Military Intelligence Program (MIP) operate a wide range of projects, many of which are highly classified, at intelligence, military, and other research facilities (ODNI n.d.-a). Some of this work also occurs on behalf of waived SAPs.

Relatedly, the National Intelligence Council (NIC)—which supports the Director of National Intelligence (DNI) in the Office of the Director of National Intelligence (ODNI)—has National Intelligence Officers (NIOs) from government, academia, and the private sector who "are the Intelligence Community's senior experts on a range of regional and functional issues" (ODNI n.d.-c). While certain NIOs may have acquired some snippets of knowledge about UAP over the course of their careers, it is highly likely that whatever information exists within the USG on UAP propulsion and/or materiel is inaccessible to the NIP, MIP, and NIOs. ODNI almost certainly also does not possess all relevant information on UAP. Some high-classified information on UAP almost certainly is held in-house by the CIA. Of note, the CIA is an independent agency of the U.S. Intelligence Community (IC) and does not come under ODNI's umbrella, even though ODNI is the umbrella for many other intelligence agencies within the USG (ODNI n.d.-b).

Information held by the USG on UAP also may exist within Independent Research and Development (IR&D, or IRAD) activities associated with defense contracting companies, where it could be organized as one or more new entities within compartments (AcqNotes 2022). What is certain is that no single entity involved with supposed UAP reverse engineering efforts holds *all* the information. Information would be scattered between entities to reduce the possibility that the directors and/or employees of any single company and/or agency could 'run with' the information to the detriment of U.S. national security.

Acronym Acrobatics

Statements emanating from DoD on December 16, 2022 (U.S. DOD 2022d) and December 17, 2022 (Lopez 2022) redefine UAP from "Unidentified Aerial Phenomena" to "Unidentified Anomalous Phenomena." As explained above, the new definition of UAP also is reflected in the language of the NDAA FY23, in which Section 1673 (c)(1) amends Section 1683 of the NDAA FY22 by replacing the word 'aerial' with 'anomalous'—which in turn means that 'Unidentified Anomalous Phenomena' replaces 'Unidentified Aerial Phenomena' as the meaning of 'UAP.' This change in terminology makes the meaning of the UAP acronym consistent with the language of the NDAA FY23, Section 1673: Unidentified <u>Anomalous</u> Phenomena Reporting Procedures. It also makes the meaning of the UAP acronym consistent with the name of the office charged with studying unidentified phenomenon deemed anomalous, namely the All-domain <u>Anomaly</u> Resolution Office (AARO) (underlining added). Interestingly, Unidentified Anomalous Phenomena is not a new term. In 2015, this term was used by a European organization, Marie Curie Alumni Association (MCAA 2015), prior to DOD's use of it in 2022, and, therefore, also prior to its use in the NDAA FY23.

Before the recent change of the acronym UAP to mean 'Unidentified Anomalous Phenomena,' the USG previously had defined UAP as Unidentified Aerial Phenomena (UAP) in two publications: ODNI's *Preliminary Assessment* from June 25, 2021; and in the NDAA FY22. But even though both publications share in common the language "Unidentified Aerial Phenomena," each publication defines this term differently. Before delineating the differences between the two definitions, one should note that in all cases, the final word, "Phenomena," is plural ('phenomenon' is the singular). Since in all its iterations, UAP already is defined in the plural, it is grammatically incorrect to put an "s" on the end of UAP when one wishes to designate more than one phenomenon. Media outlets routinely make this error by incorrectly writing "UAPs." Even NASA's webpage incorrectly states "UAPs" instead of correctly writing "UAP" to designate the plural (NASA 2022).

There is yet another acronym to consider—UAV, for Unmanned Aerial Vehicle. One example of a UAV is Northrop Grumman's RQ-180, a stealth surveillance vehicle produced for the USAF, which people colloquially refer to as the 'white bat' because of its shape. Another example of a UAV is the SR-72 hypersonic, strategic reconnaissance vehicle developed by Lockheed Martin Skunk Works. Unlike UAP, UAV is defined in the singular. Accordingly, to designate multiple Unmanned Aerial Vehicles, one must add the "s," making that acronym UAVs. Also, for the truly intrepid acronym hunters who plan to file FOIA requests, Anomalous Aerial Vehicle (AAV) is yet another term used in the literature. In a manner similar to the situation for UAP, AAVs are objects that at least theoretically could be either human or extraterrestrial in origin (Oberhaus 2018). In contrast, AAVs refer to objects that at least theoretically could be either human or extraterrestrial in origin (Oberhaus 2018), which is similar to the situation for UAP. In contrast, UAVs most definitely are of human origin.

One also should note that per Oberhaus (2018), AAVs at least theoretically could be a mix between things human and things extraterrestrial in origin.

Moving to the actual definition of UAP, it is important to note that even though the middle word of the term changed from "aerial" to "anomalous" between the NDAA FY22 and the NDAA FY23, the definition of UAP has not changed. However, the common definition of UAP found in both NDAAs is different from the common definition found in both of ODNI's documents.

ODNI's *Preliminary Assessment* (ODNI 2021, 8) and ODNI's *2022 Annual Report* (ODNA 2022, 11) both define UAP as "**Unidentified Aerial**

Phenomena (UAP): Airborne objects not immediately identifiable. The acronym UAP represents the broadest category of airborne objects reviewed for analysis." Therefore, both ODNI documents define UAP as those unidentified phenomena seen or otherwise detected in the air. In this sense, both documents define UAP in terms of what people conventionally think of as UFOs (Unidentified Flying Objects). Note, too, that the *2022 Annual Report* still defines UAP as 'Unidentified <u>Aerial</u> Phenomena' rather than as 'Unidentified <u>Anomalous</u> Phenomena,' even though by the time the *2022 Annual Report* was released in January 2023, the NDAA FY23 had redefined UAP to mean 'Unidentified Anomalous Phenomena' (NDAA FY23, 567-68).

Both ODNI documents have a narrower definition of UAP as compared to the definition of UAP found in the two NDAAs (see also Andresen 2022b, 281-82). Both NDAAs rely on the same meaning of UAP, which is defined in the NDAA FY22 and not redefined in the NDAA FY23, even though the acronym is redefined there (to make it more consistent with the definition of UAP found in the NDAA FY22). The term "transmedium" appears in the NDAA FY22 in association with UAP (NDAA FY22, 583), which is why the definition of UAP in the two NDAAs is broader than the definition of UAP in ODNI's two documents, the latter of which focus only on airborne objects. In other words, it is the transmedium feature of some (if not all) extraterrestrial UAP that may have catalyzed DOD to adopt the new term, "Unidentified Anomalous Phenomena" in the NDAA FY23. Since the transmedium nature of UAP is important, in my own writing, I use the NDAAs' broader definition of UAP as opposed to ODNI's narrower definition.

A salient issue is how to distinguish human-made UAP, some of which may be UAVs, from extraterrestrial UAP. Many human-made objects, such as the TR3-B, easily can be mistaken for extraterrestrial UAP. In fact, any UAP that appear to demonstrate plasma stealth technology easily can be mistaken as objects that are extraterrestrial in origin, even though plasma stealth technology is well within the capability of human beings (Rogoway 2019; see also BFD 2020a). Even transmedium capability should not be taken as a guarantee that one is dealing with a genuine, extraterrestrial UAP, since some human-made UAVs may have transmedium capabilities. In other words, one should not think that all anomalous objects are extraterrestrial UAP or that they are all human-made UAVs. Undoubtedly, a mixture of both exists.

Some people have suggested that a human-made Stealth version of the F-117A exists with hybrid propulsion that includes a gravity control component. I personally tend to doubt this, though it would be difficult to determine with certainty without direct access to the relevant classified projects that supposedly have developed it. The point remains, however, that classified projects in the U.S. and most likely also in China and Russia may have developed UAVs that easily can be mistaken for extraterrestrial UAP. While this may be true, however, the decades-long history of the ETI/UAP topic demonstrates that genuine *extraterrestrial* UAP also operate in our midst here on Earth.

UAP Interest in the U.S. Military

For many decades, UAP have been observed in the vicinity of military installations in the U.S. and elsewhere around the world. USN and USAF personnel observed UAP during the Vietnam War from 1955 to 1975, for example. During the 1960s and 1970s, cylindrical and Tic Tac shaped UAP commonly were seen, with one Tic Tac shaped object observed in the 1970s in the U.S. by an airplane pilot connected to the Strategic Air Command's nuclear deterrent. In 1999, a USAF serviceman witnessed a spherical UAP over the Adriatic Sea. These are just a few of the many sightings that could be mentioned.

The list of military sightings of UAP is long, which underscores the simple fact that the U.S. military has in its possession a significant amount of classified information on UAP. Military radar has detected UAP sightings by the USN, for example (BFD 2020c). On April 28, 2020, DOD verified the authenticity of three previously-released videos of UAP taken by USN personnel (U.S. DOD 2020b; see also Conte 2020). Whether or not these three videos present evidence of genuine extraterrestrial UAP or not, it is likely that they are only an exceedingly small percentage of UAP videos in DOD's possession.

UAP events have occurred in the vicinity of multiple USN vessels, including USS *Princeton*, USS *Nimitz*, USS *Ronald Reagan*, USS *Roosevelt*, USS *Russell*, USS *Omaha*, and USS *Kidd*. For example, a USN service member who functioned as a radar operator during a UAP event involving USS *Roosevelt* stated that the contact he observed was huge, and that the strength of the signal was as strong as an aircraft carrier's surface contact on the water.

In November 2004, a series of UAP events occurred on multiple days over a time frame of approximately two weeks in proximity to the USS *Nimitz*, USS *Princeton*, and other vessels comprising the Nimitz Carrier Strike Group. The aircraft carrier USS *Nimitz* is nuclear-powered (Pike 2000), as are all U.S. Navy submarines and aircraft carriers. In the past, several cruisers also were nuclear-powered, but all nuclear-powered cruisers have been retired. The Nimitz Carrier Strike Group was conducting exercises 60-70 miles off the West Coast of the U.S., in the gap between San Diego, California and Ensenada, Mexico. For example, on November 14, 2004, in conditions of excellent visibility, a Tic Tac shaped object estimated to be 40-50 feet long with no wings and no visible means of propulsion was seen and tracked on radar. The UAP demonstrated significant speed and what appeared to be instantaneous acceleration. It had no apparent constraints either on its altitude or on its ability to remain aloft for long periods of time. Testimony of radar operators, technical weapons systems operators, and pilots regarding this event is consistent, with many people with different responsibilities on the vessels and working on different platforms with different sensor systems corroborating the event. USN personnel tracked UAP using multiple high-fidelity radar systems, infrared systems, and sonar systems, which increases the credibility of the report given the redundancy of UAP detection across different systems.

USS *Princeton* radar operator Kevin Day was the first to detect the UAP, which dropped from 80,000 feet or more—possibly even from orbital altitudes—and hovered at approximately 20,000 feet. One of the objects approached the carrier battle group when it was intersected by F-18s. The F-18 pilots saw the object, and airborne command also detected the object. An Aegis class radar tracked the object, after which another F-18 was dispatched and obtained a video of the object with an infrared camera system. Day also states that the UAP he detected on radar were moving at Mach 23 and that the ship's Aegis Combat System radar data for the event was removed from the USS *Princeton* (Breslo 2020).

Of note, the USN's Aegis system can track low-frequency radar, cross-sectional data on UAP operating at extreme altitudes and velocities. Other USS *Princeton* crew members also have stated that certain individuals— perhaps from the USAF or from an entity within the IC—were brought by helicopter on board the ship shortly after the November 2004 UAP encounters occurred, and that they left a short time later, taking the ship's radar data with them—thereby corroborating Day's account mentioned above. It also has been stated that 'higher-ups' in the IC were monitoring what was happening on the ship (BFD 2020b). For a time, the radar data may have been in the possession of USAF personnel at Langley Air Force Base. Also of note, the November 2004 USS *Princeton* deck logs apparently are missing from the National Archives, with some people speculating that the USAF or another entity in the IC may have these logs.

UAP appear to be operating essentially everywhere the U.S. military has battle groups. Whereas the 2004 events recounted above occurred off the West Coast of the U.S., a 2004 event involving the USS *Ronald Reagan* occurred off the East Coast of the U.S. The USS *Ronald Reagan* is a nuclear-powered, Nimitz-class supercarrier. In this case, one of the officers on deck reportedly ordered someone to take the pages out of the deck log book, which was very unusual given that USN personnel are instructed that what is written in the log book is a legal record that should not be destroyed or otherwise tampered with (The Nimitz Encounters 2022).

In addition to the sightings mentioned above, multiple U.S. naval fighters encountered UAP off the East Cost of the U.S. in 2015. Naval pilots also are on record stating that when their carrier battle group deployed to the Middle East, they continued to encounter UAP across the Atlantic Ocean. UAP reports over Afghanistan and Syria also exist. Of note, too, government leaders in the United Arab Emirates (UAE) have an active interest in UAP and have commissioned research on this topic.

The list of UAP events in proximity to U.S. military installations, planes, and vessels at home and abroad is quite long. One can only assume that if this much UAP activity is occurring around U.S. military installations and vessels, the militaries of other nations also must be experiencing significant UAP activity. In Australia, for example, which is a member of the Five Eyes (FVEY) intelligence alliance, witnesses have reported UAP at military installations that are jointly operated by Australia and the U.S. These include the Harold E. Holt joint Australian-U.S. naval communication station on the northwest coast of Australia and, also, the Pine Gap satellite surveillance base in the Northern Territory jointly operated by Australia and the U.S. Though not jointly operated with the U.S., the RAAF Woomera Range Complex (WRC), a weapons range in South Australia, also is the site of reported UAP events.

'Discovered' Craft and Materiel

Though scrutinized (Johnson 2023), speculation exists that DOD has conducted field investigations on UAP since at least 1945, when witnesses reported that an avocado-shaped object came down near San Antonio, New Mexico (not Texas) and later was removed by the U.S. Army (Good 2007 <2006>, 53-66).

As mentioned above, in an August 1, 1952 memorandum from Edward Tauss to Philip Strong, Tauss recommends that the CIA continue to monitor the UFO situation in coordination with ATIC (G. Haines 1997 <1995>, 68, inc. fn. 15 on 80). Recall, ATIC was part of the Air Materiel Command T-2 Intelligence Department at WPAFB (NASIC n.d.). With the formation of Project Blue Book in 1952, the task of identifying and explaining UFOs fell squarely within the responsibilities of the Air Materiel Command at WPAFB (G. Haines 1997 <1995>, 68, inc. fn. 8 on 79). Indeed, because ATIC was a 'materiel' command, it would have made sense for any UFOs or parts thereof to be sent there for analysis. One also can surmise from the chronology of events discussed earlier in this chapter that by 1952—only a few years after the 1947 event purportedly involving a downed UFO in Roswell—that the CIA was deferring to ATIC, which was part of the USAF, on the topic of UAP. I am not suggesting that the CIA still defers to the USAF on UAP matters, and by now, it certainly could be the other way around.

The presence of the word "discovered" in the NDAA FY22 could be interpreted as further indication that the USG already has in its possession at least one unidentified anomalous object, or parts thereof. Section (h)(2)(K) of the NDAA FY22, states, "An update on any efforts underway on the ability to capture or exploit discovered unidentified aerial phenomena" (NDAA FY22, 581). The grammar of the passage is ambiguous, however, such that it could mean that if such an object or parts thereof *were to be discovered*, that Congress wants to know what capability is in place "to capture or exploit" such an object or parts. Of importance here, the words "capture" and "exploit" are quite offensive in the context of genuinely extraterrestrial UAP. Humans should not be attempting to do either.

Also based on the chronology of events discussed earlier in the chapter, together with USG documentation pertaining to these events, one can conclude that by 1952 at the latest, the USAF may have had one or more "discovered" unidentified aerial objects or parts thereof in its possession. Furthermore, if the USG recovered any "discovered" anomalous objects or parts thereof, then it may have sent such intact objects and/or parts to WPAFB. I am not suggesting that anything that once may have been stored at WPAFB still is there, but at one point, UAP—in part or in full—very well may have been. The USG still may have such objects and/or materiel in its possession, though not necessarily—or not necessarily exclusively—at WPAFB.

Claims also have been made that the USG and other countries possess metamaterial from extraterrestrial UAP, which causes some people to suggest that an ongoing race is occurring between the U.S., China, and Russia to understand the technology associated with such metamaterial. Ideas circulate that putative extraterrestrial metamaterial possesses unusual isotypic ratios not commonly found on Earth, and that it functions as a waveguide for high-frequency electromagnetic radiation. Certain program contractors apparently claim that metal alloys and other metamaterials have been recovered from UAP, with some reports stating that UAP debris has been tested (News18 India 2021). Another idea in circulation is that some exotic metamaterial possesses qualities similar to memory metal (Cox 2009; see also Kauffman and Mayo 1997).

While it is difficult to discern which media reports on putative extraterrestrial metamaterials are accurate, the USG is studying exotic metamaterials that some people contend came from UAP—even though it has not been confirmed that such metamaterial is extraterrestrial in origin (Tritten 2019b). In 2019, To the Stars... Academy of Arts & Science (TTSA) signed an agreement to provide UAP metamaterials to Detroit Arsenal in Warren, Michigan for analysis (Rogan 2019b). Detroit Arsenal is a manufacturing plant controlled by the U.S. Army. *Bloomberg Government* reported that on October 10, 2019, TTSA and "an Army ground vehicle research unit in Warren, Mich., signed a standard research and development agreement" (Tritten 2019b). The reporter for that story, Travis J. Tritten, confirmed to me by email on November 13, 2019, "The research group is the U.S. Army Combat Capabilities Development Command, Ground Vehicle Systems Center" (Tritten 2019a). This organization also is called the U.S. Army Futures Command (Tritten 2019b). A Cooperative Research and Development Agreement (CRADA) available online (Department of the Army 2019) provides further confirmation of the existence of this research project. The status of this research is unclear, however, since no information regarding the research has been released to the public (Rogan 2019a; Tritten 2019b). At least in part, the purpose of studying the metamaterials may be to determine if the results can be applied to making Army vehicles lighter and more resilient (Tritten 2019b).

My hope is that human beings evolve to the point at which we become interested in the potential of any exotic metamaterials—human or extraterrestrial in origin—for human health and wellbeing rather than for military applications. One wonders what the National Institute of Health (NIH) could do to benefit humankind by researching such metamaterials, for example. But what is required to make this happen is a species-level pivot from a mindset focused on 'defense' and 'national security' to one oriented towards 'health' and 'wellbeing.' If the trillions of dollars currently allocated for the military were reallocated for research on health, 'defense contractors' become 'health contractors' posthaste.

The Stakes Surrounding UAP Technology

A significant breakthrough with respect to UAP propulsion would render obsolete the current macroeconomic order and global energy structure based on fossil fuels such as oil, liquified natural gas (LNG), and coal, and, also, nuclear energy (and associated uranium mining) and renewables such as hydropower, solar, and wind. Therefore, knowledge of UAP movement technologies has the potential to upset the plans of supranational individuals and entities who currently control the world's energy resources by means of interlocking corporate and private networks, who stand to lose trillions of dollars if the current global energy structure is overturned and they do not control what replaces it. Downstream, too, people who profit from private equity investment, insider trading, and the illegal transfer of information relating to the energy sector and energy subsidies—even for renewable technologies—stand to lose enormous sums of money if the conventional energy infrastructure of the world is replaced with something else. It therefore is unsurprising that venture capitalists and individuals associated energy sector financing are attempting to obtain access to information on UAP propulsion and/or materiel in order to attempt to capitalize on it for themselves.

Any sudden and/or widespread change in the global energy market will have widespread repercussions that will unfold across human civilization relatively quickly. As an analogy yet on a smaller scale, one can consider what has unfolded in recent months as a result of the war in Ukraine. Early in the conflict, it seemed that at least one feature of global dealings in energy that was 'fueling' the conflict related to overall dynamics of the nuclear power industry and the worldwide control of uranium mining rights. One of the many subtexts of the war seems to be an attempt to reset the global energy playing board by reinvigorating the nuclear power industry in Europe and elsewhere. When the war began, Germany almost completely had eliminated its nuclear power plants because of environmental concerns and because it, like other countries in Europe, thought they could rely on the Nord Stream and Nord Stream 2 pipelines and the Nord Stream network for energy. That thinking turned out to be quite naïve after a series of explosions occurred on September 26, 2022, along the bottom of the Baltic Sea, severing three of the four Nord Stream pipelines.

During his recent address to the UN Security Council, Professor Jeffrey Sachs, Director, Center for Sustainable Development at Columbia University—a tremendously erudite scholar and genuine humanitarian—stated that an investigation by the UN Security Council into the explosion that blew up the Nord Stream natural gas pipelines is a high global priority. Prof. Sachs was a faculty member at Harvard when I was doing my doctoral work there. Later, Prof. Sachs moved to the faculty at Columbia when I was a Visiting Scholar here. Although I never have been enrolled formally as one of Prof. Sachs' students, at both Harvard and Columbia, I benefitted greatly from attending Prof. Sachs' public lectures on various topics.

As a specialist in the global economy, including global trade, finance, infrastructure, and economic statecraft (Jeffrey Sachs 2023), Prof. Sachs has an extraordinary grasp of geopolitical realities. With respect to the Nord Stream explosions, Prof. Sachs made the following excerpted remarks to the UN Security Council on February 21, 2023:

> The destruction of the Nord Stream pipelines on September 26, 2022, constitutes an act of international terrorism and represents a threat to the peace. It is the responsibility of the UN Security Council to take up the question of who might have carried out the act in order to bring the perpetrator to international justice, to pursue compensation for the damaged parties, and to prevent future such actions.

> The consequences of the destruction of the Nord Stream pipelines are enormous. They include not only the vast economic losses related to the pipelines themselves and their future potential use, but also the heightened threat to transboundary infrastructure of all kinds,

submarine internet cables, international pipelines for gas and hydrogen, transboundary power transmission, offshore wind farms and more.

The global transformation to green energy will require *considerable* transboundary infrastructure, including in international waters. Countries need to have full confidence that their infrastructure will not be destroyed by third parties. Some European countries have recently expressed concern over the safety of *their* offshore infrastructure.

For all of these reasons, the investigation by the UN Security Council of the Nord Stream explosions is a <u>high, global priority</u> (Jeffrey Sachs 2023, italics and underlining added to reflect speaker's emphasis, and paragraph divisions added for readability; see also NCTV 2023).

Prof. Sachs delivered his remarks to the UN Security Council approximately two weeks after longtime investigative journalist Seymour Hersh published his February 8, 2023 article on the U.S.' role in the destruction of the pipelines (Hersh 2023b). Hersh followed his February 8 article with an article on February 22, 2023 on the history of U.S.-Norway collaboration in covert operations (Hersh 2023a).

Despite Hersh's two outstanding articles and Sachs' cogent plea for a UN investigation into what happened, on March 27, 2023, the UN Security Council voted down a draft resolution that "would have requested the Secretary-General to establish an international, independent investigation commission to conduct a comprehensive, transparent and impartial international investigation of all aspects of the act of sabotage on the Nord Stream 1 and 2 gas pipelines — including identification of its perpetrators, sponsors, organizers and accomplices" (UN 2023). In yet another twist on this situation, the Russia government accused Germany, Denmark, and Sweden of a cover-up in their investigations into the sabotage attacks on the Nord Stream pipelines (Scahill 2023). If a U.S. covert operation is responsible for the Nord Stream explosions, I completely agree with Prof. Sachs that this constitutes "an act of international terrorism and represents a threat to the peace." Neoconservative agendas in Washington and covert action in support of these agendas should not be continued. The world must move on from such atavistic thinking and blatant disregard for international law.

With respect to shifts in the global energy market and the role or not of nuclear energy, as the war in Ukraine continued and Russian forces occupied Chernobyl and the Zaporizhzhia Nuclear Power Station—the latter of which is Europe's largest nuclear power plant—I hoped that precisely what I had warned about back in March 2022 (Andresen 2022b, 307, 315-16) would be clear to everyone. Nuclear power plants are far too dangerous to be prudent, and we need to abolish them, along with abolishing nuclear weapons and other nuclear technologies. But people responded to the situation in Ukraine with short-term rather than long-term logic. Understandably concerned about the immediate specter of a cold winter throughout Europe, people moved in the direction that I think was intended by certain individuals at the beginning of the conflict—namely towards reenergizing the nuclear power industry in Europe and elsewhere around the world. Even though the very fact that nuclear power plants can be taken hostage underscores that nuclear power installations are far too dangerous to be prudent, as the war in Ukraine continued, vested interests pushed to extend the life of nuclear power plants in Germany. This benefited German nuclear power plant operators temporarily, prior to the last nuclear power plant being switched off in Germany on April 15, 2023 (Jordans 2023).

While transforming the energy paradigm on Earth will shift wealth and power dynamics significantly, transforming energy will have an even larger impact on human psychology. If energy and transportation networks transform to a naturally abundant, nonpolluting, and free source of energy for everyone, the scarcity mindset and the concomitant dynamics of control and exploitation that arise from a scarcity mindset will disappear. At some point in the future, if *Homo sapiens sapiens* survive long enough as a species to realize this occurrence, energy abundance will provide everyone with the foundation for agricultural, manufacturing, and creative self-sufficiency. It also will improve public health by eradicating famine and by ensuring that everyone has access to ample supplies of clean water. Such welcome changes will permit people to achieve some measure of actual independence, which will give 'power' (in the sense of control over one's own destiny) to the people and take it away from the elite. This certainly will be a just result, though the path we must tread to create such a world is complex.

While access to high amounts of energy has the potential to help human beings move beyond fossil fuels and their devastating impact on the environment, the problem is that everyone racing to understand UAP propulsion today *has grossly underestimated the sophistication of UAP technology and the intense amounts of energy potentially available to be harnessed*. The downside risk that high amounts of energy could be weaponized means that if such enormous amounts of energy are unleashed before human being mature ethically and spiritually, humankind could annihilate itself by pursuing high energy utilization without proper constraints. Only when we are ethically and spiritually mature will it be safe for information on UAP propulsion and/or materiel—and UAP movement technologies more generally—to come out into the mainstream. When this time comes, humankind will have access to considerable energy at little or no cost. It also will develop sophisticated movement technologies to explore the universe.

In the meantime, we should remember that not all relationships need be mediated by technology. The most rewarding relationships are direct and genuine. Here, the most far-reaching impact of creative acculturation with ETI does not involve our energy infrastructure—it involves how we perceive and understand the nature of reality.

What we *really* need first and foremost is not enormous amounts of energy to power technology—it is learning to transform human consciousness so we are more able to manifest kindness, compassion, and love.

Chapter 5:
Bohm's Ontological Interpretation of Quantum Theory

Bohm's Contribution in the Context of Physics

Regardless of how helpful a model or theory in physics is in describing reality, it is important to recall that all models and theories are only systems of abstraction that approximate reality—they are not reality itself. The only way to access reality itself is by means of direct experience. Nevertheless, conceptually, models and theories provide helpful approximations of reality, with some abstractions being better than others in this regard.

David J. Bohm articulates a unified framework in physics that provides a causal account of reality and that reconciles general relativity and the quantum theory. Although many different interpretations of the quantum theory exist, I prefer Bohm's causal, or ontological, interpretation of the quantum theory since it reformulates the quantum theory away from a mechanistic interpretation while elucidating general relativity and the quantum theory by appealing to their deeper, shared nature—the basis common to both, which is unbroken wholeness. Conceptually, current attempts to develop a quantum theory of gravity, which have not been successful in mainstream physics, fall short of Bohm's coherent, single account of reality. Both conceptually and mathematically, Bohm's framework goes beyond a quantum theory of gravity and presents a *quantum theory of the whole*, which also is referred to as a *quantum theory of wholeness*. How Bohm reconciles the quantum theory with general relatively is a central aspect of his book aptly entitled *Wholeness and the Implicate Order* (Bohm 2005 <1980>). To unify general relativity and the quantum theory theoretically, one requires continuity across scale. Bohm articulates precisely such continuity in his quantum theory of unbroken wholeness, which sees macro and micro as part of a continuum. Noticing the continua aspect of various aspects of reality is the same method that Bohm uses to explain the inseparability of mind and matter (Bohm 1990).

Bohm goes beyond the conventional, quantum view to make alternative proposals and to provide an account of an actual process underpinning the observational results, even though the nature of this actual process is unclear. Searching for an actual, underlying process is not a return to the classical view. In some ways, it is similar to the driving force behind the present search for a generalized quantum view of gravity. Both share that they are searches for an ontology (see for example Hartle 1992) (Hiley 2002, 3).

Bohm's ontological interpretation of the quantum theory is an alternative level explanation that moves beyond interpretations of the quantum theory proposed by Niels Bohr (b. 1885) (1934; 1958) and Werner Heisenberg (b. 1901) (1949 <1930>), both of which were solely epistemological in nature. Bohm developed his causal interpretation of the quantum theory in part because he questioned Bohr's assumption that no conception of the individual quantum process is possible (Bohm 1990, 275). In contrast, Bohm wanted to account for the actual—i.e., to provide an explanation for how an ensemble of individual particles travels and arrives at the detecting instrument in a manner that accounts for the observed probability distribution. Accounting for the actual requires something *much more subtle* than a simple return to a mechanistic picture (Hiley 2002, 3, 9).

Bohm arrived at his quantum theory of wholeness because he believed— quite correctly—that *a deeper reality must exist* that is common to both the quantum theory and to general relativity. In the process of his endeavor to present a single, coherent account for reality, Bohm developed his causal, ontological interpretation of the quantum theory. Bohm himself developed the framework (Bohm 1951; 1952a; 1952b; 1953; 2005 <1980>; 1987; 1989; 1990, 276-81), with Basil J. Hiley and others making contributions later (Bohm and Hiley 1975; 1982; 1993; Bohm et al. 1987; see also Hiley and Peat 1987).

In developing his quantum theory of wholeness, Bohm was influenced by Albert Einstein (b. 1879) and by deep conversations occurring among physicists at the time with respect to the implications of the quantum theory. According to Einstein's general theory of relativity, the two aspects of reality that we ordinarily perceive as distinct—space and time—really are the same. Both are malleable entities, says Einstein, with space naturally being dynamic on a large scale. But Bohm's approach goes beyond Einstein's. With respect general relativity and the quantum theory, Bohm observes that the two theories:

> differ radically in their detailed notions of order. Thus, in relativity, movement is continuous, causally determinate and well defined, while in quantum mechanics it is discontinuous, not causally determinate and not well defined. Each theory is committed to its own notions of essentially static and fragmentary modes of existence (relativity to that of separate events, connectable by signals, and quantum mechanics to a well-defined quantum state). One thus sees that a new kind of theory is needed which drops these basic commitments and at most recovers some essential features of the older theories as abstract forms derived from a deeper reality in which what prevails is unbroken wholeness (Bohm 2005 <1980>, xvii-xviii).

Indeed, Bohm mentions the concept of unbroken wholeness many times in his descriptions of reality.

Bohm's causal interpretation of the quantum theory is the only interpretation of the quantum theory that is intelligible ontologically—i.e., it is the only interpretation of the quantum theory *that makes wholeness intelligible*. Intelligibility is very important, and Bohm's interpretation of the quantum theory accounts for ontology very clearly. Bohm points out that when confronted by an apparent lack of agreement between the fundamental concepts in general relativity and quantum mechanics—in fact, when confronted by an apparent lack of agreement between the fundamental concepts of any two theories—it is helpful to abstract to an even more fundamental level to see what both theories share. Having done this, one can begin modeling and theorizing again from that perspective. Bohm proposes that because wholeness is common to general relativity and to the quantum theory, wholeness is an appropriate basis for his theoretical model (Bohm 2005 <1980>, xvii-xviii).

Bohm's vast contributions to physics and philosophy have been significantly underappreciated. Many people, including many physicists, do not realize how far Bohm's quantum theory of wholeness extends the quantum theory. After Erwin Schrödinger (b. 1887) developed the non-relativistic wave equation in 1926, Bohm creatively extended the pilot wave theory, also referred to as the pilot wave model, that Louis de Broglie (b. 1892) initially presented in his dissertation in 1924 and articulated in more detail at the October 1927 Fifth Solvay International Conference (see Bohm and Sheldrake 1986, 111-12 and Fiscaletti 2018). The Solvay Conference series was founded in 1912 by Belgian industrialist Ernest Solvay, and presentations there played an important role in the early development of the academic discipline of physics. Because of the conceptual overlap between de Broglie's and Bohm's frameworks, some people describe this general line of inquiry as 'the de Broglie-Bohm theory,' though Bohm extended the general framework significantly beyond de Broglie's early work.

Mainstream quantum theory, which attempts to describe the realm of the very small, often is referred to as quantum field theory (QFT). Quantum processes were discovered about a hundred years ago when scientists were studying the kind of light that glows from gases when they are heated in a glass tube. They realized that such light forms distinct lines or pencil-like beams of color as opposed to a continuous spectrum that one sees when a piece of cut glass projects light on a table and is observed through a prism. The reason light forms distinct, pencil-like beams of color is because when it is heated or otherwise excited, electrons leap from one orbit to another around the nucleus of an atom instead of moving gradually between orbits. This 'quantum leap' describes the property of electrons to move from one orbit to another seemingly without moving through the space between the orbits. Since electrons only can occupy discrete, specific orbits, electrons and atoms possess energy that comes in discrete chunks that cannot be subdivided. These specific, minimum quantities are called 'quanta.' Bohr observes that when atoms are de-excited, they make downward leaps in their specific orbits and emit light in specific wavelengths that produce specific colors. Despite clarity on some features of quantum theory, however, questions remain regarding probability, the act of observation, and measurement. As Bohr observes, measurement changes everything, since the particle's location is uncertain before the particle is detected or measured by instruments. When observation or measurement occur, however, uncertainty regarding the particle's location disappears, since the act of observation or measurement forces a particle to relinquish all the possible places it could have been and select one location where it is observed or measured (PBS 2011).

Bohm states that physicists should concern themselves with questions about self-reference, movement, process, and description of sequential moments. This contrasts with how physicists often focus on arriving at numbers that agree with experimental findings and then leave it at that, thereby resulting in a domain of physics that has become overly narrow. But physics was not always like this. As recently as the 1920s and 1930s, physicists concerned themselves with a much larger domain of interest. Scientists such as Heisenberg, Bohr, Sir Arthur Eddington, and Einstein considered much broader questions, though now, because very few physicists ask truly broad questions, the teaching of physics has degenerated and become more dogmatic and mechanical. Profound, philosophical questions that were vivid in the 1930s, for example, and which underlie the entire approach of physics, have faded. Now, students are presented with formulae and told, "This is quantum mechanics," observes Bohm (1986a, 97-98).

It is curious why more scientists, especially physicists, have not recognized the significance of Bohm's ontological interpretation of the quantum theory. Bohm himself cites to the history physics as an academic discipline to explain why. Bohm refers to de Broglie's 1927 contribution at the Fifth Solvay Congress, mentioned above, noting that de Broglie's work provides the beginnings of an ontological interpretation of the quantum theory. Because de Broglie's model went against the positivist leanings of most physicists of the day, however, his views encountered significant resistance. Many physicists also thought de Broglie's model was too conceptual, with "too many pictures in it that were unverifiable," says Bohm. Bohm also acknowledges that de Broglie was not in the position to respond to important objections posed by physicist Wolfgang Pauli with respect to what is referred to as the many-body problem, i.e., physical issues that arise with regard to the properties of microscopic systems that are comprised of many particles. Einstein, who obviously was a potent political force in the history of science, also did not like de Broglie's model because it was nonlocal. Lacking support in the academic discipline of physics, de Broglie's model went by the wayside, and even de Broglie himself lost interest in the project (Bohm and Sheldrake 1986, 111-12).

One reason some physicists have considered it impossible to arrive at a physical interpretation of quantum theory—which is what Bohm presents—is because of the problem of multidimensional space and what a multidimensional space means physically. Bohm views multidimensional space as a field of information, which is helpful conceptually since information is organized in any number of dimensions. Accordingly, there is no reason why the information field around the electron cannot be said to be in multidimensional space (Bohm and Sheldrake 1986, 110-11).

Unfortunately, negative reception of de Broglie's work also created a negative reception for Bohm's ideas. Most physicists today therefore are unaware of how robust Bohm's causal interpretation of the quantum theory actually is. This accords with what Thomas Kuhn (1962) claims in *The Structure of Scientific Revolutions*, namely that scientific paradigms are accepted or rejected according to the interests and beliefs prevailing among scientists at a given time. Once an idea or paradigm is accepted by the consensus, one instantly responds to the situation according to conditioning, which is why many physicists did not explore Bohm's conceptual framework and ontological interpretation after Bohr's physical interpretation of the quantum theory was more or less accepted. In contrast to Bohm's framework, Bohr's view is that 'quantum mechanics' cannot discuss what is since it deals with nothing but phenomena and only provides statistical rules for connecting phenomena (Bohm and Sheldrake 1986, 112; see also Andresen 1999).

The academic discipline of physics historically has been influenced by very strong personalities. As mentioned above, de Broglie's and Bohm's ideas unfortunately were sidelined in part because Einstein had personal difficulty accepting the concept of nonlocality. Another reason that Bohm's views have not been treated fairly is because J. Robert Oppenheimer (b. 1904), who was deeply jealous of Bohm, unscrupulously and unprofessionally maligned Bohm's views. This is movingly documented in the film *Infinite Potential: The Life & Ideas of David Bohm* (Howard 2020). As mentioned in the film, Oppenheimer was Bohm's doctoral advisor. For political reasons, Bohm himself was not permitted to work directly on the development of the atomic bomb. Nevertheless, Oppenheimer relied on some of Bohm's mathematical calculations for some of Oppenheimer's own work at Los Alamos. Instead of acknowledging Bohm's physics and mathematical genius, however, Oppenheimer blackballed Bohm across the scientific community. Oppenheimer's academic assault on Bohm was so pronounced and Oppenheimer's standing in the field was so high that many of the physicists dismissed Bohm's ideas out of hand without reading Bohm's papers for themselves.

Oppenheimer's attempt to destroy Bohm's career reminds me of Sigmund Freud's attacks against his student Carl G. Jung. Even after Jung had achieved faculty status—or perhaps because of it—Freud undertook a vicious campaign to undercut Jung's career. This type of intellectual jealousy is common in academia and elsewhere in human society. In the case of Bohm and Oppenheimer, it has set the discipline of physics and commensurate human understanding back decades.

Many students studying physics do not even read Bohm's work. For example, when I took physics and other science and mathematics courses as part of my undergraduate engineering curriculum at Princeton University, we studied Einstein, Bohr, Schrödinger, and Heisenberg—but we never studied Bohm. Bohm was never even mentioned. I came across Bohm's writings some years later when I was a graduate student studying Indo-Tibetan Buddhism. At that time, I read *Dialogues with Scientists and Sages: The Search for Unity*, in which Renée Weber (1986), then Professor of

Philosophy at Rutgers University, interviews many people, including Bohm, the Dalai Lama, Rupert Sheldrake, and others. After reading Bohm's insightful interviews in that collection, I went back and read Bohm's physics and other academic publications and immediately realized the tremendous value of Bohm's work.

Bohm states that physics and mathematics go hand in hand, and he emphasizes that having a good physics intuition is important to any undertaking. One reason that intuition is so valuable is that culture has a bad habit of getting in the way. This is true even in science, where much thinking is mediated unduly by cultural factors. Historical research has shown clearly that themes in science often arise in conjunction with political and other social movements that are part of the fabric of our history as human beings. Having arisen, these themes then inform scientific models and theories. For example, from the late nineteenth century to the end of the Nazi regime, German science invoked holism as a reaction to mechanism and reductionism and employed a 'holistic' view to attempt to heal a world fragmented, disenchanted, and torn apart by industralization, urbanization, and war. Also in Nazi Germany, however, invoking the philosophical concept of holism unfortunately became a way to manipulate society conceptually into supporting fascist, authoritarian, and totalitarian political agendas (Harrington 1996; see also Macrakis 1998). The extent to which Nazism influenced German science underscores how all human knowledge systems—be they scientific, religious, sociological, etc.—are culturally conditioned. It is particularly easy to forget that about science, however, since scientific theories often are taken as 'fact' when they are nothing more than explanations that demonstrate an approximate fit to context in any one historical time frame.

Although Bohm's framework is constrained by human mathematical and linguistic conventions, his focus on ontology moves us past a stagnant fixation on algorithms and the repetitive exercise of grasping after them because they 'work' under certain conditions. This makes Bohm's framework less historically and sociologically conditioned than it otherwise might be. Even further, Bohm's ontological notion of wholeness is in many ways conceptually the opposite of how 'holism' is understood in the historical and sociological context of Nazism. In contrast to German understanding of holism, which is an ideology that favors the centralization of power and authority and elite control over others, Bohm's ontology of wholeness is one that values the creativity of each facet of the whole. Bohm's concept of unbroken wholeness leads to the manifestation of multifaceted instances of creativity in society, which is the opposite of any pull towards 'one world government,' 'one world currency,' or 'one world' anything. Said more generally, Bohm's ontology of wholeness is consistent with decentralization and creativity rather than with centralization and stagnation.

Not Bohmian '**Mechanics**'

When I refer to 'the quantum theory' rather than to 'quantum mechanics,' I specifically am following Bohm, whose elucidation of the quantum theory is decidedly not mechanical or mechanistic at all.

Bohm's 'nonmechanics' has many obvious advantages over mainstream quantum mechanics. Bohm did not think that the quantum formalism suggested a mechanistic interpretation or that the quantum theory reduces to mere scientific materialism. In a section of *Quantum Theory* entitled "The Need for a Nonmechanical Description," Bohm writes, "This means that the term quantum mechanics is very much a misnomer. It should, perhaps, be called quantum nonmechanics" (Hiley 2002, 4, citing Bohm 1951, 167).

Bohm maintained his 'nonmechanical' description of the quantum theory in later papers (Bohm 1952a; 1952b). Bohm's book *Causality and Chance* (Bohm 1984 <1957>) also includes many arguments against a mechanistic outlook in physics. Indeed, *Causality and Chance* presents the foundation of Bohm's thinking, which he summarizes later in *Wholeness and the Implicate Order* (Bohm 2005 <1980>). Some of Bohm's ideas from his 1952 proposals also are found in *The Undivided Universe* by Bohm and Hiley (1993). There, Bohm and Hiley move beyond John Bell's (1987) discussion of differences between the mechanistic view and the standard approach to quantum mechanics one finds in mainstream physics (see also Bell 1966). As Hiley puts it, "There are strong indications that something much more radical was needed." Bohm and Hiley "decided to explore the structure more deeply to see if it was telling us anything more. We found it was," writes Hiley (2002, 4-5).

Explicate and Implicate, Etc., Levels of Order

Bohm articulates his ontological interpretation of the quantum theory in terms of levels of order, two of which are explicate order and implicate order. As I mention in chapter 1, Bohm extends the quantum theory beyond these two levels of order to postulate *a series of levels of order of increasing subtlety*—from implicate, to super-implicate, super-super-implicate, etc.— potentially to infinity. Bohm also considers the relationship between these levels of order (Bohm 1986a, 96-98).

Explicate order is what is 'unfolded.' It is what conventional perception renders to our conventional five senses all day, every day, and it is what many people mistake for the entirety of reality. Ordinary experience and classical, Newtonian physics focus on explicate order. Even though explicate order *appears* to stand by itself, this appearance only demonstrates the *relative*—not absolute—independence of things. But one cannot fully understand explicate order apart from its ground in the primary reality of the implicate order (Bohm 1990, 273).

Because explicate order describes what is manifest, one could describe it as the realm of apparent forms. Human beings often mistake what manifests in explicate order for reality in its entirety, without consciously perceiving the implicate, super-implicate, super-super-implicate, etc., levels of order from which explicate order phenomena manifest. While physical objects possess some degree of coarse level of matter, they are generated by a higher level of order, i.e., implicate order. In this sense, what we conventionally think of as a single object actually is not static. Physical objects' appearance of continuity in spacetime is nothing more than an epiphenomenon of perception. In reality, a different instance of each object is created in each instant.

Super-implicate Order and Beyond

One can begin to understand the super-implicate order according to a model that proposes that enfoldment occurs on two levels. The first is an enfolded order of the vacuum with ripples on it that unfold. The second is a super-information field of the whole universe, which is a super-implicate order that organizes the first level, i.e., implicate order, into various structures. The super-implicate order is capable of tremendous development of structure. This contrasts with the holographic theory, for example, which has an implicate order but does not have anything to organize it (i.e., a super-implicate order). In the holographic theory, the implicate order is linear, merely passing through itself and diffusing around. Although special devices can unfold it, it does not have an intrinsic capacity to unfold an order. By way of contrast, in Bohm's framework, the super-implicate order is the higher field since the implicate order is a wave function and the super-implicate order is *a function of the wave function* — i.e., a higher order, or super-wave function, if you will. The super-implicate order makes the implicate order non-linear and organizes it into relatively stable forms with complex structures (Bohm 1986b, 33).

A super-super-implicate order also may exist, and there may be an [super-super-super-]implicate order even beyond that one. Bohm (1986b, 33) states:

> I'd like to propose that we are making a series of abstractions and any level of thought must cut off somewhere. Even if we put more in, there is still more left out. It's inherent in thought that it is not going to grasp the actual totality. But the holistic part of thought would be thought which does not make a break, thought which is unbroken.

Asked by Weber if this is a continuum of ordering principles, Bohm replies, "That's right. Even when we say that we have made a break, we realize that it really shades off into the unknown. That's essential for this quality of wholeness."

Holomovement

In contrast to explicate order, which is 'unfolded,' implicate, super-implicate, super-super-implicate, etc., levels of order are 'enfolded.' These relatively subtle levels of order comprise the enfolded ground from which explicate order manifests. According to Bohm, existence is the reciprocal flow back and forth between explicate and implicate, etc., levels of order. He calls this reciprocal flow 'holomovement.' Bohm states, "Because the implicate order is not static but basically dynamic in nature, in a constant process of change and development, I called its most general form the *holomovement*" (Bohm 1990, 273). Everything is in flow in holomovement. While language gives the impression that implicate and explicate levels of order are dichotomous, they are not. It is more accurate to visualize implicate and explicate levels of order as holomovement's background and foreground, with no sharp demarcation between them.

Bohm describes holomovement in ontological terms as the ground of what is manifest. The basic movement of holomovement is folding and unfolding. What is manifest is abstracted and floating in the holomovement. According to Bohm, existence as a whole is a holomovement that manifests in relatively stable form. It is the flux coming into balance for the time being, i.e., coming into relevant closure similarly to a vortex, which closes on itself even though it always is moving. The coming into temporary balance of the flux results in more stable forms of matter as compared to a more dynamic manifestation of the flux itself. For example, a cloud holds a relatively stable form, so even a cloud can be regarded as a manifestation of the movement of the wind. One similarly can regard matter as forming clouds within the holomovement, which is what manifests the holomovement to our ordinary senses and thought. All entities such as cells, atoms, and human and other types of intelligent beings are forms in the holomovement (Bohm 1986b, 26-27).

Unfoldment from implicate, super-implicate, etc., levels of order begins to provide a good account for what is meant by the mathematics of the quantum theory. The unitary transformation, which is the basic mathematical description of movement in the quantum theory, is the mathematical description of holomovement. According to present quantum theory, the vacuum contains an enormous amount of energy—so much that it is virtually unlimited from a human perspective—even though physicists generally ignore this energy because human-made instruments cannot measure it. The tacit philosophical assumption underlying this stance is that only what can be measured by an instrument can be considered real. But this is inconsistent, since physicists also accept that particles exist that cannot be detected by instruments at all. What one can say is that the present state of theoretical physics implies that empty space has all this energy, and that matter is a slight increase of the energy. Accordingly, matter is like a small ripple on this tremendous ocean of energy, having some relative stability and being manifest. Bohm suggests that implicate order implies a reality immensely beyond what we call matter. Matter itself is merely a ripple in this background, which one can think of as an ocean of energy. The ocean is not primarily in space and time at all—it is primarily in implicate order, which is to say that it is unmanifest. While it may manifest in this little bit of matter, i.e., the ripple, the source is the implicate level of order, which is an ocean of energy that is unmanifest and, at least by humans, untapped. Beyond that ocean are even larger oceans, which in Bohm's framework refer to the super-implicate, super-super-implicate, etc., levels of order, potentially extending to infinity (Bohm 1986b, 26-28).

It is important to keep in mind that explicate, implicate, etc., levels of order are not absolutes—they merely are abstractions that are useful for explanatory purposes. Existence itself is a dynamic flow between levels of order—the unceasing manifesting and unmanifesting from implicate to explicate, back to implicate order, etc. I tend to use the word 'manifesting' to refer to the movement from implicate to explicate and the word 'unmanifesting' to refer to the movement from implicate to explicate. Bohm (1983, 120) uses the terms 'creation' and 'annihilation':

> What we have here is a kind of universal process of constant creation and annihilation, determined through the super-quantum potential so as to give rise to a world of form and structure, in which all

manifest features are only relatively constant, recurrent and stable aspects of this whole.

This describes the process of manifesting and unmanifesting that is so central to Bohm's overall framework.

Bohm proposes that the implicate order can be defined by the quantum potential, Q, which is the field consisting of an infinite number of pilot waves. The overlapping of the waves generates the explicate order of particles and forces, and, ultimately, of space and time (Scaruffi 2000). The original field, the quantum potential, or Q, is the first implicate order, while the super-quantum potential is the second implicate order, or super-implicate order (Pylkkänen 2007, 178; see also Fiscaletti 2018, 57-61).

Another important aspect of the quantum potential is its geometrodynamic nature. As Fiscaletti (2018, 13) writes:

Taking account of the geometrodynamic nature of the quantum potential, one can also say that the quantum potential expresses the geometric properties of space which determine the behaviour of the particles and are derived just from a more fundamental geometrodynamic background, namely the implicate order.

This underscores the fundamental role of geometry as an integral aspect (pardon the pun) of the nature of reality. What is particularly interesting about Fiscaletti's proposal is that one can view geometry not only as an aspect of explicate order, but also as an aspect of implicate order.

An important concept in Bohm's thinking is that *the essential qualities of fields exist only in their movement*. The conceptual picture is not one of a field with some essential qualities, which may or may not move. Instead, it is movement that gives rise to the essential qualities of fields. One could say that movement is more fundamental, while the essential qualities—be they of fields or of particles—are derivative. Holomovement refers to the totality of movement, which is assumed to be the most fundamental nature of existence currently known to us and what gives rise to the essential qualities of fields and of particles. Although fields can give rise to particle-like manifestations via certain recurrent unfoldment and enfoldment, if reality is more fundamentally *movement*, then the notion of a permanently existent entity with a given identity—even at the level of a particle—is incorrect (Pylkkänen 2007, 25).

Interpreting the Wave Function

In Bohm's framework, one can interpret the wave function "in terms of possibilities, or better still, in terms of *potentialities*" (Hiley 2002, 3).

In classical physics, a potential is a force field generated by an outside agency that describes the potential evolution of the particle starting from a given position (Hiley 2002, 16). In contrast, to understand potentialities in the quantum context, one refers to the quantum potential, or Q. Bohm originally intended to provide a description of the evolution, understood as the development over time, of an actual quantum process. For such a description, mere potentialities are insufficient. But the Bohm approach uses the wave function to obtain equation #17 (see Hiley 2002, 12 for equation #17), which means that the equation contains information about potentialities (Hiley 2002, 15).

According to Bohr, the wave function is simply a term in an algorithm from which one can calculate the probable outcome of any given experiment. On Bohr's view, the wave function does not say anything about how a particle leaving a source arrives at the detecting instrument. Somewhat poetically, or at least metaphorically, Miller and Wheeler (1984) therefore liken a quantum process to a "smoky dragon." Bohm pushes the ball (or the smoky dragon, as the case may be) down the court from here. Bohm's trajectory provides an explanation of how an ensemble of individual particles can travel and arrive at the detecting instrument in a way that accounts for the observed probability distribution. This is necessary to help understand the kind of underlying process that could produce observed probability distributions in general without recourse to smoky dragons and other fantastical beasts, so to speak. Although it is well known that one cannot 'see' the particles travelling along a trajectory, in the absence of any information to the contrary, one can imagine that the particles do travel along such paths, even if one is not entirely sure whether they do or do not (Hiley 2002, 3).

Even though Bohm's approach clarifies many things, the quantum world seems to insist on retaining some degree of murkiness. For example, it makes sense to ask if one can claim that a particle does not travel along a trajectory, though Hiley argues that it is better to adopt a position that understands a particle to travel along such a trajectory unless it leads to a contradiction rather than saying that one has no idea how a particle gets from A to B. A more fundamental question is whether localized, classical-type particles exist at the quantum level at all—or whether the quantum behavior is a pointer to a subtler form of motion that has more to do with process than with particles/fields in motion in spacetime. This in fact may be the case (see for example Brown and Hiley 2000 and Hiley 2001) (Hiley 2002, 4).

If, for example, classical-type particles do not actually exist at the quantum level, this would have significant implications for our understanding of ETI/UAP. ETI itself may understand a subtler form of motion that depends on process more than on the notions of particles/fields in motion in spacetime.

Nonlocality and Entanglement

Nonlocality and entanglement refer to the way in which one aspect of a system participates in another without a local connection between them (Bohm and Hiley 1975). Bohm makes a characteristically astute and insightful remark:

> In the implicate order we not only always deal with the whole (which the [i.e., Einstein's] field theory also does), but we also say that the connections of the whole have nothing to do with locality in space and time but have to do with *an entirely different quality, namely enfoldment* (Bohm 1986b, 26, italics added).

In other words, *enfoldment explains nonlocality.*

Although the concepts of nonlocality and entanglement are very similar, they are not strictly equivalent since some mixed, entangled states do not generate nonlocal correlations (Massar and Pironio 2012). Putting this slightly differently, in certain circumstances but not in all circumstances, nonlocality occurs because of entanglement. Subtle differences exist between the concepts of entanglement and nonlocality, which nevertheless also are closely connected. Entanglement refers to when particles are permanently correlated such that they depend upon the states and properties of one another and behave like a single entity. Einstein referred to this as "spooky action at a distance" (Cofield 2017), since two entangled particles demonstrate a nonlocal connection by behaving as if they 'know' something about one another even when they are separated by a great distance.

Bohm's causal interpretation of the quantum theory helps illuminate the nature of entanglement and nonlocality. First, Bohm's interpretation assumes that the electron is a particle following a well-defined trajectory, somewhat like a planet orbiting the Sun. Second, Bohm's interpretation proposes that the electron always is accompanied by a new kind of quantum field. Although one tends to think of a field as something that is spread out over space, such as a magnetic field, the quantum field is different from this conventional concept. The quantum field possesses qualitatively new features that imply a radical departure from Newtonian physics. Speaking generally, one can represent fields mathematically in physics by using certain expressions called potentials. Bohm (1990, 276) writes, "In physics, a potential describes a field in terms of a possibility or potentiality that is present at each point of space for giving rise to action on a particle which is at that point." Similarly, as with electric and magnetic fields, the quantum field also can be represented in terms of a potential, which as mentioned above, Bohm calls the quantum potential, or Q. Such a notion is fundamentally different from classical Newtonian ideas. Whereas in Newtonian physics, the effect of the potential on a particle is always proportional to the intensity of the field, as for example is the case with electric and magnetic potentials, *the quantum potential depends only on the form and not on the intensity of the quantum field*. This means that even a very weak quantum field strongly can affect the particle, which in turn implies that even distant features of the environment strongly can affect the particle.

The two slit interference experiment elucidates the nature of entanglement and nonlocality very clearly. The quantum potential, which consists of a series of plateaus separated by deep valleys, is present in front of the slits. When an electron crosses one of these valleys, it is accelerated sharply. The quantum potential deflects the electrons, even in the empty space in front of the slits, and this deflection still may be large even far away from the slits. In its movement, the electron is affected by the quantum potential, which is determined by the wave that in general precedes the particle. However, if one follows the entire set of trajectories, which represents an initially random distribution of particles, one observes that they are bunched systematically into a fringe-like pattern. This becomes apparent after many electrons have arrived at the screen in front of the slits. This demonstrates how the electron can be a particle yet also can manifest characteristic wave-like properties statistically. *For this explanation to work, the quantum potential must depend only on the form of the wave, so that it can be strong even when the wave intensity is weak.* As Bohm writes, "what is basically new here is the feature that we have called non-locality, i.e. the ability for distant parts of the environment (such as the slit system) to affect the motion of the particle in a significant way (in this case through its effect on the quantum field)" (Bohm 1990, 276-79).

The implications of entanglement and nonlocality are profound. As Fiscaletti (2018, 14) writes:

> In Bohm's view the non-locality is a characteristic subtended of space-time and the particles are seen as vibration modes of the global field which is the dynamical expression of the fundamental level, of the deep geometrical structure. One can thus say that the geometrodynamic properties of space determine a non-local global field, the implicate order and that the subatomic particles emerge from this fundamental background.

Emphasized here is that geometry is central to our understanding of the nature of reality, and that the geometrodynamic properties of space determine a nonlocal global field, which is implicate order.

An extremely important aspect of the nature of reality is that everything that exists does so by means of participation rather than interaction. At low temperatures, wholeness, nonlocality, and the organization of movement by means of common pools of information accounts for the basic behavior of matter. At higher temperatures (Bohm and Hiley 1987), the quantum potential tends to take the form of independent parts, which implies that the particles move with an *apparent* independence. Whereas classical physics treats the whole merely as a convenient way of thinking about what many people erroneously assume is nothing more than a collection of independent parts involved in mechanical interaction with one another, Bohm's view is much different. He observes that the notion of participation rather than interaction is a more accurate way of thinking about what is occurring, and he emphasizes that "the possibilities for wholeness in the quantum theory have an objective significance" (Bohm 1990, 280-81).

Reality is not comprised of 'individual' entities interacting like billiard balls, i.e., colliding, moving apart, and colliding again. Instead, everything participates in everything else. Quantum wholeness means that in an observation carried out to a quantum theoretical level of accuracy, the observing apparatus and the observed system cannot be regarded as separate. Rather, each *participates* in the other to such an extent that it is impossible to attribute the observed result of their interaction unambiguously to the observed system alone" (Bohm 1990, 275).

About nonlocality, Bohm observes that "under certain conditions, particles that are at macroscopic orders of distance from each other appear to be able, in some sense, to affect each other, even though there is no known means by which they could be connected" (Bohm 1990, 274). In the case of a single particle, a nonlocal connection is demonstrated experimentally because the quantum potential does not necessarily fall off to a negligible value when the particles are separated even by macroscopic orders of distance (Bohm 1989, 8). Despite such experimental verification of nonlocality, human beings' rigid insistence on thinking in a linear manner and in terms of localized interactions between 'objects' and inherently separate 'beings' causes humankind to go terribly wrong. To survive as a species, human beings must learn to perceive the wholeness of reality and to become much more aware of nonlocality and entanglement.

Focusing on the quantum potential, or Q, leads one to see the quantum significance of entanglement in an entirely new way, namely its striking implication for quantum nonlocality (see Bohm and Hiley 1975). By reflecting on the significance of quantum entanglement for quantum nonlocality, Bell (1987) derived his famous inequalities. The calculations of Dewdney, Holland and Kyprianidis (1987) for the case of two coupled spin-half systems in an entangled state demonstrate "how the dramatic appearance of the non-local was mediated by the quantum potential." Normally, calculations explaining an experimental result would be taken as an indication of the success of the theory. However, in the instance of the calculations of Dewdney, Holland and Kyprianidis, the politics of science intervened negatively such that some people regard the calculations as meaningless even though they are not (Hiley 2002, 9).

Despite experimentally confirmed violations of the Bell inequality, some people find the appearance of nonlocality a weakness of Bohm's framework. Really, it is a strength. Bohm's theory provides a detailed explanation of the Einstein-Podolsky-Rosen (EPR) paradox rather than leaving it unexplained. In fact, Bohm's explanation helped Hiley (1995) make more sense of Bohr's answer to the EPR paradox (Hiley 2002, 9). EPR is a physics conjecture that two entangled particles are connected by a wormhole, also referred to as an Einstein-Rosen bridge. In keeping with the notion that all things in the universe are connected, a theoretical link has been proposed between quantum entanglement and wormholes. Decades ago, scientists stated that it was not ruled out theoretically that by threading exotic forms of energy through a wormhole, it may be possible to travel through the wormhole to a part of the universe that at that moment was inaccessible along a linear trajectory (Moskvitch 2013; Lindley 2005).

Scientists continue to make new, unexpected discoveries with respect to entanglement. Brookhaven National Laboratory recently uncovered a new kind of quantum entanglement that resolves a longstanding mystery with respect to the protons and neutrons present in the nuclei of atoms. People are hopeful that this discovery will help with many topics ranging from quantum computing to astrophysics (Ferreira 2023).

Many-Particle Systems

When one considers a system consisting of many particles in contrast to a system consisting of a single electron, several new concepts emerge that more fully illuminate the meaning of wholeness. First, as stated above, even when two or more particles are separated by long distances, they can affect each other strongly by means of the quantum potential. Second, in a many-particle system, one can think of the participation of the particles as depending on a common pool of information that belongs to the system as a whole in such a way that is not analyzable in terms of pre-assigned relationships between individual particles (Bohm 1990, 280).

Since matter depends on a common pool of information as soon as it enters a many-particle system, matter is being informed constantly, which means that whatever would determine matter as one particle no longer determines it in a many-particle system. Furthermore, the many-particle system can become part of an even larger system, which has its own pool of information, etc., which means there is no final determination—instead, a relative determination exists in accordance with context. According to the quantum theory, the structure of a molecule depends on the pool of information that is in the wave function of the whole system, such that one structure or another manifests in accordance with the pool of information. Said differently, molecular structures are determined by the wave function of the whole system, which is the whole system's pool of information (Bohm 1989, 15-16).

An intuitive picture of the meaning of the quantum theory is provided by Bohm's observation that in a superconducting state, the wave function is similar to the score of a ballet. The meaning of the fields around the electrons determines the interrelationships of the particles, with the implication being that things that are quite a distance from one another remain connected. Another implication is that the interrelation of particles depends on the state of the whole in a way that is not a pre-assigned property of the parts. Before the parts manifest, the whole has a certain prior meaning, which is something similar to the score of a dance. As mentioned above, in a superconducting state, electrons move in a regular, coordinated way without scattering, whereas in any ordinary state, they act like a disorganized crowd of people (Bohm and Sheldrake 1986, 111).

The phenomenon of superconductivity exemplifies how in a many-particle system, the participation of the particles depends on a common pool of information that belongs to the system as a whole in such a way that it is not analyzable in terms of pre-assigned relationships between individual particles, as mentioned above At ordinary temperatures, electrons moving inside a metal are scattered in a random way by obstacles and irregularities in the metal, resulting in a resistance to the flow of electric current. In contrast, at low temperatures, electrons move together in an organized way, including going around obstacles and irregularities to re-form their pattern of orderly movement together. Since the electrons are not scattered, the current can flow indefinitely without resistance. More detailed analysis shows that the quantum potential for the whole system constitutes a nonlocal connection that brings about the organized and orderly pattern of electrons moving together without scattering (Bohm 1990, 280).

As mentioned immediately above, in the superconducting state, which may arise at very low temperatures, electric current flows indefinitely without friction because electrons are not scattered by irregularities or obstacles in the metal in which they are flowing. Accordingly, the superconducting state of elections in a metal demonstrates the activity of a common pool of information in a mechanical context. Bohm states, "In terms of this model, one sees that in the superconducting state, the common part of information induces an organized and coordinated movement of electrons resembling a ballet dance, in which the particles go around irregularities and obstacle without being scattered" (Bohm 1989, 8).

Describing how the wave function is similar to the score of a dance, Bohm states:

Now if you compare this to the ballet, you could say that in the super-conducting state, the wave function is like the score—it's a kind of information—and the dance is the meaning of the score. It is possible for the wave function to have a special form—mathematically called *factorize*—where the various dancers are independent. Generally they are not. This explains why at higher temperatures we get independence and classical behavior. But the lower the temperature, the less the independence. There are other conditions which favor quantum mechanical behavior, but in general high temperature favors independence of the various dancers, so the system falls into independent parts. The meaning of the whole score is such that it determines how many independent dances are going on and what they are. To make this ballet dance analogy better, let's say the wave function score is not fixed, but it's a score which depends on the initial configuration of the particle. The dance would vary according to the initial configurations of the dancer (Bohm and Sheldrake 1986, 111).

In the quote above, Bohm is describing a situation that is more similar to improvization as compared to choreography. In this analogy, the 'score' evolves according to Schrödinger's equation, as if there were rules by which the dance itself is constantly changing (Bohm and Sheldrake 1986, 111). Bohm states:

That gives you a fairly good intuitive picture of what the meaning of quantum mechanics [i.e., 'quantum nonmechanics'] is. We can say that the basic properties of matter—which determine chemistry, super-conductivity, DNA, life—come from this. They don't come from the mechanics; they come from this. There would be no chemistry and no molecules were it not for this property (Bohm and Sheldrake 1986, 111).

The aspect of the 'score,' which evolves according to Schrödinger's equation, has been missed in mechanistic interpretations of the quantum theory. Per Bohr, such interpretations view the wave function simply as a term in an algorithm from which one can calculate the probable outcome of any given experiment (Hiley 2002, 3).

Either one can take Bohr's mainstream view of the quantum theory and say it is nothing more than a system of calculations that people use only for the purpose of working out the mathematics, or, alternatively, one can try to interpret the quantum theory and to find the meaning, as Bohm does. This becomes more evident when one moves from physics to chemistry. For example, Bohm proposes that one can understand chemistry in terms of something such as information and its meaning, even though this proposal is not commonly known or accepted. Nevertheless, the quantum theory currently is considered the best theory available to explain what is occurring in many-particle systems such as chemistry. Applying Bohm's interpretation to chemistry, molecular structures are determined by the wave function, which is the pool of information (Bohm 1989, 15-16).

In essence, then, there are many different ways to reflect on and interpret what is occurring in reality. A physics approach starts with looking at things primarily with the available evidence. Here, one can say that there is an energy that is relatively unformed amd that is organized by the field, and that this gives rise to the particles. In other words, the chemical molecule now has a field, namely the Schrödinger field, which organizes it such that, as Bohm observes, "there is not such a sharp distinction between the field and the particle." He continues, "The organizing field [i.e., implicate order] is present everywhere and I'm proposing that there may be super-organizing fields [super-implicate, super-super-implicate, etc.] which organize the wave function and that might shade over into all sorts of other fields, so that we are looking at two different areas in which this organization is taking place" (Bohm and Sheldrake 1986, 117).

Quantum Potential Carries Active Information

In Bohm's framework, particles respond to information. The difference between an electron and a proton is largely the different way in which each responds to information in the wave function (Bohm 1989, 11). The electron responds with its own energy to the form and not to the intensity of the wave, however strong or weak the intensity of the wave may be. This response strongly can reflect distant features of the environment, which implies a certain new quality of wholeness of the electron with the environment that is not present in classical physics. Each particle depends on the field of information in the Schrödinger wave, and the whole environment is reflected in information in the form of the wave. The electron, for example, may have an inner complexity comparable to that of a radio set or a vessel guided by an automatic pilot. This idea goes against the general approach in physics over the past few centuries, which assumes that as one analyzes matter into smaller and smaller units, its behavior becomes simpler and simpler. The implication here is that in the quantum domain, such an approach is no longer adequate. Rather, the situation is more like what is encountered in a large population of human beings. Whereas some relatively simple statistical analyses can be made with respect to a large mass of people, when it comes to the individual, one discovers a complexity and subtlety beyond analysis (Bohm 1989, 7-8).

According to Bohm's ontological interpretation of the quantum theory, the quantum potential represents *active information*. Bohm theorizes that the quantum potential contains active information, which is information that is potentially active everywhere but actually active only where and when there is a particle. Active information may be organized in pools of sizes that vary according to the conditions. This implies that the current quantum laws are only simplifications and abstractions from a vast totality, and, also, that the current quantum laws describe reality in somewhat of a superficial manner. In physical experiments and observations carried out thus far, deeper levels of the vast totality have not yet revealed themselves (Bohm 1989, 6, 9).

Bohm uses the new notion of active information to elucidate further the feature of nonlocality. In Bohm's framework, nonlocality is the ability of distant parts of the environment such as the slit system to affect the motion of the particle in a significant way by means of its effects on the quantum field. It is by means of the relation of active information to the quantum potential that the notion of wholeness becomes intelligible. Here, one takes the idea of "in-form," meaning to put form into, literally, which is different than the technical meaning of negentropy found in information theory. Although one may think of the electron as moving under its own energy, the quantum potential acts to put form into its motion. This form is related to the form of the wave from which the quantum potential is derived. The notion of active information therefore implies the possibility of a certain kind of wholeness of the electron with distant features of its environment. Bohm describes this idea by saying that "the possibility of a certain kind of wholeness of the electron with distant features of its environment. ...can be understood in terms of the concept of a particle whose motion is guided by active information." Although in certain ways Bohm's view is similar to Bohr's notion of wholeness, Bohm's view is different in that it invokes the idea that active information guides the motion of the particle as the basis for understanding the wholeness of the electron with distant features of the environment. In contrast, in Bohr's approach, there is no corresponding way to make such wholeness intelligible. Bohm's view is that intelligibility is important, and his framework relating to wholeness helpfully moves beyond Bohr's thinking by employing the notion of active information to provide for intelligibility (Bohm 1990, 278-80).

Hiley (2002, 1) discusses some background with respect to the notion of active information for the purpose of accounting for quantum processes. He writes:

> To appreciate the full significance of this new notion, we show why it is essential to distinguish our approach from the approach that goes under the name 'Bohmian mechanics'. We then show for the first time how the quantum potential emerges from the Heisenberg picture thus providing a new perspective to the whole approach. This enables us to clarify the role of the energy associated with the quantum potential and the status of active information. We conclude with some remarks on the relation between active information and Shannon information.

Furthermore, focusing on the quantum potential illuminates the significance of quantum entanglement in a new way by helping to demonstrate its striking implication for quantum nonlocality (Hiley 2002, 9; see also Bohm and Hiley 1975).

It is important to understand the quantum potential and its origins as clearly as possible. In fact, the notion of the quantum potential and its origins even confuses some physicists who generally support Bohm's interpretation. While the quantum potential emerges from the Schrödinger picture, it is important to ask where the quantum potential is in the Heisenberg picture. Since the two pictures are equivalent when one does not consider quantum fields, something equivalent to the quantum potential also must be present somewhere in the algebraic picture of Heisenberg. Monk and Hiley (1998) and Brown and Hiley (2000) suggest that one should take a more general point of view according to which one takes the algebraic structure of the quantum formalism as more basic and the representation as a secondary feature of the theory. This approach may be helpful particularly when one's primary objective is to illuminate how the quantum potential emerges from the Heisenberg picture (Hiley 2002, 10-15).

By working through the math, one arrives at a simple derivation of the Aharonov-Bohm effect and the Berry phase (Brown and Hiley 2000). According to this approach, the quantum potential only takes on an explicit form when one goes to a representation. Both the guidance condition and the quantum potential *only appear when we choose a representation for the operators*. In other words, only when one projects operators into a representation space do these two conditions arise. Furthermore, in this representation, two sets of trajectories exist for the same system. This in turn means that in effect, there are two phase spaces. This outcome arises when one assumes the algebraic elements and takes the algebraic formalism of the quantum theory seriously without bringing in conventional prejudices with respect to classical or quantum physics. If one assumes that the algebraic elements describe some new notion of 'process' per se, one can ask what the structure of this process tells us about the nature of quantum processes. In both cases, a guidance condition and a quantum potential both make an appearance in the same set of equations, namely those that arise from the specific representation of the anti-commutator (Hiley 2002, 10-15).

One immediate advantage of the algebraic approach is that it restores the symmetry between the r-representation (Belinfante 1973) and the p-representation (see Brown and Hiley 2000)]. Lack of apparent symmetry was a feature that Heisenberg had criticized. But there is no lack of symmetry—indeed, there is completely symmetry reflecting the fact that the symplectic (canonical) symmetry that is present in both the classical domain and in the standard interpretation also is present in the Bohm approach, as it must be. Nowhere does the quantum potential or the guidance condition appear in the algebra of operators. Both only appear when a particular representation is chosen, and then they both appear together in the same equation. Therefore, if the Schrödinger equation is assumed to be valid, there is no need to establish the guidance condition as a primitive condition. With respect to the *internal energy* of the particle evolving in a given experimental setup, it is essential that one regard this energy as internal since the quantum potential has no external source. There also is the possibility of distinguishing between the actual and the potential, such that we can describe the actual kinetic energy of the particle (Hiley 2002, 13-15).

The symplectic symmetry is important, and canonical relations are carried over to the quantum domain. One equation can be viewed as a generalized expression for the conservation of energy when one regards Q—quantum potential energy, which appears as a necessary consequence of the symplectic symmetry—as a new form of potential energy that is negligible in the classical world and is apparent only in quantum systems. One should not think of Q as the source of a new force to be added into the Newtonian equations of motion, for example. Also, while the symplectic symmetry is important, the double cover of the metaplectic group plays a key role in the quantum theory (for an account of this structure, see de Gosson 2001). Even though nothing mathematically new has been added and one is working from the assumption of the validity of the Schrödinger equation, what is novel is for what physical reason Q appears and what it could represent. Even though it is not particularly radical to ask this type of question in physics, when Bohm posed these questions, the politics of the discipline was such that it was seen as a radical departure from what others were doing at the time (Hiley 2002, 6). Nevertheless, one must take seriously the new form of quantum potential energy Q, which only acts in the quantum domain, since without it, energy would not be conserved. Furthermore, without understanding this energy, the form of the trajectories is merely another quantum mystery (Hiley 2002, 8).

To consider a possible interpretation of the quantum potential, Q, in terms of information, one can consider the role of information in physical theories. Here, it is helpful to distinguish how the term 'information' is used in Bohm's framework as compared to its more commonplace usage as Shannon information. In classical physics, the notion of a potential refers to a force field generated by an outside agency, and it describes the potential evolution of the particle starting from a given position. Since particles in different positions experience different effects, one can think of the potential as being revealed by means of the movements of an ensemble of particles. In contrast, one can think about the quantum potential as describing a field of energy, though it is problematic if one thinks that it can be regarded as producing a force on the particle. Firstly, as stated above, the quantum potential has no external source, so there is nothing for the particle to push against, so to speak. Since the energy is internal, something more subtle is involved. It is more akin to the role the gravitational field plays in general relativity, in which gravitational energy is understood to curve spacetime itself. Secondly, the quantum potential does not arise directly from the Hamiltonian and it therefore does not appear explicitly in two algebraic equations (see Hiley 2002, 10-11, equations #8 and #9). It only appears when one projects equation #9 into a particular representation space. This is even more like gravitation, in which the 'force' appears only when one projects the geodesics into a Euclidean space. Only in a Euclidean space does one see the deflected trajectories revealing the presence of the gravitational force (Hiley 2002, 16). Thirdly, the form of the quantum potential in equation #17 (as mentioned above, see Hiley 2002, 12 for equation #17) shows that it depends on the amplitude of the quantum 'wave,' which is a second derivative. Not only this, but the amplitude also appears in the denominator, which implies that the potential is no longer proportional to the amplitude of a wave, which is what one would expect from classical physics (Hiley 2002, 16-17).

Mathematically, the key to the quantum potential energy lies in the second spatial derivative, which indicates that the shape or form of the wave is more important than its magnitude. Accordingly, *a small change in the form of the wave function can produce large effects in the development of the system.* Also, the quantum potential produces a law of force that does not necessarily fall off with distance. This means that *the quantum potential can produce large effects between systems that are separated by large distances* (Hiley and Pylkkänen 2005, 19).

In accordance with the discussion above, the quantum potential gives rise to effects that are totally different from those expected from a classical wave. In the classical case, the force produced by a wave is directly proportional to its amplitude—which one knows directly from swimming, for example. The appearance of the amplitude in the denominator also explains why the quantum potential can produce strong, long-range effects that do not necessarily fall off with distance. These are properties of typical entangled wave functions. Accordingly, even though the function spreads out, the effect of the quantum potential need not necessarily decrease, which is precisely the type of behavior that explains the EPR paradox. If for example one examines the expression of the quantum potential in the two-slit experiment, one finds that it depends on the width of the slits, the distance between them, and the momentum of the particle—in other words, it contains information about the overall experimental arrangement. This is unsurprising mathematically, since the wave function is a solution of the Schrödinger equation, which necessarily must reflect the boundary conditions (Hiley 2002, 16-17).

From the discussion above, it becomes clear that one can treat the quantum potential as an information potential (Hiley 2002, 18, citing to Bohm and Hiley 1987; 1993; and to Hiley 1995; 1999). Hiley (2002, 17) writes:

> [W]e want to suggest that we can think of the process in terms of a local particle being fed this information locally through the information contained in the potential field as the particle evolves along its path.

But how are we to understand these puzzling features physically? Because there is nothing to push against we should not regard the quantum potential as giving rise to an efficient cause ('pushing and pulling') but it should be regarded more in the spirit of providing an example of Aristotle's formative cause. That is the quantum potential gives new form to the evolution of the trajectories, in a way that is very reminiscent of the morphogenetic fields proposed by Waddington (1956) and Thom (1975[<1972>]) in biology. The form is provided from within but it is, of course, shaped by the environment. Thus the quantum potential reflects the experimental conditions. Close one slit and the quantum potential changes and the subsequent evolution of the particle is different There seems to be a kind of 'self-organisation' involved.

Self-organization requires the notion of information to be *active*. In a biological system, this information is provided by the environment, soil conditions, amount of moisture, etc. In a quantum system, one also can suggest that the information is provided by the experimental conditions, which is the quantum system's 'environment.' But this information is not passive. It is active and it causes the internal energy to be redistributed between the kinetic (pB) and potential (Q) parts. Literally, then, *the quantum process is formed from within* (Hiley 2002, 17).

The notion of objective, active information opens new connections between information and meaning, some of which Hiley (2002) discusses. The proposal is that active information at the quantum level organizes the dynamical evolution of the system itself. As Hiley and Pylkkänen (2005, 19) state:

> To make this notion more concrete we can consider the root of the word information, which literally means to put form into a process. As we remarked earlier, the quantum potential appears to be some kind of internal energy which carries information about the environment. Therefore we consider the idea that the whole process, particle plus active environment (which in terms of a measurement requires the specification of experimental conditions), is being

formed partly from within. This suggests that the process is more organic than mechanical. In this respect the system does not require an external force to solely determine its future behavior, thus clearly enriching the notion of causality. It looks as if the classical potential can be regarded as a 'push-pull' potential while the quantum potential involves something subtler such as information.

After a detailed examination of this potential in many different experimental situations, Bohm and Hiley (1993) suggested that it acts by producing a change in the form of the process. In a somewhat Aristotelian fashion, a 'formative cause' is present in addition to 'material' and 'efficient' causes.

About formative cause, Bohm states that it basically is very similar to meaning. Meaning operates in a human being as a formative cause inasmuch as it provides an end towards which the human being is moving. It permeates the person's attention and gives form to his or her activities in a manner that tends to realize that end (Weber 1987; see also Sheldrake 1981). In addition, the notion of 'formative cause' may be another way of interpreting the wave function, in addition to interpreting it in terms of potentialities. As Bohm states, "I would have an explanation of the electron in the following terms: it is constantly forming and dissolving in a similar way, and what is behind it is this formative cause; that is, a formal and a final cause constantly tending to form." The present wave function is the description of a very small part of the formative field of the electron, or the system. This relates to meaning, since one can interpret the wave function as representing a description of the formative cause but also as having a type of meaning that would connect to things such as nonlocal correlation. In this manner, there would be a meaning connected with the whole system, and, also, with any one part (Weber 1987).

Whereas in a mechanical explanation all influences are mechanical, the framework developed by Bohm and Hiley suggests a more organic picture in which information plays a dynamic and active role, making it so-called active information. In this sense, Bohm's interpretation clarifies Bohr's interpretation rather than contradicting it (Hiley 2002, 18). Active information may determine the unfolding of more general processes that are not necessarily governed by the Schrödinger equation but, instead, involve deeper factors (Hiley and Pylkkänen 2005, 20).

The quantum potential energy does not behave like an additional, classical form of energy. It is a form of internal energy that is split off from the kinetic energy, and it has no external source. When one examines the quantum potential energy in mathematical detail in the context of traditional quantum mechanical problems, one finds that it contains information about the experimental environment in which the particle finds itself. This provides the route to its possible role as an information potential. Another way of describing this is that the quantum potential feeds energy into each individual particle as it proceeds through the experimental apparatus (Hiley and Pylkkänen 2005, 11; see Brown and Hiley 2000).

Viewing an electron as moving along a trajectory under the influence of the quantum potential resolves many well-known paradoxes and ambiguities of quantum theory (Bohm and Hiley 1993). It also opens an entirely new way of looking at the role of energy and information in physics in general, which can have far-reaching implications in other domains such as biology, psychology, and, as described above, also in chemistry. For example, one of the traditional puzzles in quantum mechanics is so-called quantum tunneling, namely how particles can penetrate a potential barrier when their kinetic energy is less than the barrier height. The standard approach leaves one with the account that in some instances the particle goes through the barrier while in other instances it does not. All that one can know are the probabilities of this happening in an ensemble of particles. Furthermore, many people assume that whatever happens to the individual particle is genuinely indeterministic. In contrast, in the Bohm theory, *"whether a particle goes through a barrier or not is determined by the position of the particle in the initial wave function"* (Hiley and Pylkkänen 2005, 12, italics added).

The particles can penetrate the barrier without violating the conservation of energy because of redistribution between the various qualities of energy—kinetic energy, the quantum potential energy, and the classical potential energy—even though the total energy is strictly conserved (Hiley and Pylkkänen 2005, 14). As the particle from a beam of particles travels along one path, it is subjected to the quantum potential calculated from the wave function associated with that path, which means that the quantum potential carries active information (Hiley and Pylkkänen 2005, 17-18).

In the explanation of active information above, the notion of information is not used in its usual sense. Instead, its meaning is understood in a literal sense. As Miller (1987) points out, the etymological origin of the word 'information' stresses the *active* role of information. For example, 'in-form' literally means to form from within. Miller writes, "The central stem ({\em forma}) carries the primary meaning of visible form, outward appearance, shape or outline. So *informare* signifies the action of forming, fashioning or bringing a certain shape or order into something." Accordingly, this is not information for oneself as the experimenter, but, rather, it refers to the activity occurring in the system itself. The notion of active information is that information *for the particle* is objective. This is different from the idea of information that we commonly think of as information *for human beings* (Hiley 2002, 18). Active information is not 'information for us,' but instead it is objective information for the particle, which therefore means that it measures some form of objective information. In this way, information plays an objective and active role in all quantum processes—which fits well with Bohr's (1935) original answer to the EPR paradox. There, Bohr argues that one should not regard the coupling between entangled pairs as arising from a mechanical force. Instead, Bohr refers to *"an influence on the very conditions which define the possible types of prediction regarding the future behaviour of the system."* Realizing that this was a key idea, Bohr italicized the passage in his paper from 1935, "Can Quantum-Mechanical Description of Physical Reality be Considered Complete?" and repeated it word for word in his "Discussions with Einstein on Epistemological Problems in Atomic Physics" (Bohr 1949) (Hiley 2002, 18, 22).

A common yet erroneous view is that when an observer plays an active role such as looking at the detector, this is what causes the wave function to collapse. In reality, however, the role of the observer in the quantum theory is no different from that of an observer in classical physics. The observer merely sets up the apparatus initially. Irreversibility, which renders certain information permanently inactive, also occurs in specific quantum scenarios. Furthermore, in certain quantum scenarios, a particle can behave as if the wave function collapsed while no process of wave function collapse occurs. What occurs instead is a redistribution of quantum potential energy, so from that point on, the system behaves as if it were a mixed state described by the density matrix, with the entropy of the system having increased. Considering this is more heuristic terms, in a Stern-Gerlach magnet experiment, the magnet increases the number of accessible quantum states available to the system, which means the entropy of the system must have increased. This increase in entropy means that the information capacity of the system with respect to Shannon information has increased. For example, one can consider the case in which the original particle with its spin along the x-axis is fired into the Stern-Gerlach magnet aligned along the z-axis. This produces two possible output states, such that two qubits have been created. According to the Bohm approach, "the two bits of information have their physical origins in the activity in one of the two possible quantum potentials" (Hiley 2002, 18-21).

At a more nuanced level, it is important to note that the issue of 'information for the particle' or 'information for us' is not an either-or issue. Whereas one view is to consider the two quantum spin states and to argue that the information is for us because we do not know which of the two spin states a particular particle is in, it also is true that information for us must be carried by some physical process. What is important here is that quantum processes provide a novel way of carrying active information that have fascinating consequences. One can explore this in terms of cognitive processes, for example (Khrennikov 2000). It also is important to note that if one is considering physics in terms of information in the sense of Wheeler's (1991) speculations, then one is discussing objective information rather than subjective information. Another interesting area of investigation is how one can use the idea of the transfer of active information to explain quantum teleportation (Maroney and Hiley 1999) (Hiley 2002, 21, inc. fn. 1).

To summarize certain aspects of the discussion above, the ontological interpretation of quantum theory suggests that active information connected to the quantum potential energy plays a key role in quantum physical processes (Hiley and Pylkkänen 2005, 7). Accordingly, one can use the notion of active information to explain quantum effects (Hiley and Pylkkänen 2005, 16). Thinking about quantum physics from the perspective of active information helps one understand how information plays an active role in physical processes. Furthermore, "[a]n important possibility is that this information is not only present in the traditional domain treated by quantum theory, but it could also have implications on the macroscopic scale" (Hiley and Pylkkänen 2005, 9). This last point is particularly relevant to the discussion of ETI/UAP, as discussed later in the book.

Untangling the Physics Literature discussing Bohm's Framework

One confusing aspect of the physics literature regarding Bohm's framework is the fact that Dürr, Goldstein, and Zanghi (2003 <1992>, henceforth DGZ), who are referred to as the quintessential 'Bohmians,' and, also, some other physicists who also espouse 'Bohmian mechanics' — articulate a view that is different from Bohm's view in important ways. Although these theorists think of themselves as 'Bohmians,' in many respects, their views do not reflect Bohm's actual framework. For example, even though Bohm's own approach is not mechanical in the least, DGZ use the term 'Bohmian mechanics' to describe their own mechanistic minimalist approach to Bohm formalism. In contrast to Bohm's approach, DGZ try to retain as many of the traditional features of a mechanistic view of physics as possible by introducing the minimum number of assumptions that seem necessary to generate the formalism (Hiley 2002, 4).

Although DGZ claim they are developing a radically new theory, they appear to be trying to remain as close as possible to classical physics notions. In sharp contrast to Bohm, DGZ think the quantum potential is artificial and should be avoided. Taking such a view implicitly follows Heisenberg (1959 <1958>), who states that Bohm's framework requires "some strange quantum potentials introduced *ad hoc* by Bohm." Heisenberg did not elaborate on what was "ad hoc" about Bohm's math, and it is difficult to understand how a term that arises naturally from a basic equation should be called "ad hoc." Nevertheless, Dürr et al. (1996) take this position and state that "the artificiality suggested by the quantum potential is the price one pays" for putting a non-classical theory into a classical framework (Dürr et al. 1996). Despite such an unclear statement, Goldstein (1998b, 40) (see also Goldstein 1998a) similarly writes, "In particular, Bohm's invocation of the 'quantum potential' made his theory seem artificial and obscured its essential structure" (Hiley 2002, 7-8). However, when one discusses potentialities by means of the concept of a potential—namely the quantum potential, or Q, one realizes that Q is not 'ad hoc,' as Heisenberg contends, since it must be present to conserve the total energy of the system. Thus, far from obscuring the theory's essential structure, the quantum potential is essential in the formalism (Hiley 2002, 15-16). Fiscaletti (2018, 73) agrees, writing "the quantum potential … cannot be considered ad hoc but is entirely within the structure of the quantum world."

The quantum potential, Q, which is so important in Bohm's framework, emerges straightforwardly from the mathematics. It is neither 'ad hoc' nor is it 'invoked.' Bohm and Hiley simply give it a name and then explore the consequence of this additional energy. Indeed, it is a normal practice in physics to attempt to attach physical meanings to terms that arise naturally in fundamental equations. Like Bohm, DGZ also assume the validity of the Schrödinger equation and the Born probability postulate. Instead of examining the contents of the Schrödinger equation in term of its real and imaginary parts, however, they argue that one must add a new idea, namely that the movement of an individual particle is described by the guidance condition. The approach of DGZ also differs from Bohm's in other ways. For example, they have a statistical theory with one fundamental equation of motion, namely Schrödinger's equation, and nothing else, which is why Hiley refers to the approach as "mechanical minimalism" (Hiley 2002, 7-8).

Starting from a more general perspective, Bohm's approach first assumes that one accepts the Born probability postulate, which also is referred to as the Born rule, or Born's rule. This is a postulate of the quantum theory that provides the probability that a measurement of a quantum system will yield a given result. Bohm (1951) then interprets the wave function in terms of potentialities. Heisenberg (1959 <1958>, 53) also favors the use of potentialities. Bohm argues that the potentialities *are latent in the particle* and only can be brought out more fully by means of interaction with a classical measuring apparatus. In essence, this is the conventional, quantum view. But Bohm thinks one should inquire more deeply, otherwise there is a complete absence of an account of the actual in the standard quantum formalism, according to which nothing seems to happen unless and until an interaction with a measuring apparatus occurs. In other words, no actualization occurs until some form of instrument is triggered. Bohm does not leave it at that. He asks why the measuring instrument is so different, given that it is merely another collection of physical processes governed by the same laws of physics. Besides, it seems that something surely must trigger the measuring instrument. To resolve these and other issues, Bohm goes beyond the conventional, quantum view to make alternative proposals and to provide an account of an actual process underpinning the observational results, even though the nature of this actual process is unclear. Searching for an actual, underlying process is not a return to the classical view. In some ways, it is similar to the driving force behind the present search for a generalized quantum view of gravity. Both share that they are searches for an ontology (see for example Hartle 1992) (Hiley 2002, 3).

A commonality between Bohm's approach and that of DGZ is that both calculate the trajectories and interpret them in precisely the same way. This means that both approaches provide a way to discuss the actual evolution of individual particles in a fundamental statistical theory. However, Bohm contends that one must go beyond the mechanistic picture in order to obtain a better understanding of this behavior. Accordingly, although DGZ embrace the imaginary part of the Schrödinger equation and give it meaning as an expression of the conservation of probability, as do Bohm and Hiley, when it comes to the real part of the same equation, DGZ dismiss it because it seems "artificial." This is an odd stance, since it implies that one has some idea of how quantum particles *should* behave, and that when the trajectories do not fit this preconceived framework, things become 'artificial' and 'obscure.' But the fact that behavior is apparently bizarre is not a reason to ignore it—to the contrary, paying attention when things are seemingly bizarre helps one elucidate the quantum theory (Hiley 2002, 7-8).

Others theorists also reject aspects of Bohm's approach because these aspects seem too strange. For example, Scully (1998) and Aharonov and Vaidman (1996) employ similarly dismissive arguments, oddly against the trajectories themselves. They call the trajectories "surreal" apparently because the trajectories do not conform to what one would expect on classical grounds. But such criticisms cannot be sustained (Hiley et al. 2001). When compared to intuitive ideas based on classical theories, quantum phenomena indeed are bizarre, and one could argue that it would be quite surprising if the quantum potential and the trajectories and possessed classical properties. Quantum theory is a highly non-classical theory, and one should expect surprises. When this occurs, one should use all available approaches to explore in precisely what ways the behavior is non-classical (Hiley 2002, 8-9).

While 'Bohmian mechanics' and Bohm's own framework agree that an account of the actual is needed, the two approaches differ crucially with respect to what form an account of the actual should take. There is no dispute with respect to the form of the equations, yet each group nevertheless employs different physical explanations that it finds acceptable. In contrast to many advocates of so-called Bohmian mechanics, Bohm himself and in his collaboration with Hiley take a different attitude towards the formalism. Having considered the formalism and reflected on how it works, they conclude that rather than a simple return to a mechanistic picture, something much more subtle is involved (Hiley 2002, 9).

In standard quantum mechanics, one is taught that particles sometimes behave like waves and sometimes like particles. The standard view also includes the measurement problem, Schrödinger's paradoxical cat, Wigner's friend thought experiment, and the idea that no phenomenon *is* a phenomenon until it is an observed phenomenon. Some physicists do not feel comfortable with these seemingly surreal aspects of standard quantum mechanics, perhaps in part because physicists use completely different mathematical structures to describe classical and quantum phenomena. Hiley sets about the task of bringing both classical and quantum descriptions to a common form in order to determine the precise differences between the two domains. Hiley cites Bohm's method of handling the Schrödinger equation as one possibility. Bohm splits the Schrödinger equation into its real and imaginary parts under polar decomposition of the wave function. The imaginary part gives the conservation of probability, while the real part possesses a striking resemblance to the classical Hamilton-Jacobi equation. Hiley acknowledges that others have taken a different approach, namely reformulating classical physics in the Hilbert space formalism. To review the development of the Bohm formalism, Zeh (1999) correctly observes that Bohm and Hiley (1987; 1993) assume that the Schrödinger equation is valid and that one does not need to add any new mathematical terms to this equation. The Born probability postulate also is assumed. The derivation of the two equations that form the basis of the Bohm approach is straightforward. With the wave function expressed in polar form, the equation breaks into imaginary and real parts. The imaginary part of the Schrödinger equation is the conservation of probability equation, and the real part of the Schrödinger equation, which caught Bohm's attention, has a close similarity to the classical Hamilton-Jacobi equation. Although the reappearance of what strongly resembles the Hamilton-Jacobi equation in the real part of Schrödinger's equation could be a coincidence, it most likely occurs because Schrödinger's (1926) original 'derivation' of his equation begins by assuming the classical Hamilton-Jacobi equation. Hiley explores this formalism without presumed metaphysical assumptions and without resorting to the preconceptions of the classical view, even leaving behind some common, deeply ingrained quantum preconceptions. He takes this approach to order to gain new insights into quantum phenomena by considering things from a different perspective (Hiley 2002, 2, 5-6).

Superquantum Potential and Why the 'Laws' of Physics Must Change

Since all theories are ways of looking at a situation that may be correct in a given domain but are likely to be incorrect in some more general domain

(Hiley and Pylkkänen 2005, fn. 1 on 12), no single, fundamental level of reality exists (Bohm 1984 <1957>; Schaffer 2003) (Hiley and Pylkkänen 2005, fn. 1 on 12).

Bohm notes that one can extend the quantum theory by improving the analogy between mental processes and quantum processes, and he explains how two processes—the relationship of various levels of subtlety in mind and the relationship of various levels of subtlety in quantum processes—essentially are the same. Bohm proposes that the quantum behavior of a system of electrons is similar to the behavior of mind. Bohm also explains the analogy between mental processes and quantum processes by considering that both mental processes and quantum processes may extend "to indefinitely great levels of subtlety" (Bohm, 1990, 283).

Bohm uses the notion of the quantum potential to explain how mental and material are interwoven with one another. The quantum potential is information, and its activity guides "the 'dance' of the electrons." The idea is that just as the quantum potential constitutes active information that can give form to the movements of the particles, there also exists a *superquantum potential* that can give form to the unfoldment and development of the first order quantum potential. Importantly, this view provides insight into *how one's understanding of physics must transform, since the first order quantum potential no longer would satisfy the laws of the current quantum theory.* The laws of the current quantum theory are an approximation that only works when the action of the superquantum potential can be neglected (Bohm, 1990, 283).

One can suppose a series of orders of superquantum potentials. Each order constitutes information that gives form to the activity of the next lower order, which is less subtle. In this way, one arrives at a process that is very similar to one regarding the relationship of various levels of subtlety in mind. Accordingly, both processes—the relationship of various levels of subtlety in mind and the relationship of various levels of subtlety in quantum processes—essentially are the same. What we experience as mind, in its movement through various levels of subtlety, in a natural way ultimately moves the body by reaching the level of the quantum potential and the 'dance' of the particles. No unbridgeable gap or barrier exists between any of these levels—rather, at each stage, some kind of information is the bridge. The implication is "that the quantum potential acting on atomic particles, for example, represents only one stage in the process." Since the content of human consciousness is part of this overall process, the implication is that "in some sense a rudimentary mind-like quality is present even at the level of particle physics, and that as we go to subtler levels, this mind-like quality becomes stronger and more developed. Each kind and level of mind may have a relative autonomy and stability" (Bohm, 1990, 283).

Because any scientific theory, including any physics theory, is a way of looking at a situation that may be correct in a given domain but is likely to be incorrect in some more general domain, one should not assume, for example, that the model of an electron moving along trajectories is a complete and final description of a purported 'fundamental' level of reality. Although this model of an electron may be correct and may provide insight in its own domain, it may turn out to be incorrect in a broader domain. For example, at the level of Planck time (10^{-43} seconds), movement appears discrete, making the notion of particles moving continuously along trajectories inapplicable. At this level, it is more helpful to think in terms of the notion of a series of converging and diverging quantum waves that give rise to particle-like manifestations in a discrete fashion (Bohm and Hiley 1993, 367-68; Hiley and Pylkkänen 2005, fn. 1 on 12).

Regarding the origin of the universe, physics and cosmology raise questions with respect to the tremendous evolution that occurred within the first few fractions of a second of the universe. During this time, particles, space, time, and everything else we know in conventional reality had yet to manifest. Then somehow it all developed. Bohm states:

> People take it for granted that during this whole process the laws of quantum mechanics were the same, though everything else changes. They assume that, even though quantum mechanics is supposed to refer to nothing but the results of measurements, although

measurements would have been impossible without anything to measure (Bohm and Sheldrake 1986, 117).

In other words, it simply is illogical to think that the quantum theory only refers to the results of measurements.

With respect to whether natural laws are eternally given or whether they are built up gradually, Bohm's position is that in view of the implicate order, the notion of formative fields gradually becomes necessary. Even modern physics points to the idea of formative fields by saying there was a time prior to the Big Bang before any units such as quarks, atoms, and molecules even existed. The claim that certain fixed, everlasting laws of atoms and molecules exist cannot hold if one traces back to the time before the atoms and molecules existed, since it is clear that physics can say nothing about that (Bohm and Sheldrake 1986, 117-18). Bohm notes:

> It [i.e., physics] can say only that there was a formation of these particles at a certain stage. So there would have to be an actual development in which the necessity in a certain field grew more and more fixed. You can even see that happening as you cool down a substance that liquefies; at first you get little clumps of liquid which are transient, and then they get bigger and more determinate. Now physicists explain all this by saying that the laws of the molecules are eternal; molecules are merely consequences of those laws, or derived from those laws. But if you follow that back and ask, 'Where were molecules? Well, they were originally protons and electrons, which were originally quarks, which were originally sub-quarks. And it goes right back to a stage where none of the units we know even existed, so the whole schema sort of fades out. It's then open to you to say that, in general, fields of necessity are not eternal; they are constantly forming and developing (Bohm and Sheldrake 1986, 118).

In other words, there is no basis for the view that the so-called laws of physics are fixed and unchanging.

Despite the recognition by some scientists and historians and philosophers of science that no physics theory will be correct across all domains, many academics and especially scientists stubbornly hold onto the idea of fixed, immutable 'physical laws.' Some of this stubbornness may be an epiphenomenon of the history of science, which began with a kind of neo-Platonic, neo-Pythagorean idea of timeless laws—a view that some scientists have taken for granted for years. When it was proposed, the evolutionary theory in biology catalyzed some change in this stance, since it brought with it an evolutionary view of reality regarding animals and plants. Nevertheless, some people continue to think that a timeless background exists for the physical world of atoms and molecules. But the cosmology of the Big Bang supports the idea of a radically evolutionary universe, and the very idea of a historical Big Bang calls into question the idea of timeless laws that have existed foverever and somehow pervade space and time. The idea of timeless laws ceases to have much meaning because it raises the problem of where the laws were or how they may have existed before the Big Bang. One could carry this back to a point at which present quantum theory was just coming into being. Indeed, the entire formative structure of the quantum theory may have emerged from some other state of affairs and so on, as a field of meaning in a participatory universe in which everything is the observer and everything is the observed. Whereas physics has not paid much attention to how the general meaning field has evolved, biology discusses the evolution of this formative field. In physics, the most similar comparison is thinking about the origin of the universe (Bohm and Sheldrake 1986, 118-19, 122).

Putting this section's discussion very straightforwardly, a universe organized according to fixed, immutable laws is not the universe we inhabit. Our reality is one in which not only *do* the laws of physics change, they *must* change.

Morphogenesis and the Origin of Form

Bohm discusses how form develops from that which is beyond form. Bohm's view is that the implicate order can account for the origin of form. The notion of the implicate order is a way of discussing the origin of form out of the formless, via the process of explication, or unfolding. The *genesis* of forms is in implicate order, which can be understood as the creative source of new forms—which in turn are replicated. As a form projects out in different places in explicate order, it has certain tendencies that constantly are being built up by means of repetition. This is what gives the appearance of the 'causation' of the present by the past. To begin with, the form is enfolded, and then it unfolds. The forms are not completely pre-existing in the implicate order, but instead, they develop by means of a process—which means there is a developmental process in the implicate order, namely a development of form (Bohm 1986a, 93-4).

Sheldrake (1995 <1981>; 1995 <1988>) develops the notion of morphogenetic fields based on a framework in which memory is postulated to have two aspects—individual, and species collective. Individual memories are stored in dependence on resonance with ourselves in the past. Morphic resonance is the collective memory of habits/regularities of nature and the transfer of information across space and time by a process of resonance. Morphogenetic fields are "[f]ields that play a causal role in morphogenesis," which is understood as "the coming into being of form." Relatedly, morphic resonance is "[t]he influence of previous structures of activity on subsequent similar structures of activity organized by morphic fields" (Sheldrake 1995 <1988>, 371).

According to Bohm, Sheldrake makes the point that Sheldrake himself differs from Plato in thinking that the forms are not completely pre-existing but rather that they develop. Bohm agrees and adds the idea "that a form develops through the process of projection and injection, re-projection, re-injection, and so on." This refers to the development of form in the implicate order. This also relates to the notion of creativity and how it relates to projection, injection, and re-protection. As Bohm states:

> As far as the implicate order is concerned, every new moment could, in principle, be entirely unrelated to the previous one—it could be totally creative. You could say that creativity is fundamental in the implicate order, and what we really have to explain are the processes that are *not* creative. You see, usually we believe that in life the rule

is uncreativity, and occasionally a little burst of creativity comes in that requires explanation. But the implicate order turns all that around and says creativity is the basis and it is repetition that has to be explained That's where the morphogenetic field theory comes in.

Morphogenetic fields are explained in the implicate order theory as due to the constant repetition produced by projection and injection, re-projection and re-injection, and so on. A form emerges or is creatively projected from the whole, then it influences the whole, or is "injected" back into it; in the implicate order it "resonates" with similar forms and then is re-projected back into the explicate order. This whole process—forms ceaselessly emerging and then being reabsorbed—accounts for the influence of past forms on present ones, and also allows for the emergence of new or creative forms. A morphogenetic field is simply a field that, through constant repetition, has become relatively stable (Bohm 1986a, 94).

The maladaptive repetition of past thought-forms, habit-patterns, conditioning, memories, etc.—especially negative ones—obstructs genuine intelligence. This is particularly the case when such repetition becomes obsessive to the point that it blocks awareness of the present. Bohm states:

Well, the point is that we must give the past its due. The past is intrinsically neither good nor bad. But it's necessary. We must have some form—we can't live entirely in the implicate order. The problem that [Jiddu] Krishnamurti addresses is that the past is overemphasized and becomes dominant, thus resisting creativity where it would otherwise be natural and appropriate. The past … tends to hold on, and that is the trouble. But the past itself, if properly addressed, is useful enough. You see, the past can also have a part in creativity. If you had absolute creativity—absolute novelty with no past—then nothing would ever exist because it would all vanish at the very moment of creation. Nothing could last if everything were entirely new. Therefore, it's a dialectical movement which requires both sides, creativity and stability, the creative present and the relatively fixed past (Bohm 1986a, 94-95).

When asked by Weber what the proper way is to relate to the past, Bohm states, "Through intelligence. We want neither to ignore the past nor to become attached to it. If something from the past begins to cause conflict and confusion, then that has to be dropped" (Bohm 1986a, 95).

Bohm emphasizes that forms are not pre-existing models. Instead, forms may develop out of that which is beyond form. According to Bohm, one can think of this as the universe experimenting and trying out various forms. Natural selection explains the way things survive after they emerge or appear, but it does not explain why so many forms *have* appeared. What Bohm finds significant is the tendency to produce structure and form, which is intrinsically creative, whereas survival or natural selection is merely the mechanism that selects which forms are going to remain. According to Bohm, the implication of this is that the universe itself is learning (Bohm 1986a, 96).

Bohm relates his thinking on morphogenesis to Sheldrake's notion of morphogenetic fields. A morphogenetic field is the governing form of any organism or entity. Nature is full of a multitude of forms, so naturally, a corresponding multiplicity of morphogenetic fields also exists. One can see this diversity in a more unified way by considering the possibility that these morphogenetic fields are arranged in a hierarchical order—the more specific fields are included in more general fields, which themselves are subsets of even more inclusive and general fields, and so on. Even though the diversity is real, a larger, underlying unity always exists (Bohm 1986a, 98).

Regarding the status of morphogenetic fields and whether they are actual and concrete forces or merely nominal ways of referring to what happens, Bohm states that they are not mere names. He uses an analogy to the radio receiver. A radio wave is sent out from a radio station. The wave has a form, e.g., music, and this form is carried by a very weak electrical wave that is picked up by the antennae of a particular radio set. When music emanates from the radio set, almost all its energy comes from the power plug in the wall socket—but its form comes from the very weak electrical wave picked up by the antennae. In other words, a very subtle energy, which is picked up by the antennae, molds a denser energy that is coming from the wall socket. Similarly, one can picture a formative or morphic field as a very subtle aspect of the implicate order that can impress itself on denser and explicate energies and ride on it. The example of a seed provides another way of thinking about this. The energy and nutrients come from the Sun, air, soil, water, and wind, and the seed itself has very little energy. Nevertheless, somehow the seed possesses the form of the plant, and that tiny energy or form impresses itself on all those other factors and thus produces the plant. Somehow, that tiny bit of energy governs all subsequent growth so that the whole system is made to produce a plant instead of producing something else (Bohm 1986a, 99).

Whereas Sheldrake's view is that the morphic field has no energy at all, Bohm thinks that it has a very weak energy, in a manner analogous to the radio wave. This is consistent with Bohm's view of the implicate order, which is that the formative field is, or has, a weak, or subtle energy, i.e., subtle by comparison with the ordinary standards of mechanical energy (Bohm 1986a, 99).

Bohm's Framework Helps Explain ETI/UAP

Many different aspects of Bohm's ontological approach to the quantum theory shed significant light on the ETI/UAP issue. Here I consider twelve.

First, discussion of the ETI/UAP topic must occur within the context of a broad discussion of ontology—for example how ontology is understood within the context of the quantum theory. Here, Bohm's interpretation of the quantum theory, which is ontological by design, helps guide us towards the clearer ontological view that ETI already possesses.

ETI is trying to help humankind cross the threshold from only perceiving the explicate order aspects of reality to a perspective that includes subtler implicate, super-implicate, super-super-implicate, etc., aspects of reality. *We simply must perceive the subtler aspects of reality in order to survive as a species.* If we remain constrained by a view of reality that only focuses on explicate order, our trajectory as a species will continue to be coarse, vicious, and violent, to the point that we become extinct because of our own shortsightedness.

<u>Second</u>, the behavior of extraterrestrial UAP shows that the ETI responsible for them clearly perceives, understands, and operates in accordance with the inseparability of mind and matter and the wholeness of existence. ETI also understands how to account for the actual process underlying quantum observational results and to operationalize this understanding in its UAP displays.

The behavior of ETI's UAP indicates that ETI clearly has transcended the Cartesian split between mind and matter—or it simply evolved in such a way that this split was never part of its cognition. As Bohm notes, the erroneous view that mind and matter are separate is the basis for so much that is wrong in current human knowledge and society. In a prior publication (Andresen 2022b, 300, 317), I gloss how one moves from the inseparability of mind and matter to how one overcomes subject-object dualism. I unpack that argument in considerable detail in a forthcoming publication (Andresen forthcoming). Here, I want to emphasize how understanding the inseparability of mind and matter may help ETI in many ways, for example by facilitating ETI-human direct encounters involving telepathy (Turing 1950, 453-54), the reality of which also underscores the inseparability of mind and matter (Andresen in preparation).

<u>Third</u>, Bohm's explanation of participation in whole existence has tremendous implications for human society and for how humankind relates to ETI. In the sense that everything participates in everything else, extraterrestrials are we and we are they. In other words, creative acculturation between humankind and ETI is no more or less participatory than are interactions between human beings. When one realizes that beings one habitually perceives as independent and autonomous really are interdependent in nature, one also realizes it is futile to seek personal happiness over and apart from the wellbeing of the whole. This is a major reason why we can rest assured that ETI is well intentioned towards humankind.

<u>Fourth</u>, Bohm's concept of holomovement helps us understand ETI/UAP. As almost all witness reports attest, UAP move and maneuver in a very advanced manner. Aspects of ETI/UAP movement can be described as

post-technological and, also, trans-technological inasmuch as these movement technologies demonstrate ETI's deep understanding of the nature of reality. ETI clearly understands quantum entanglement and nonlocality very clearly. As more human beings begin to perceive these aspects of reality more clearly, they too will begin to move and maneuver more fluidly themselves.

Fifth, Bohm's understanding of explicate and implicate levels of order moves us beyond the task of attempting to categorize or pigeonhole 'types' of UAP. Considering the English words 'explicit' and 'implicit,' one can correlate these two words with Bohm's view of explicate and implicate order and, also, with descriptions of UAP along the defined to amorphous continuum I propose elsewhere (Andresen 2022b, 301-4) and describe briefly in chapter 3 of this book. Here, I add the idea that relatively defined UAP signal and communicate relatively more explicitly, while more amorphous UAP signal and communicate relatively more implicitly. For example, spheres, cubes, and triangles that all possess approximately the same degree of definition may be conveying a consistent message with respect to the nature of reality. In other words, at the broadest level, what may be most important is the *degree of definition* of various UAP, not their shapes—even if UAP with different shapes sometimes do perform different functions. As compared to more defined UAP, more amorphous UAP such as orbs may convey something more implicit and subtle regarding the nature of reality. The shapes of orbs and other amorphous UAP also may relate to function, but perhaps not to the same degree as is the case with more defined and often geometrical UAP.

Sixth, ETI/UAP clearly understands what Bohm refers to as 'pools of information.' The traditional view is that entanglement, nonlocality, and the organization of movement by means of common pools of information is relevant at low temperatures, whereas at higher temperatures, the quantum potential tends to take the form of independent parts—which in turn implies that the particles move with an *apparent* independence (Bohm et al. 1987; Bohm 1990, 280-81). However, the incredible coherence witnessed when multiple UAP move together in seemingly organic, biological, and swarm-like ways suggests that *UAP can manifest quantum-like behavior at a macroscopic scale*. In other words, UAP displays are helping human beings move beyond the idea that general relativity applies at the macro level and that the quantum theory applies at the micro level, since even at the macro level, something about UAP displays that visually whoops "quantum." ETI therefore is using UAP displays to show human beings *how quantum realities are not merely an aspect of micro-level reality, but they also manifest at the macro level.* Comprehending this will help humankind tremendously in areas such as gravity control and, also, in terms of human awareness, perception, and realization.

Seventh, UAP display entanglement inasmuch as they appear to possess nonlocal elements in which each object recognizes a certain wholeness with distant features of its environment. The *coherence* evident in UAP displays has both mental and physical aspects, which of course is natural given the inseparability of mind and matter. UAP coherence in motion and intent can teach human beings more about how the inseparability of mind and matter and creativity more generally create fluid motion in spacetime.

Eighth, ETI clearly perceives what Bohm refers to as levels of order of increasingly subtlety and can operationalize these subtle levels of order by creating intentional manifestations that consciously employ them.

Ninth, ETI/UAP can manifest and unmanifest (i.e., appear and disappear) at will from detection by human observers and their instruments. My view is that in certain circumstances, this involves manifesting from implicate to explicate order and unmanifesting from explicate to implicate order. In other words, ETI/UAP sometimes make *a conscious and intentional transition back and forth between implicate and explicate levels of order.* The implications of this are profound. At a minimum, it means that ETI understands the genesis and constitution of form and the process and dynamics of manifesting and unmanifesting. Essentially, then, ETI/UAP can reconstitute themselves at will at different points in spacetime. Literally, ETI/UAP are demonstrating how the infinite becomes finite and how the finite becomes infinite.

Tenth, ETI creates pedagogical demonstrations including those involving UAP to show human beings how everything comprises a fluid, dynamic, and relational holomovement with constant movement between infinity/continuity on the one hand and the apparent discreteness/ discontinuity of form on the other.

<u>Eleventh</u>, with respect to the temporal aspect of ETI/UAP, it is important to recall that what happens in space coincidently happens in time. As stated above, Bohm's view is that the implicate order is a ground beyond time, a totality out of which each moment is projected into explicate order (Bohm 1986a, 93). This idea can help us understand how the ETI operating in our midst can utilize time to advantage, either in its own appearances and disappearances, or in encounter experiences with human beings (Andresen in preparation).

One should reiterate that the concepts of 'space,' 'time,' and 'spacetime' are epiphenomenal abstractions that arise in human consciousness. Since human beings are embodied for a portion of their existence, these abstractions have adaptive value in helping people navigate. Even at an intraspecies level, however, human beings in different cultural contexts understand and experience time differently. Mainstream culture divides time numerically, which results in many people thinking about time in a linear manner. But this is not uniform across human beings—think for example of Aborigines and their understanding and experience of dreamtime, which is not strictly linear. In nonlinear cognitive states, people access something profound—since as Bohm (1986a, 93) notes, at the level of implicate order, linear time has no meaning, with *things that are similar resonating together*. ETI/UAP can operationalize this understanding, which also suggests that they may be able to move between human time trajectories at will.

<u>Twelfth</u>, Bohm is quite clear that *the so-called laws of physics are not fixed*— they can and must change. This is analogous to the situation in society in which human laws can and must change as society develops. Given that the laws of physics are not fixed, scientific investigation should be open to moving beyond the standard model in physics, which even considers Bohm's ontological interpretation of the quantum theory novel, in order to achieve a meaningful understanding of and relationship with ETI/UAP.

Chapter 6:
The Precariousness of the Nuclear Situation

Precariousness Quotient (PQ) Assesses Nuclear Risks

The risks associated with nuclear weapons and other nuclear technologies are enormous. A nuclear war would devastate humankind, other species, and Earth itself (Diaz-Maurin 2022b).

Humankind has struggled with the nuclear technologies since the first nuclear weapon was created. On July 16, 1945, one week after White Sands Missile Range (WSMR) had been established, a plutonium implosion device, which was the world's first atomic bomb in recorded history, was detonated in the north-central portion of the Range, approximately sixty miles north of White Sands National Monument. If we should have learned anything from almost seventy-eight years of the nuclear age, it is that the existence of nuclear weapons has shaped almost every aspect of geopolitical relations and foreign policy since they were invented. All dialogue between nation-states ultimately capitulates to the reality of nuclear weapons—who has them, who does not have them, who is trying to develop or otherwise obtain them, who may use them, and who is unlikely to use them, even if provoked.

'Trinity' was the code name for the first test of the atomic bomb, and the detonation occurred at what is referred to as the Trinity Test Site. Known in Spanish as Jornada del Muerto, the site is at the northern end of the Alamogordo Bombing Range, in Socorro County, near the towns of Carrizozo and San Antonia in New Mexico. Given that the area includes an almost waterless ninety-mile (~140 km) trail beginning north of Las Cruces, New Mexico and ending south of Socorro, New Mexico, Spanish conquistadors named the desert basin Jornada del Muerto. Translated into English, this means "Dead Man's Journey," or "Route of the Dead Man." Historically, the trail was part of the Camino Real de Teirra

Adentro, which led northward from central colonial New Spain, which is part of present-day Mexico, to the farthest reaches of the viceroyalty in northern Nuevo México Province, which is the area around the upper valley of the Rio Grande. After testing the first atomic bomb at the Trinity Site, the U.S. detonated atomic bombs over Hiroshima and Nagasaki, Japan, on August 6 and 9, 1945, respectively.

Ever since 1945, humankind has continued to struggle with nuclear technologies. The U.S. engaged in the disgraceful detonation of twenty-three nuclear weapons on seven test sites in the Marshall Islands in Micronesia between 1946 and 1958, with horrific repercussions for people living there (Pilger 2016). In 1962, there was the Cuban Missile Crisis, which was followed in 1986 by the nuclear reactor accident at Chernobyl. By 2022, many nuclear risks coincided. In Ukraine, military attacks targeted nuclear reactors and nuclear facilities. Furthermore, Russian leadership stated that it was contemplating using nuclear weapons on the battlefield, thereby threatening to undo the 77-year tradition of nuclear weapons' non-use during situations of conflict. North Korea test-launched more ballistic missiles than ever before in a single year and appears to be preparing for one or more nuclear weapons tests (Diaz-Maurin 2022a). North Korea has conducted six nuclear weapons tests previously, in 2006, 2009, 2013, twice in 2016, and in 2017.

In addition, and, also, by 2022, Iran had resumed construction of its underground nuclear complex, disconnected International Atomic Energy Agency (IAEA) surveillance cameras, and accelerated its uranium enrichment program. Saudi Arabia took further steps towards enriching uranium and refused IAEA inspections meant to ensure that it does not conduct covert activities related to nuclear weapons. In June, nuclear non-proliferation and disarmament efforts failed when participants in the first meeting of states parties of the Treaty on the Prohibition of Nuclear Weapons (TPNW)—also known as the Nuclear Weapon Ban Treaty—did not reach agreement with respect to calling out Russia's nuclear threats and rhetoric amidst the war in Ukraine. Then, in August, EU-mediated talks between the U.S. and Iran failed to revive the 2015 Joint Comprehensive Plan of Action (JCPOA) intended to limit Iran's nuclear program, which former President Donald Trump abandoned in 2018 (Diaz-Maurin 2022a).

The TPNW is not the only treaty regarding nuclear weapons to have encounter challenges. The usefulness of the Treaty on the Non-Proliferation of Nuclear Weapons (NPT)—also referred to as the Nuclear Non-Proliferation Treaty, or simply the Non-Proliferation Treaty—also has been called into significant question. On August 26, 2022, Russia blocked the NPT Review Conference from reaching consensus on a substantive outcome document (Hernández and Kimball 2022). After a monthlong UN review of the document (Foltynova 2022), Russia's stated rationale in blocking consensus was differences in accounts of what caused the nuclear safety crisis at the Zaporizhzhia Nuclear Power Station in Ukraine—even though many countries agreed that the crisis was caused by the Russian occupation of the plant (Hernández and Kimball 2022). The review conference of the parties to the NPT ended without an agreement after Russia refused to sign off on a final outcome document, saying that it that lacked "balance" because it criticized Russia's occupation of the Zaporizhzhia plant (Foltynova 2022).

In late November 2022, hopes that on-site inspections under the New Strategic Arms Reduction Treaty (New START) would resume fell through after Russia postponed a meeting of the Bilateral Consultative Commission (BCC), which is the Treaty's implementing body (Diaz-Maurin 2022a). The session of the BCC on the Treaty slated to meet in Egypt in late November abruptly was called off. The U.S. blamed Russia for this postponement, with a State Department spokesperson saying Russia made the decision "unilaterally" (Chernova et al. 2023).

Originally signed on April 8, 2010, New START went into force on February 5, 2011. In 2021, New START was extended another five years until February 5, 2026. New START is the only bilateral nuclear arms control treaty between the U.S. and Russia, which are the two countries in the world with the largest nuclear arsenals (Diaz-Maurin 2022a). New START regulates the U.S. and Russia primarily by limiting the number of deployed intercontinental-range nuclear weapons that both countries can have to 1,550 each (Chernova et al. 2023; see U.S. DOS n.d. for a description of the features of the Treaty).

On February 21, 2023, Russian President Vladimir Putin announced during his state of the nation address that Russa was suspending its participation in New START—though Putin also stated that Russia not withdrawing from the agreement per se. Under the Treaty, both the U.S. and Russia are permitted to conduct inspections of one another's weapons sites, although inspections have been halted since 2020 because of the Covid-19 pandemic, and U.S. officials contend that Russia continually has refused to allow inspections of its nuclear facilities. In January 2023, a U.S. State Department spokesperson stated, "Russia is not complying with its obligation under the New START Treaty to facilitate inspection activities on its territory," adding that "Russia's refusal to facilitate inspection activities prevents the United States from exercising important rights under the treaty and threatens the viability of U.S.-Russian nuclear arms control." Given that in early 2021, New START only was extended until 2026, both the U.S. and Russia will need to begin negotiating another arms control agreement soon (Chernova et al. 2023). In addition, on March 2, 2023, during a speech to the Russian parliament, Putin stated that Russia was ready to resume nuclear weapons tests if the U.S. conducted one first (Messmer 2023).

The result of the discussion above is that the risk of a catastrophic nuclear detonation is higher today than it was even when the U.S. detonated two atomic bombs over Japan in 1945. In 2020, Ernest J. Moniz, former United States Secretary of Energy, wrote, "as I understood while serving as U.S. secretary of energy, responsible for our nation's nuclear stockpile and key nonproliferation programs, there have been more close calls along the way than most people realize." He also observed, "Nearly 30 years after the Cold War, the risk of a nuclear weapon's being used is higher than at any other time since the U.S. and the Soviets came to the brink of nuclear war during the Cuban Missile Crisis" (Moniz 2020). Since Moniz wrote those words, the potential for a catastrophic nuclear war has increased even more following Russia's recent invasion of Ukraine on February 24, 2022.

On October 27, 2022, the administration of U.S. President Joe Biden released the unclassified and long-delayed version of its Nuclear Posture Review

(NPR), entitled *2022 Nuclear Posture Review* (U.S. DOD 2022a; 2022b). This occurred alongside the administration's release of the unclassified version of the National Defense Strategy (NDS) and the Missile Defense Review (MDR) (U.S. DOD 2022b). As the Biden administration's NPR was being drafted, DOD abruptly shifted leadership of the effort, causing concern among arms control advocates that the NPR would reflect the Cold War era's overreliance on nuclear weapons (Smith 2021). In fact, reception to the NPR was quite mixed, with the document's articulation of an "integrated deterrence" strategy seen as particularly outdated (Grieco et al. 2022). The NPR specifically disappointed experts inasmuch as it maintained the nuclear status quo instead of articulating how to reduce the role of nuclear weapons in U.S. security strategy (Diaz-Maurin 2022a).

Previous efforts to reduce nuclear arsenals and the role that nuclear weapons play have been subdued by renewed strategic competition abroad and by opposition from defense hawks at home. Nevertheless, the NPR does resist some efforts by defense hawks and nuclear lobbyists to add nuclear weapons to the U.S. arsenal and to delay the retirement of older types. Instead, the NPR adjusts the existing force posture and increases integration of conventional and nuclear planning. Even so, the NPR concludes it still may be possible to reduce the role nuclear weapons play in scenarios in which nuclear use is not credible. What is disappointing, however, is the fact that although Biden spoke strongly in favor of adopting policies of No First Use (NFU) and Sole Purpose (SP) during his presidential election campaign, the NPR explicitly rejects both. This makes the document disappointing from both arms control and risk reduction perspectives (Kristensen and Korda 2022a).

Even in its most stringent formulation, a SP declaration is not equivalent to a pledge of NFU. NFU is a statement about when the U.S. would or would not *use* nuclear weapons. In other words, NFU is an explicit constraint with respect to employing nuclear weapons, and it commits a state *not* to use nuclear weapons except in retaliation for nuclear attacks. In contrast, sole purpose, as its name implies, is a statement about *why* the United States possesses the nuclear arsenal that it does, not how it will use it. Sole purpose does not impose constraints on employing nuclear weapons. Instead, it explicitly de-emphasizes the role of nuclear weapons in U.S. national security strategy (Panda and Narang 2021; see also Mount 2021).

To capture the breadth and scale of the risks facing humankind from nuclear weapons and other nuclear technologies, I use the term *precariousness quotient* (PQ)—the antithesis and nemesis of the commonly known intelligence quotient (IQ). The pages below describe some of the numerous elements that increase the PQ associated with nuclear weapons and other nuclear technologies: 1) number of countries with nuclear capabilities; 2) complex geometry of nuclear-armed rivals; 3) scope of nuclear capabilities; 4) global stockpiles of fissile materials; 5) depleted uranium weapons; 6) lethality of nuclear weapons; 7) destabilizing impact of engineering modifications to nuclear weapons; 8) nuclear weapons kept on hair-trigger alert; 9) widespread famine from soot injection into the atmosphere; 10) excessive reliance on deterrence strategies; 11) arms control "non-solutions;" 12) nuclear risks from terrorist groups and other hostile actors; 13) nuclear weapons pointed at population centers and infrastructure; 14) nuclear risks amplified by planetary dynamics; 15) decades-old nuclear power infrastructure susceptible to accidents; and 16) risks of blackmail involving nuclear weapons and nuclear power. I describe some of these topics elsewhere (Andresen 2022b, 305-307).

Number of Countries with Nuclear Capabilities

Almost certainly, ten countries already possess nuclear weapons. Eight countries are overtly nuclear—Russia, the U.S., China, France, the U.K., Pakistan, India, and North Korea. The first seven of these are listed in order from greatest to least total number of deployed plus stored warheads. North Korea's position on the list refers to the number of warheads that it could build with the amount of fissile material people estimate it has produced (Kristensen et al. 2021, 2).

In addition to the eight overtly nuclear countries, Israel and Iran almost certainly have nuclear weapons but do not claim to possess them. It has been estimated that Israel possesses a nuclear arsenal with approximately ninety warheads, together with significant capacity involving nuclear-capable aircraft, land-based ballistic missiles, and sea-based missiles and submarines. Over decades, however, Israel's policy has been one of nuclear 'ambiguity,' or 'opacity' (Hebrew *amimut*), according to which numerous Israeli government officials have made purposefully ambiguous statements regarding whether Israel possesses nuclear weapons (Kristensen and Korda 2022b; see also Trevithick 2019; Gross 2021).

Even as far back as May 1989, the Director of Central Intelligence (DCI) in the U.S., who at the time was Mr. William H. Webster (CNN 2021), indicated that Israel may have been trying to construct a thermonuclear weapon (Cordesman 2005, 49). In 2012, experts in Israel and Germany confirmed that nuclear-tipped missiles were deployed on the submarines that Germany exported to Israel (SPIEGEL 2012). It therefore is logical that despite Israel's posture of nuclear ambiguity, the country does possess nuclear weapons.

Iran also almost certainly possesses nuclear weapons. Woolsey et al. (2016; 2021) argue that Iran deceptively claims it has no nuclear weapons program while it nonetheless operates a clandestine military nuclear program underground. They also argue that Iran most likely has nuclear warheads for its Shahab-III medium-range missile, which it already has tested to make electromagnetic pulse (EMP) attacks (see also Trevithick 2019).

Woolsey at al. (2016) state that Iran is creating a large, deployable missile force that could use a satellite to launch a surprise nuclear EMP (NEMP) attack against the U.S. They also note that North Korea already possesses this capability. A NEMP attack could lead to a protracted nationwide blackout with the potential to kill millions of people. Woolsey et al. (2021) contend that not becoming overtly nuclear permits Iran to build enough nuclear missiles in secret to create an irreversible capability. Woolsey et al. (2016; 2021) further contend that given the risk of an EMP attack, or, worse, an NEMP attack, it is imperative that critical infrastructure, e.g., electrical grids, etc., in the U.S. be safeguarded and hardened.

On November 15, 2022, Biden signed a $1.2 trillion infrastructure bill into law that allocates millions of dollars for studies on how to protect critical infrastructure from the effects of an EMP. An EMP could be induced by the explosion of a nuclear weapon or by other means. Many people are concerned that new nuclear weapons could employ intense EMP fields to surmount the EMP hardening currently in place in the U.S. Such new nuclear weapons also could devastate civilian power transmission and interrupt the provision of electricity to U.S. military bases that rely on unhardened commercial power plants (Pompeo 2022).

Complex Geometry of Nuclear-Armed Rivals

A complex geometry of nuclear-armed rivals exists today (Andresen 2022b, 305-306). This geometry is made even more complex because a country with nuclear weapons can give nuclear-capable missiles and nuclear weapons to a proxy, thereby creating an extraordinarily dangerous free-for-all. This last point is not merely theoretical, either, since Belarus recently claimed that nuclear-capable Iskander missile systems given to the country by Russia are ready for use (EURACTIV 2022). More alarmingly, on March 25, 2023, Putin announced that Russia will station tactical nuclear weapons in Belarus (AP 2023).

According to Michael Krepon, who co-founded Stimson Center (Stimson) and who also shaped the debate on arms controls while working to reduce the role of nuclear weapons in security strategies (Diaz-Maurin 2022a), the current geometry of nuclear competition (Krepon 2021a) includes four pair of nuclear-armed rivals, and, also, the U.K. and France. According to Krepon, the four pair of nuclear-armed rivals are the U.S. and China, the U.S. and Russia, China and India, and India and Pakistan (Krepon 2021b). But Krepon's geometry is incomplete, since an accurate geometry of nuclear-armed rivals also includes the U.S. and Iran and Israel and Iran (Woolsey et al. 2016; 2021).

If we are not careful, the geometry of nuclear-armed rivals will become even more complex if Saudi Arabia, potentially with Chinese assistance, begins to manufacture its own ballistic missiles (Cohen 2021). If this occurs, it will become necessary to add Israel and Saudi Arabia, and, also, Iran and Saudi Arabia, to the current geometry of nuclear-armed rivals—thereby greatly complexifying what already is a complex geometry of nuclear competition on Earth. The increased risk this would pose to humankind is difficult to calculate, since, as Krepon (2021a) notes, adding any new pairings of nuclear-armed rivals increases the chances of nuclear catastrophe. The potential for nuclear conflagration increases even more during times of strained geopolitical relations and escalating dangerous military practices, which is the situation we face in the world now. Krepon singles out military practices between the U.S. and China, the U.S. and Russia, India and China, and Pakistan and India as particularly dangerous. One must add to this list of dangerous military practices what is occurring in Ukraine, the Taiwan Strait, and the South China Sea, which also includes the U.S.' escalatory encirclement of China (Wilpert 2021; see also Pilger 2016).

While arms control has limited Russian and U.S. strategic nuclear forces somewhat, during the past decade, China has expanded both its strategic and intermediate-range nuclear forces, thereby demonstrating the limitations of bilateral treaties that only limit the nuclear forces of Russia and the U.S. China is building hundreds of ICBM emplacements, and it possesses a formidable ballistic missile submarine, or sub-surface ballistic nuclear (SSBN) force. China also is developing a stealth bomber that is expected to be able to carry nuclear missiles (Pompeo 2022).

Scope of Nuclear Capabilities

Russia has the most nuclear weapons of any country on Earth (6,257 total, of which 1,458 are active, 3,039 are available, and 1,760 are retired). The U.S. is next (5,550 total, of which 1,389 are active, 2,361 are available, and 1,800 are retired). China has 350 available nuclear weapons, but it is expanding its nuclear arsenal rapidly. India, which like the U.S., China, and Russia, bases its nuclear forces on the concept of a strategic triad, as discussed below, has 156 available nuclear weapons (WPR 2023).

Most nuclear weapons in the U.S. arsenal are not deployed but instead are held in reserve or are awaiting dismantlement. However, most deployed weapons are on ballistic missiles, followed by a few hundred located at strategic bomber bases in the U.S. An additional, estimated 100-200 tactical bombs are deployed at European bases. As of 2019, remaining warheads were in storage at an estimated twenty-four geographical locations in eleven states and five European countries. The large Kirtland Underground Munitions and Maintenance Storage Complex (KUMMSC) south of Albuquerque, New Mexico had the largest inventory of nuclear weapons in the U.S., with most of these being retired weapons awaiting shipment for dismantlement at the Pantex Plant in Texas. The Strategic Weapons Facility, Pacific (SWFPAC) located in Bangor, Washington had the second largest inventory of nuclear weapons in the U.S. Of the five nuclear weapons storage locations in Europe, Incirlik Air Base in Turkey at least at one point in time stored the most nuclear weapons, about one-third of those in Europe, although unconfirmed rumors suggest that these weapons may have been withdrawn (Kristensen and Korda 2019, 122; see also Kristensen 2019).

The nuclear forces of the U.S., China, Russia, and India all are based on the concept of the strategic triad—land, sea, and air. The former Soviet Union demonstrated multidimensional capability first, with the U.S. nuclear triad becoming operational in 1959. In the U.S., the nuclear triad is composed of ICBMs, ballistic missile submarines, and bombers (Pompeo 2022). Land-based ICBMs are operated by the USAF. Next-generation ICBMs—previously referred to as the Ground Based Strategic Deterrent (GBSD) (Northrop Grumman 2022) and now called the LGM-35A Sentinel ICBM system—is the modernization program of the land-based leg of the U.S. nuclear triad (Air Force NWC n.d.).

The USN operates nuclear-powered ballistic missile submarines. The USAF operates strategic bombers, which are organized into nine bomb squadrons in five bomb wings at three bases—Minot Air Force Base in North Dakota; Barksdale Air Force Base in Louisiana; and Whiteman Air Force Base in Missouri. While some weapons are deployed at bomber bases, many weapons are kept in central storage at KUMMSC. With respect to nonstrategic nuclear weapons, the U.S. has one type, the B61 gravity bomb. All versions remain in stockpile, though some are thought to be deployed at six bases in five European countries: Aviano and Ghedi in Italy; Büchel in Germany; Incirlik in Turkey; Kleine Brogel in Belgium; and Volkel in the Netherlands (Kristensen and Korda 2019, 127-31).

Global Stockpiles of Fissile Materials

Global stockpiles of HEU and separated plutonium have reached horrific levels (IPFM 2022a; IPFM 2015, 2). India, Pakistan, Israel, and North Korea produce fissile materials (IPFM 2022a). In addition, Iran began stockpiling enriched uranium in 2019 (IAEA 2021; Iran International 2021).

Experts in the area of nuclear disarmament have proposed a Fissile Material Cut-off Treaty (FMCT) to prohibit the production of highly enriched uranium (HEU) and plutonium, which are the two main components of nuclear weapons. Discussions of a FMCT have occurred at the United Nations Conference on Disarmament (CD). Of note, this is a different organization than the UN Disarmament Commission (UNDC). The CD was established as the sole multilateral negotiating forum on disarmament, and it is comprised of sixty-five member nations (UNODA n.d.-b). Because the CD operates by consensus, however, it often is stagnant, which has impeded progress on a FMCT (Kimball 2018). Nevertheless, the CD currently is discussing a FMCT (NTI 2023b).

Nations that joined the NPT as non-weapon states already are prohibited from producing or acquiring fissile material for weapons. A FMCT would provide new restrictions for the five recognized nuclear weapon states (NWS), namely the U.S., Russia, the U.K., France, and China, and, also, for the four nations that are not NPT members, namely Israel, India, Pakistan, and North Korea (Kimball 2018).

Without tangible progress on a FMCT, humankind's peril resulting from nuclear weapons will continue to worsen. Producing and stockpiling fissile materials is reckless, since the very fact they are being produced and stockpiled increases the likelihood that they will be used. This is because stockpiling removes the threshold barrier of not using fissile materials, namely investing the resources and industrial effort required to produce and stockpile them in the first place. When fissile materials are sitting in stockpiles, it naturally catalyzes some people to start thinking about how to acquire these materials by means of purchase, bribery, and/or theft, and, also, to think about how to use them. These would be nonissues if the materials had not been produced and stockpiled in the first place. The same kind of dynamic applies to weapons generally, but of course the risks are greater with respect to nuclear weapons and other WMDs. When stockpiles of weapons reach a certain level—i.e., a tipping point—the balance can tip towards using them rather than permitting them to continue to accumulate. One can see this dynamic playing out now in Ukraine, where both the U.S. and Russia are drawing from stockpiles of old, military equipment and ammunition to fuel the conflict.

Furthermore, the MAD (mutually assured destruction) argument is inapplicable to WMDs, which are not merely accessible to nation-states, but which also are within reach of nefarious, rogue actors who often are motivated by irrational ideologies. Of particular concern are ideologies that promise eschatological rewards to those who act in a manner that is destructive to the whole, themselves included.

Both military and civilian stocks of fissile materials exist. On the military side, the nuclear-weapon states have enough fissile materials in their weapon stockpiles for tens of thousands of nuclear weapons. On the civilian side, enough plutonium has been separated to make a similarly large number of weapons. HEU is used in civilian reactor fuel in more than one hundred locations, and the total amount used for this purpose is sufficient to make about one thousand atomic bombs of the type used in Hiroshima. The design of a hydrogen bomb (H-bomb) also is well within the technical capabilities of terrorist groups (IPFM 2022b).

Specific details with respect to the amount of fissile material in existence are shocking. At the beginning of 2021, the global stockpile HEU was estimated to be about 1,255 metric tons. The global stockpile of separated plutonium was approximately 545 metric tons, of which approximately 405 metric tons were produced outside of weapon programs, were covered by obligations that the material not be used in weapons, or were not directly suitable for weapons—leaving approximately 140 metric tons of plutonium in weapons or available for weapons. Russia tops the list of weapons-usable fissile materials, with 678 metric tons of HEU and 88 metric tons of separated plutonium available for weapons. The U.S. is next in both categories. Countries also on the list are the U.K., France, China, Pakistan,

India, Israel, North Korea, and others. In addition, production of military fissile materials continues in many countries—India produces plutonium for weapons and HEU for naval propulsion; Pakistan produces plutonium and HEU for weapons; Israel is believed to produce plutonium; and North Korea has the capability to produce weapons-grade plutonium and HEU. Moreover, France, Russia, the U.K., Japan, and India operate civilian reprocessing facilities that separate plutonium from spent fuel of power reactors; China is operating a pilot civilian reprocessing facility; and thirteen countries—Russia, the U.S., France, the U.K., Germany, the Netherlands, China, Japan, Argentina, Brazil, India, Pakistan, and Iran—operate uranium enrichment plants, including countries in the Urenco consortium. North Korea also is believed to have an operational uranium enrichment plant (IPFM 2022a). Of note, the Urenco Group is a British-German-Dutch nuclear fuel consortium operating several uranium enrichment plants in Germany, the Netherlands, the U.S., and the U.K.

The International Panel on Fissile Materials (IPFM) fosters initiatives to reduce stocks of HEU and plutonium and to end the production of these materials. Founded in January 2006, the IPFM is comprised of an independent group of arms control and nonproliferation experts from both nuclear weapon and non-nuclear weapon states. The IPFM's mission is to analyze the technical basis for practical and achievable policy initiatives to secure, consolidate, and reduce stockpiles of HEU and plutonium. Since these fissile materials are the key components in nuclear weapons, controlling them is central for nuclear weapons disarmament. Controlling fissile materials also is necessary to halt the proliferation of nuclear weapons and to ensure that terrorists do not acquire nuclear weapons (IPFM 2022b).

The members of the IPFM include nuclear experts from seventeen countries—Brazil, Canada, China, France, Germany, India, Iran, Japan, Mexico, Norway, Pakistan, South Korea, Russia, South Africa, Sweden, the U.K., and the U.S. IPFM research and reports are shared with international organizations, national governments, and non-governmental groups. IPFM's full panel meets twice a year at capitals around the world, and it also holds specialist workshops throughout the year. These meetings and workshops often occur in conjunction with international conferences at which IPFM panels and experts make presentations (IPFM 2022b).

Since 2015, the IPFM has been co-chaired by Professor Alexander Glaser and Dr. Zia Mian of Princeton University and Professor Tatsujiro Suzuki of Nagasaki University. Previously, it was co-chaired by Professor José Goldemberg of the University of São Paulo, Brazil (2006-2007); Professor Emeritus Ramamurti Rajaraman (2007-2014) of Jawaharlal Nehru University, New Delhi, India; and Professor Frank von Hippel of Princeton University (2006-2014) (IPFM 2022b).

As an undergraduate at Princeton University, I had the immense privilege of studying directly with Prof. Frank von Hippel. He is a remarkable theoretical physicist and an even more remarkable human being. Among the many brilliant academics whom I have had the fortune of meeting and often studying with over the years, Prof. von Hippel stands out, not merely for his intellectual acumen, but also because of his kindness and genuine concern for all humankind.

Prof. von Hippel is one of the world's leading voices on the risks posed by nuclear weapons. He is a Professor and Co-Director of the Program on Science and Global Security at Princeton University, which he co-founded, and Professor of Public and International Affairs, Emeritus, at Princeton School of Public and International Affairs, where I studied with him. Before coming to Princeton, Prof. von Hippel worked for ten years in theoretical elementary-particle physics. Among numerous accomplishments, he served as Chairman of the Federation of American Scientists in the 1980s and as Assistant Director for National Security in the White House Office of Science and Technology Policy in the early 1990s. He remains a member of the IPFM, which he previously co-chaired (Princeton University 2023). Prof. von Hippel also serves as a Member of the National Advisory Board of the Center for Arms Control and Non-Proliferation, which is the research branch of the Council for a Livable World (CLW). CLW is a non-profit, advocacy organization based in Washington, D.C. dedicated to eliminating the U.S. arsenal of nuclear weapons. CLW was founded in 1962 by Hungarian nuclear physicist Leó Szilárd as the Council for Abolishing War.

With colleagues, Prof. von Hippel has worked on fissile material policy issues for the past thirty years. His policy research areas focus on nuclear arms control and nonproliferation, nuclear power and energy issues, and checks and balances in policymaking for technology. He also has played a major role in developing cooperative programs to increase the security of Russian nuclear-weapons-usable materials. For example, Prof. von Hippel contributed to ending the U.S. program to foster the commercialization of plutonium breeder reactors, and he convinced former President of the Soviet Union Mikhail Gorbachev to embrace the idea of a Fissile Material Production Cutoff Treaty. This catalyzed an era of significant U.S.-Russian cooperation on nuclear materials protection, introduced a control and accounting program, and broadened efforts to eliminate the use of HEU in civilian reactors worldwide.

Depleted Uranium Weapons

Nuclear weapons and the fissile materials on which they depend are not the only problem. Depleted uranium (DU) weapons are horrific and have devastating effects on human beings, other species, and our planet. DU weapons also are potential agents of omnicide, since they could make the entire planetary environment radioactive to such an extent that it no longer would be a viable host for humanity or for many other species. Such radioactive poisoning of the environment already has occurred in some areas of Central Asia, notably Afghanistan, and West Asia, notably Iraq (Moret 2004).

In an appalling and immoral development, the U.K. announced on March 20, 2023 that it will be sending DU rounds to Ukraine. These rounds are to be sent to Ukraine together with Challenger 2 tanks. This horrific development portends negative health consequences for the Ukrainians themselves. DU, which is a byproduct of the enrichment process used to make reactor-grade uranium, is a chemically toxic and radioactive heavy metal. Use of DU rounds causes toxic, radioactive dust to be released into the atmosphere, which when inhaled and/or ingested in other ways, results in debilitating health consequences. The U.S. and the U.K. used DU munitions on a large scale during the Gulf War in 1991 and in Iraq in 2003, with incident rates for cancer increasing sharply in areas in which DU rounds were used. DU rounds also have been implicated in a rise in birth defects from areas adjacent to the main Gulf War battlefields. Other health problems associated with DU include kidney failure, nervous system disorders, lung disease, and reproductive problems (Hudson 2023).

NATO forces' use of DU weapons in the former Yugoslavia in 1995 and 1999 also resulted in negative consequences such as those mentioned above. Still, the U.S. continued to use DU weapons in Syria in 2015. The negative impacts of these weapons also have not been confined to local populations. Troops involved in or close to the use of DU, and, also, military clean-up teams sent to deal with the impact of the DU, regularly experience negative health impacts. In 2006, the European Parliament strengthened its previous calls for a moratorium on DU by calling for an introduction of a total ban, classifying the use of DU, along with white phosphorus, as inhumane. Since 2007, repeated UN General Assembly resolutions have highlighted serious concerns over the use of DU weapons. However, the U.S., the U.K., France, and Israel consistently have voted against the resolutions, being the only states to do so. The situation in Ukraine is horrific already, and DU weapons will make the situation worse for everyone, including for Ukrainians (Hudson 2023; see also Copp 2023).

Lethality of Nuclear Weapons

Some nuclear weapons today are at least 3,000 times more powerful than the bomb dropped over Hiroshima, which is estimated to have killed between 90,000 and 166,000 people (Bennett 2020).

People often make a distinction between 'strategic' and 'tactical' nuclear weapons, though it is unclear whether that distinction is relevant today. So-called strategic nuclear weapons are those with yields ranging from 100 kilotons to the low megaton range. They are designed for use against buried hard targets such as missile silos, or against wide area targets including: larger bomber or naval bases; military command centers; industrial centers for arms design and manufacturing; transportation, economic, and energy infrastructure; and heavily populated areas such as cities and towns that contain these targets (Baker 2017). Tactical nuclear weapons sometimes are referred to as 'nonstrategic' nuclear weapons. Tactical nuclear weapons have less explosive energy as compared to strategic weapons, though the word 'tactical' refers to the fact that such nuclear weapons do in fact possess a large amount of explosive energy.

Russia possesses a total of approximately 4,500 active and available nuclear warheads, 2,500 of which are strategic and 2,000 of which are tactical. Russia's strategic nuclear weapons, i.e., those with the largest yield, are deployed on submarines, bombers, and ICBMs. Russia developed its tactical nuclear weapons to be used against troops and installations in a small area or in a limited engagement. The Iskander ballistic missile, which has a range of about 500 kilometers and which has been used by Russia in Ukraine, is the same short-range missile that has the capability to launch tactical nuclear weapons. The U.S. also has tactical nuclear weapons stationed around Europe, i.e., approximately 100 nuclear gravity bombs with less sophisticated guidance as compared to the U.S.' strategic nuclear weapons. Today, most nuclear weapons are variable-yield, also referred to as dial-a-yield, which means that they provide a set amount of explosive energy that can range from fractions of a kiloton to multiples of a megaton. Whereas the atomic bomb used in Hiroshima was about 15 kilotons, the newest version of the B61 nuclear bomb in the U.S. arsenal can release variable amounts of explosive energy, i.e., 0.3, 1.5, 10, or 50 kilotons (Tannenwald 2022). Furthermore, the existence of so-called suitcase or backpack nuclear devices adds yet another, alarming layer of complexity to the situation involving nuclear weapons (Crux 2022).

Tactical nuclear weapons, which are smaller and which have more precise targeting, exist because people think they make nuclear weapons more 'usable.' In contrast, the sheer destructiveness of higher-yield strategic nuclear weapons is a more obvious deterrent to their use. While the existence of tactical nuclear weapons makes deterrence threats more credible, it also makes first use of tactical weapons more tempting because a party to a conflict may want to use one or more tactical nuclear weapons instead of reserving them for retaliation. However, the destructive power of a thermonuclear explosion of any size means that using even a single tactical nuclear weapon makes no sense. Even at the 0.3 kilotons of the smallest yield nuclear weapon, the immediate damage produced would be far more than that of a conventional explosion. In addition to the initial fireball, the follow-on effects would produce shock waves and deadly radiation that would cause long-term health damage in survivors. Radioactive fallout would contaminate air, soil, water, and the food supply (Tannenwald 2022).

As discussed below, in the event of a nuclear explosion, the food supply is particularly vulnerable to soot injection into the atmosphere (Xia et al. 2022; Ray 2022). Even worse, using a tactical nuclear weapon could trigger full-scale nuclear war, since escalation could arise regardless of what kind of nuclear weapon—tactical or strategic—had been used first. In other words, the immensity of the moment could render moot any imagined distinction between 'tactical' and 'strategic' nuclear weapons (Tannenwald 2022).

Destabilizing Impact of Engineering Modifications to Nuclear Weapons

Although in 2010, former President Barack Obama committed to not developing new nuclear weapons or modifying them in any significant way in support of new military missions, the U.S. has not adhered to that commitment. The many significant engineering modifications that have occurred to the U.S. nuclear arsenal since Obama's 2010 statement have resulted for all intents and purposes in a new type of nuclear weapon. Even though modifying the U.S. nuclear arsenal is destabilizing and escalatory, members of Congress raised little if any controversy when the engineering modifications were proposed, nor have they done so as the new fuzing system is being installed on a widespread basis. Instead, both Democratic and Republican administrations have depicted the fuzing system as a slight modernization of a single component, claiming that this does not violate Obama's 2010 promise to foreswear the development of new nuclear weapons or to modify nuclear weapons in support of new military missions (Smith 2021).

The numerous engineering modifications that have been made to nuclear weapons in the U.S. arsenal have made these weapons much more destructive and riskier. Plans are to apply the engineering modifications to hundreds of U.S. ballistic missiles, at least, specifically to the most powerful warheads. The modifications involve a sophisticated electronic sensor buried in hardened metal shells at the tip. Wires, sensors, batteries, and computing gear provide these weapons with an increased ability to detonate with more precise timing over particularly challenging targets. The new components are being paired with other engineering enhancements that collectively increase what military planners refer to as the "hard target kill capability" of individual warheads. This refers to an improved ability to destroy Russian and Chinese nuclear-tipped missiles and command posts in hardened silos or mountain sanctuaries, and, also, to obliterate hardened military command and storage bunkers in North Korea, which are additional, potential U.S. nuclear targets (Smith 2021).

The engineering modifications to the U.S.' nuclear weapons have not raised controversy in Congress because of money, not because of what the newly designed fuzing system does or does not do. Congressional representatives in the U.S. rely on large donations to finance their campaigns for election and re-election, and people associated with the defense sector often are the ones making particularly large donations to Congressional campaigns. Accordingly, the decisions that Congressional representatives make often are driven by political and financial motivations. Even more problematic, many Congressional representatives invest financially in the very defense companies that make money when multi-billion-dollar contracts are granted to them for the design, redesign, and manufacture of weapons (Freeman and Hartung 2023)—exactly the sort of program that resulted in the new fuzing system.

Development of the new fuzing system is part of a major effort over two decades in the U.S. to modernize major components in five types of existing nuclear warheads—*at a cost exceeding $40 billion*. The U.S. simultaneously is modernizing almost all the launchers for these warheads, including elements in the U.S. missile, bomber, and submarine forces. This requires investing *$634 billion* over the next decade, which brings the complete cost of modifying the nuclear arsenal to *roughly $1.2 trillion* over the next three decades. But because the U.S. nuclear force now has improved capabilities with enhanced nuclear killing power as a result of the engineering modifications, it makes no general sense whatsoever to spend over a trillion dollars on the overall modernization, operation, and maintenance of the U.S. nuclear arsenal, nor does it make sense to keep so many warheads in the stockpile (Smith 2021). These actions only make limited sense if one's motivation is to make money—*a lot of it*—and if one happens to work for one of the companies involved in the modernization program—or if one benefits from large donations to a Congressional campaign. Next, when individuals win elections, they often have access to the kind of information necessary to make their own private investments in the very defense contractors they voted should receive new defense contracts. To be sure, the entire cycle reflects the actions of a self-interested circle of people with personal, vested, financial interests rather than those who act in the long-term, best interests of the United States and the rest of the world—not to mention those who act in the long-term, best interest of *Homo sapiens sapiens* as a species.

Many officials and nuclear weapons experts have raised significant concerns with respect to the engineering modifications that have been applied to the U.S.' nuclear arsenal since Obama's 2010 pronouncement. Despite what both Republican and Democratic administrations have said, experts who do not rely on large campaign donations to have jobs are concerned that the warhead improvements look like new designs. Hans Kristensen, who monitors technological efforts for the Federation of American Scientists (FAS), a nonprofit group in Washington, D.C., observes that as the U.S. arsenal has shrunk by roughly one-third as a result of arms agreements over the past two decades, engineers and weaponeers have tried to enhance the capabilities of the remaining weapons. The problem, states Kristensen, is that the results are so far removed from the Obama era's limitation that they are one step short of a new nuclear weapon (Smith 2021).

The National Nuclear Security Administration (NNSA) in Washington, D.C. funded the work on the engineering modifications, and the U.S. government-owned Sandia National Laboratories employed a team of several hundred people on its New Mexico campus during the initial years of the technology's development. Sandia possesses some of the most advanced research facilities, tools, and equipment in the world. Sandia's design subsequently was turned into hardware by Honeywell, which not only manages the lab but also operates the federally-owned nuclear weapons production facility in Kansas City, Missouri that is producing hundreds of the new systems (Smith 2021).

The warhead fuzes' accompanying sensors and computers are embedded in a stubby capsule about two feet high and a foot wide. It is so compact that it was carried to a congressional hearing in 2014 to show representatives. Plans are to install the warhead fuze and its accompanying sensors and computers on three new types of warheads atop land and sea-based missiles, and, also and in part, on a warhead to be carried by U.S. F-16 and F-35 warplanes deployed in Europe. Production of the first of many high-yield nuclear warheads containing the components was developed over the past decade at a cost of billions of dollars and was completed for installation on missiles aboard Navy submarines. This follows the development and installation of similar fuzes, also designed by Sandia, on hundreds of smaller-yield submarine warheads. The USAF also plans to install some of the same technology aboard new land-based missiles slated for deployment by the end of the decade. If this occurs, the technology will be deployed on more than 1,300 warheads in the U.S. arsenal (Smith 2021).

Publicly, DoD describes the components of the warhead fuze as a routine engineering improvement that provides no substantial new military capabilities. But that statement is inaccurate given that the new warheads are significantly more damaging than previous, similar weapons. The increased destructiveness of the new warheads means that in some cases fewer weapons could be needed to ensure that all the objectives in the U.S.' nuclear targeting plans are met fully. Although this could open a path to future shrinkage of the size of the U.S. nuclear arsenal, it also could create an escalatory spiral in which other countries try to outpace one another's engineering modifications while simultaneously invoking rhetoric to reduce the overall number of arms. Precisely for this reason, many people express the view that the engineering modifications create security perils. Georgetown University Professor Keir Lieber and Dartmouth University Associate Professor Daryl Press, the latter of whom is a consultant to DOD, estimate that the fuzes roughly double the destructive power of the U.S. submarine fleet alone. Officials and nuclear weapons experts contend that this shift in weapons capabilities has military and political consequences. The leaders of target countries, knowing that U.S. nuclear strikes are more certain to be effective in destroying weapons in target countries, might be more deterred from taking provocative actions that could draw a U.S. nuclear attack. The opposite scenario also is possible. Leaders of target countries, knowing that many of their protected, land-based weapons and associated command posts could not escape destruction in the event of a nuclear attack, might be more prone to order the use of their own weapons early in a conflict or crisis to ensure that these weapons are not destroyed by incoming warheads. This could promote a hair-trigger launch policy that could escalate cataclysmically (Smith 2021).

Officials state that one of the Sandia team's objectives when developing the engineering modifications was to make the fuze of U.S. warheads more resilient in the face of electronic jamming efforts. Electronic jamming typically is intended to make a fuze malfunction or detonate too early. The Sandia team also wanted to make the fuze more resilient in case a potential nuclear blast created high radiation levels, for example if Russia deliberately set off a nuclear blast at high altitude of approximately thirty miles over key command centers as part of that country's missile defense system (Smith 2021). High-altitude nuclear explosions, which previously were performed by the U.S. and the former Soviet Union between 1958 and 1962, result from nuclear weapons testing within the upper layers of the Earth's atmosphere and in outer space. For example, the Starfish Prime high-altitude nuclear test conducted by the U.S. in 1962 was the largest nuclear test conducted in outer space. It had devastating effects. Interestingly, sixteen years later, the site of the launch, Johnston Atoll, became the site of an oft-discussed UFO encounter.

Dr. James Acton, a physicist who co-directs the nuclear policy program at the Carnegie Endowment for International Peace (CEIP) and who has written extensively about the need to avert unnecessary conflicts, thinks that efforts to modernize the nuclear arsenal should be more focused on ensuring the weapons' safety, security, and reliability and less about improving their accuracy. He states:

> If China or Russia believe in a conflict or a crisis that we are going to attack or destroy their nuclear forces and command posts, that gives them an incentive to use nuclear weapons first, or to threaten their use. They have strong incentives to take steps that would further escalate the crisis and create new dangers.

Acton also expresses concern that distributing impressive accuracy improvements throughout the U.S. arsenal will raise concerns in China and Russia about targeting the leadership of these countries. Acton's position—with which I agree—is that the U.S. should be more restrained in making engineering modifications to its nuclear arsenal, since the additional escalation risks outweigh a relatively modest increase in utility against targets such as deeply-buried command posts. In fact, the U.S. already can hold at risk many, though possibly not all, key Chinese and Russian targets (Smith 2021).

Like Acton, physicist Dr. John R. Harvey argues that making significant engineering modifications to a nuclear arsenal heightens the risk of a preemptive attack (Smith 2021). From 2009 to 2013, Harvey was the Principal Deputy to the Assistant to the Secretary of Defense for Nuclear and Chemical and Biological Defense Programs, which is DOD's top nuclear weapons authority; from 2001 to 2009, he was Director, Policy Planning Staff of the National Nuclear Security Administration (NNSA); from 1995 to 2001, he was Deputy Assistant Secretary of Defense for Nuclear Forces and Missile Defense Policy; and from 1989 to 1995, he was Director of the Science Program at the Standard University Center for International Security and Arms Control (Harvey n.d.). Harvey therefore is in an excellent position to assess the risks associated with nuclear weapons. Harvey's concern is that the deployment of additional capabilities to target key enemy warheads could put "the adversary in a posture that would generate a rapid response, which could conceivably be the result of misinterpretation." An adversary therefore might launch weapons based on the erroneous sensing of an attack and/or based on inaccurate fears regarding a potential imminent attack (Smith 2021).

Kristensen raises another key problem associated with modifying nuclear weapons, namely that some weapons designers have accepted the idea that building weapons that are more accurate and destructive while producing lower collateral damage "makes them easier to use." An argument that has been broached is that lower-yield, more accurate weapons are better deterrents because if an adversary regards them as more usable, their deterrence effect is greater, which makes the overall likelihood of their use lower rather than higher (Smith 2021). I disagree with this assessment. My view is that the very existence of lower-yield weapons increases rather than diminishes the likelihood that a nuclear weapon will be used, making lower-yield weapons more destabilizing than higher-yield weapons. A similar logic applies to engineering modifications of the type being discussed here—if an adversary perceives that increased accuracy makes modified nuclear weapons easier to use, then the engineering modifications are destabilizing.

By upgrading the "hard-target kill capability" of nuclear weapons, it was thought that DOD would agree that the U.S.' nuclear force could be smaller. Former President Obama voiced this view in June 2013 after a comprehensive, classified review. Obama stated that even after the New START limitations were met in full, it was possible to ensure the security of the U.S. and its allies and partners and to maintain a strong and credible strategic deterrent while pursuing up to a one-third reduction in deployed nuclear weapons beyond what New START required. Despite what Obama said, however, his administration did not implement reductions in the size of the U.S. nuclear arsenal. Jon Wolfsthal, former White House Adviser on Arms Control and Nuclear Issues who now is a senior adviser to Global Zero, an advocacy group working towards the eventual elimination of nuclear arms, states that the military's support for a one-third reduction of warheads from the size of the current arsenal—estimated by Smith (2021) at around 500 warheads—was not conditioned during the Obama administration's review on a requirement that similar reductions be taken by Russia. Instead, it was decided based on an assessment of what the U.S. needed to be able to hold key Russian targets at risk. This was determined in part by a recognition that the U.S. arsenal was becoming more accurate and in part by the fact that the total number of targets that the U.S. needed to destroy in Russia had declined. Wolfsthal thinks that Obama did not implement approved reductions in the size of the U.S. arsenal for political rather than military reasons. According to Wolfsthal, Obama and Biden opposed acting unilaterally because they hoped to persuade Russia to act similarly, which they thought would contribute to an overall reduction in global nuclear risks. Regardless, Wolfsthal states that he thinks it is "crazy" to be spending so much on new weapons when "we could live with a smaller force" (Smith 2021).

Andrew Weber, the former Assistant Secretary of Defense for Nuclear, Chemical, and Biological Defense Programs from 2009 to 2014, and former Staff Director of the Nuclear Weapons Council that approved development of the fuzes, thinks that the deployment of the fuzes reduces the need to keep developing smaller nuclear weapons slated for deployment in the next decade. Weber think that new air- and sea-launched cruise missiles are not necessary, and that they have the potential to undermine deterrence because they are stealthy, surprise-attack weapons that will make opponents nervous enough to adopt hair-trigger launch policies, a significant downside risk, as discussed below. Since new air- and sea-launched cruise missiles can be deployed with both conventional and nuclear warheads, and since it is impossible for adversaries to tell the difference, their use could cause unintentional escalation from a conventional to a nuclear war. Weber states that it is time to stop "replacing everything mindlessly" (Smith 2021).

Programs proposed to create new air- and sea-launched cruise missiles are estimated to cost tens of billions of dollars (CRS 2021b). In May 2021, the Congressional Budget Office (CBO) estimated that the U.S. will spend a total of $634 billion over the next ten years to sustain and modernize its nuclear arsenal, which is 28% higher than the previous ten-year projection released in 2019. This estimate in planned spending is projected to consume 6.0-8.5% of projected total spending on national defense during fiscal years 2021-2023. Over the next thirty years, the total sustainment and modernization costs of U.S. nuclear forces could reach \$2 trillion (Bugos 2022)—unless, of course, this insanity is checked.

So far, the Biden administration has said little about its larger plans for the nuclear arsenal besides affirming in budget plans that it intends no major change in the modernization programs that were created by Obama and that were continued if not expanded by former President Donald Trump (Smith 2021). This is not a solution. We must find a way to abolish nuclear weapons. Continuing to spend trillions of dollars 'modernizing' and replacing nuclear weapons wastes money and further destabilizes the world. We need another approach, namely unilateral restraint, to end this exorbitantly expensive and utterly insane cycle of nuclear posturing—a cycle that benefits a few individuals financially while risking longevity of *Homo sapiens sapiens* and many other species on Earth.

Nuclear Weapons Kept on Hair-Trigger Alert

Many nuclear weapons are kept on hair-trigger alert, which makes time pressure on decision-making very pronounced. This increases the possibility that a nuclear weapon could be launched accidentally because of faulty sensor data, a computer glitch, and/or some other form of incorrect information (UCS 2015). If launch of a nuclear weapon based on misinformation were misinterpreted as intentional hostility, it theoretically could trigger even further exchange of nuclear weapons.

Widespread Famine from Soot Injection into the Atmosphere

A follow-on consequence of a nuclear exchange, particularly of high-yield weapons, is that the spread of soot ejected into the stratosphere would change global weather patterns, thereby triggering global famine (Ray 2022, 567).

Atmospheric soot loading from the detonation of a nuclear weapon would disrupt Earth's climate such that it would limit terrestrial and aquatic food production severely. A recent study uses climate, crop, and fishery models to estimate the impacts of six scenarios of stratospheric soot injection. According to these models, soot injections of sufficient scale would lead to mass food shortages, with livestock and aquatic food production unable to compensate for reduced crop output. *An estimated more than two billion people could die from nuclear war between India and Pakistan, and more than five billion people could die from a nuclear war between the U.S. and Russia.* As should be completely obvious to everyone, all human beings urgently must work together to catalyze the global cooperation necessary to prevent nuclear war (Xia et al. 2022, 586).

Excessive Reliance on Deterrence Strategies

Political scientists and government officials traditionally mention two approaches to preventing nuclear war—deterrence, and arms control. This section discusses the major shortcomings of deterrence, while the next section explains how arms control also has fallen short.

Michael R. Pompeo is a vocal proponent of deterrence. During the Trump administration, Pompeo held the positions of Director of the CIA and, later, Secretary of State. He states the following:

> Deterrence arises from strength and never from weakness because weakness invites belligerency. Unilateral restraint in the maintenance and disposition of forces does not support deterrence, for deterrence cannot arise from unilateral restraint if such restraint circumscribes power in the face of burgeoning threats. Unilateral restraint can signal weakness, which may begin a dangerous cascade of responses by nations that believe they are unbound (Pompeo 2022).

While Pompeo believes that unilateral restraint is a sign of weakness, I argue the opposite later in this book. In point of historical fact, unilateral restraint by the U.S. resulted in rapid reductions of the nuclear arsenal of the former Soviet Union (National Security Archive 2016). As a species, we have no more time for posturing, particularly posturing that involves nuclear weapons. If we want *Homo sapiens sapiens* to survive, we must move past utilitarian thinking and take a deontological step forward—which is precisely what unilateral restraint is.

On the opposite side of the continuum from unilateral restraint, deterrence is the idea that possessing nuclear weapons protects a nation from attack because of the threat of overwhelming retaliation (Tannenwald 2022). Nuclear-armed competitors increasingly are relying on turning back to deterrence since arms control diplomacy is not producing results—so much so that many people now hold the opinion that arms control is passé (Krepon 2021a). The problem is, like arms control, deterrence also does not work.

As mentioned above, the U.S., China, Russia, and India currently rely on a nuclear triad with land-, sea-, and air-launch capabilities. Deterrence advocates argue that deploying forces on an array of different platforms by land, sea, and air supports deterrence because one can structure each arm of the triad to make a disabling first strike all but impossible. In the U.S., deterrence advocates argue that the nuclear triad secures deterrence against WMDs, including nuclear, chemical, biological, and electromagnetic pulse (EMP) weapons. Deterrence advocates do not favor comprehensive bilateral (the U.S. and Russia) or trilateral (the U.S., Russia, and China) nuclear talks to reduce nuclear systems. Concerned that the ICBM force is almost fifty years old, they want to procure the Minuteman III ICBM replacement, since they see the ICBM force and submarines as the most survivable deterrent. Deterrence advocates also do not want to agree with Russia to eliminate all fixed, ground-based strategic missiles, which would cancel the Minuteman's successor, previously called the Ground-Based Strategic Deterrent (GBSD), and now referred to as the LGM-35A Sentinel ICBM system. Deterrence advocates also want to develop and deploy the Long-Range Standoff Cruise Missile (LRSO), which they see as a hedge against a defensive breakthrough that could impair U.S. stealth bombers from reaching their targets undetected. Deterrence advocates also do not want the U.S. to move towards a two-legged dyad consisting of ballistic missile submarine, or sub-surface ballistic nuclear (SSBNs) submarines and bombers, since they assume China and Russia would retain strategic triads and, also, HGVs (Pompeo 2022).

Furthermore, deterrence advocates do not want the U.S. to announce a unilateral moratorium on the development of strategic, as opposed to tactical, anti-ballistic missile (ABM) systems, while requesting China and

Russia to do the same. They believe this will never happen. Many people also think the difference between strategic and tactical ABM systems is no longer clearly demarcated because of technology and operational doctrines. In addition, deterrence advocates do not want the U.S. bomber fleet to face any constraints, such as an announcement that the new B-21 Raider bomber will be incapable of carrying nuclear weapons and/or that the B-52 re-engineering program will be canceled (Pompeo 2022).

Given indications of Russia's and China's non-compliance with the Comprehensive Nuclear-Test-Ban Treaty (CTBT), deterrence advocates do not want the U.S. to remain committed to this Treaty without preconditions, since they believe this will inhibit the ability of the U.S. to deploy next-generation nuclear weapons. Deterrence advocates argue in favor of next-generation weapons by saying that such weapons would have enhanced reliability without requiring recurring, underground testing. Finally, deterrence advocates do not want the U.S. to declare a No First Use (NFU) policy regarding nuclear weapons, since they think NFU would undermine and degrade the ability of the U.S. to respond to a range of threats (Pompeo 2022). While it may be true that next-generation nuclear weapons would not require recurring, underground testing, this is not an argument per se for developing such weapons in the first place. We need to walk back from the precipice, not design and manufacture new weapons that help us walk over it.

Having laid out the argument for deterrence, above, I spend the remainder of this sections deconstructing the kind of faulty logic on which deterrence thinking is based. I realize that all nuclear-armed countries are turning towards deterrence because of disappointment with arms control, but relying on deterrence in this era of great power competition and escalating, dangerous military practices is unwise. Excessive reliance on deterrence strategies has increased the risks associated with nuclear weapons, not reduced them. Because renewed great power struggles negatively impact international security so severely (CRS 2021a), it is unreasonable to think that any combination of countries could fight a nuclear war without cataclysmic repercussions for the entire world.

Even planning for nuclear war is highly destabilizing, since indirect signaling between nation-states is ambiguous and easily can be misinterpreted. Yet as DOD refocuses its efforts from the so-called war on terror to great power competition, it recently used the newest war plan of the U.S. Strategic Command (USSTRATCOM), which is called STRATCOM CONPLAN [Concept Plan] 0810-12, to practice nuclear war fighting during the midst of the war in Ukraine. Practicing nuclear war fighting at any time, let alone now during the war in Ukraine, poses a significant risk to strategic stability for the simple reason that one's perceived adversary has an imprecise ability to differentiate deterrent moves from actual threats (Arkin and Ambinder 2022).

In addition to being atavistic overall, U.S. nuclear deterrence theory is woefully out of date. Many military and other government officials in the U.S. who stubbornly hold to the idea that deterrence works nevertheless acknowledge that U.S. nuclear deterrence theory is outdated. Officials at USSTRATCOM therefore recently began writing a new deterrence theory that simultaneously faces Russia and China while considering how these threats changed in 2022. Until recently, the U.S. had not faced two peer, nuclear-capable opponents at the same time. Adding to the complexity of this exercise is the view that Russia and China must be deterred differently. Since operational deterrence expertise is not what it was at the end of the Cold War, some people think that rewriting deterrence theory will reinvigorate operational deterrence. USSTRATCOM leadership wants to evolve past the traditional nuclear deterrence theory of mutually assured destruction (MAD), which posits that any use of nuclear weapons would result in retaliatory use and total annihilation of all parties. While some people think that MAD should be credited with preventing nuclear war since 1945, others are concerned that rather than starting a global thermonuclear war, an actor such Putin could use smaller, tactical, shorter range nuclear weapons in limited numbers and directed to specific targets (Copp 2022). A valid concern is that Russia may perceive and try to exploit a so-called deterrence gap—i.e., a threshold below which an actor such as Putin mistakenly believes it is possible to employ nuclear weapons.

The very fact that countries continued to produce nuclear weapons after 1945 suggests that leadership in these nations consider nuclear weapons intrinsically useable. Certainly, there is sufficient evidence available in the public domain to show that planning for the use of nuclear weapons in circumstances that fall short of a global nuclear conflict has been an element of strategy since nuclear weapons first were deployed in substantial numbers. Today, if Putin views nuclear weapons as useable, with the perceived benefit of war-time use exceeding the perceived risk of escalation to all-out nuclear war, this makes MAD a myth. In any event, the topic should be of significant interest to the public, which should discuss it carefully (Rogers 2023a).

Even though national security hawk Mike Pompeo recognizes that tactical nuclear exchange, at whatever level, dramatically would increase the potential for global nuclear war by escalation or miscalculation, he wants to carve out space to justify potential U.S. use of a tactical nuclear weapon by blaming the U.S.' adversaries for their development of next-generation weapons. Pompeo is disingenuous here, since he does not acknowledge that the U.S. also is proceeding full tilt ahead developing its own next-generation weapons. Pompeo argues that adversaries' development of next-generation weapons may reduce the threshold below which the U.S. could employ tactical nuclear weapons of reduced yield (Pompeo 2022). Such exceedingly dangerous thinking must be challenged.

At a global level, China and Russia theoretically can escalate unilaterally to any level of violence in any domain, with any instrument of national power. Of course, so can the U.S. and other members of NATO. From both perspectives, then, that of NATO and that of the countries that comprise the BRICS+, nation states are facing a competitive, confrontational geopolitical landscape at scale. A disconcerting scenario is that countries such as China and Russia combine their ambitions to force the U.S to face simultaneous nuclear threats (Copp 2022). From the perspective of BRICS, however, a disconcerting scenario is that NATO grows even larger and the countries that comprise BRICS face nuclear threats from even more nations than they do now. The question is whether overhauling deterrence theory in the U.S., or elsewhere, will reduce the global risk that all human beings face from nuclear weapons, or whether a new approach is needed. I think it should be completely evident to everyone that a new approach is long overdue.

Unfortunately, instead of implementing a new approach, the U.S. for its part continues along the long, unsuccessful road of revising its deterrence theory—now by invoking new concepts that it hopes will be effective in making deterrence work. For the two-party scenario, for example, the U.S. has moved beyond classic deterrence theory to incorporate ideas such as nonlinearity, linkages, chaotic behavior, and inability to predict into its updated deterrence models (Copp 2022). While the idea of incorporating such concepts may make limited sense in theory, it has little practical significance. No amount of nonlinear modeling of scenarios makes a whit of difference if Vladimir Putin is in a bad mood. The sorts of models that USSTRATCOM increasingly is relying upon have no way of calculating the actions of exceedingly emotional human actors such as Putin—or, exceedingly calculating ones, if one interprets Putin's meetings with Xi in that light. What is missing from USSTRATCOM's approach is the simple reality that world leaders are human beings, and that human beings— *especially* world leaders—are notoriously complex emotionally and cognitively. To have any hope of achieving real peace, we need an approach that transcends both deterrence theory and arms control. We need to move beyond algorithms and mathematical modeling to cut through the ontological and spiritual impasses that cause people such as Putin and many other world leaders to order ruthlessly destructive actions in the first place.

We also need to start learning from our mistakes instead of mindlessly repeating them. Deterrence has not worked, regardless of the methods used and the numerous forms of deterrence theory and strategies that have been employed. Actions to sharpen deterrence tend to push nuclear-armed rivals towards the next crisis. To date, nuclear-armed rivals already have engaged in two border wars: the Soviet Union and China in 1969; and India and Pakistan in 1999. In addition, regular clashes occur along disputed borders between India and China and between India and Pakistan. Furthermore, deterrence has resulted in many negative effects: it has depleted resources that otherwise could have been spent addressing social and environmental issues; it has resulted in five-digit-sized U.S. and Soviet nuclear arsenals; it has led to almost 2,000 nuclear tests, including over 400 in the atmosphere; and it has generated dangerous military practices, such as maintaining delivery vehicles for nuclear weapons on a state of high alert (Krepon 2021a).

Since deterrence strategies historically have led to both sides in a conflict escalating rather than de-escalating their weapons buildup and their engagement in dangerous military practices, we cannot count on deterrence strategies to keep us safe. In addition, following a strategy to increase nuclear deterrence while simultaneously engaging in dangerous military practices heads directly off a cliff. The ideology underlying deterrence also is deeply problematic given that it is based on the quest to seek advantage while seeking to avoid disadvantage—with both these impulses reflecting little more than tactical and strategic thinking applied to the simpleminded pursuit of dominance.

The problem we must address is that many people advocate deterrence, not because it works—since it doesn't—and not because it is the best option on the table—which it isn't. The real problem is greed. Defense industry players and Congressional representatives benefit financially from deterrence. Deterrence is profitable because it requires that defense contractors make things, whether that be new fuzes, new hypersonic delivery systems, etc. Deterrence generates contracts, and *contracts generate money*—lots of it—for people who are motivated by greed. This is true everywhere in the world, not only in the U.S. Defense contractors—which often are international—wield tremendous power globally. Despite the trillions of dollars spent on nuclear arsenals, however, nuclear deterrence brings with it tremendous risks to our species, to other species on Earth, and to Earth itself. Nobody is sleeping soundly under a nuclear umbrella, and the world clearly would be better off without nuclear weapons (Tannenwald 2022).

Another major problem is that the rhetoric surrounding nuclear deterrence can be manipulated in situations of actual conflict. This has become very clear during the current war in Ukraine. Russia's invasion of Ukraine on February 24, 2022 demonstrates how little actual protection nuclear weapons provide and the risks that they pose. Everyone can see that the war in Ukraine is upending the entire security order of Europe and the West more generally. At strategic moments during this war, Putin has made various nuclear threats and has given orders to increase the alert level of Russia's nuclear forces. By increasing the alert level, however, Putin increases the risk of nuclear use from desperation, miscalculation, or from an accident that could occur during what people refer to as 'the fog of war.' Previously, during Russia's invasion of Crimea in 2014, Russian leaders also talked openly about putting nuclear weapons on alert. In 2015, Russia threatened Danish warships with nuclear weapons if Denmark joined NATO's missile defense system. All these actions appear to be Putin's way of reminding the West that Russia still is a so-called great power (Tannenwald 2022). It is eerily similar to Kim Jong-un brandishing his country's nuclear weapons to try to prove to the world and to himself that North Korea is a new 'great power.'

While the war in Ukraine is a humanitarian nightmare, it also has exposed the limits of the West's reliance on nuclear deterrence. If human hostilities escalate and/or if human intentions are misinterpreted because of an accidental firing of a nuclear weapon, humankind could obliterate itself using the very weapons systems it has designed, manufactured, and deployed. Although deterrence is widely credited for helping prevent armed conflict between the U.S. and the former Soviet Union during the Cold War, the downside risk of deterrence has become abundantly clear for all to see since Russia's recent invasion of Ukraine in 2022. Now, Russia is using the threat of using nuclear weapons to attempt to deter the West from intervening with conventional military forces to defend Ukraine. Although it has not worked at a broad level, the point is that this is a radically different situation from a potential scenario in which Putin might use nuclear deterrence to protect Russia. Throughout the war, Putin and his hardline allies in Russia repeatedly have signaled that if the U.S. and its NATO allies continue to arm Ukraine, Putin 'might' reach for a tactical nuclear weapon. The result has been that although NATO's nuclear weapons presumably deter Russia from expanding the war to NATO countries such as Poland, Romania, or the Baltic states, and although the nuclear balance of terror may deter wider war in Europe, Putin's use of nuclear rhetoric has left Ukraine struggling with insufficient support to win the conflict outright and to retake its territory. Here, "NATO states do not seem very reassured by their vaunted nuclear deterrence," since NATO members continue to be preoccupied by the possibility of a nuclear attack by Russia, either within Ukraine itself or beyond it (Tannenwald 2022).

Concern that Russia may use a tactical nuclear weapon in Ukraine has grown as Russia has suffered heavy losses on the battlefield and as Ukraine has begun to take back its territory. It is intimating first use of nuclear weapons supposedly on grounds of 'defense,' After the 'referenda' in four Ukrainian oblasts—Luhansk, Donetsk, Zaporizhzhia, and Kherson—and the subsequent 'annexation' of these regions by Russia on September 30, 2022, many people assumed that Putin might ramp up his nuclear rhetoric by claiming that Russia's military doctrine would validate a decision by him to use nuclear weapons to 'defend' the annexed regions. In such a scenario, these regions would be construed by Kremlin leadership as 'part of Russia.' Although Putin has not done this, the underlying threat of him doing so has succeeded in terrifying the entire world—including many Russians themselves. At the same time, Russian leaders have made it clear that they would view any nuclear attack as the start of an all-out nuclear war. that if the West were to use nuclear weapons first, then it would interpret this as an all-out nuclear war and would respond by using nuclear weapons itself.

Days after three separate drone strikes in early December 2022 hit bases hundreds of miles within Russia's borders—an attack that Moscow blamed on Ukraine—Putin suggested on December 8, 2022 that he was considering making changes to Russia's nuclear deterrence doctrine to include a first-strike policy. He referred to this as a "disarming strike" (McFall 2022). Putin stated that Russia could adopt such a preemptive military strike doctrine, noting that the country has the hypersonic weapons capable of carrying out such a strike intended to knock out an adversary's command facilities (AP 2022). Putin also stated that such a strategy would be motivated by current U.S. deterrence policy (McFall 2022). Here, he specifically noted that the U.S. has not ruled out the first use of nuclear weapons. In fact, for years, Russian military doctrine has stated that Russia reserves the right to first use of a nuclear weapon in response to largescale military aggression (AP 2022). In addition, both the U.S. and Russia already have nuclear doctrine permitting them to use a nuclear weapon in response to non-nuclear threats rather than only in retaliation to a nuclear attack. In his recent speech, however, Putin clearly was invoking "strategic messaging" to signal that Russia may use nuclear weapons in Ukraine if the US/NATO continue to provide weapons to Ukraine—especially the type of weapons that enable Ukraine to strike deep into Russia itself (McFall 2022).

Russia's current nuclear doctrine is based on the so-called launch on warning concept, which envisions the use of nuclear weapons in the face of an imminent nuclear attack detected by its early warning systems. On December 8, 2022, Putin made the following comments:

> When the early warning system receives a signal about a missile attack, we launch hundreds of missiles that are impossible to stop. Enemy missile warheads would inevitably reach the territory of the Russian Federation. But nothing would be left of the enemy too, because it's impossible to intercept hundreds of missiles. And this, of course, is a factor of deterrence (translated from Russia).

Russia's nuclear doctrine also states the country can use nuclear weapons if it comes under a nuclear strike or if it faces an attack *with conventional weapons* that threatens "the very existence" of the Russian state. For example, the Kremlin has expressed concern for years with respect to U.S. efforts to develop Prompt Global Strike (PGS) capability that envisions hitting an adversary's strategic targets anywhere in the world *within one hour* using precision-guided conventional weapons. Despite such plans, the U.S. argues rhetorically that nuclear powers should avoid provocative behavior and lower the risk of proliferation to prevent escalation and nuclear war (AP 2022). Given the PGS agenda, it is questionable how sincere the U.S. argument is.

Understanding the thinking of current leaders in Russia with respect to nuclear use is challenging. In June 2020, Putin made a presentation, "Foundations of State Policy of the Russian Federation in the Area of Nuclear Deterrence" (CNA 2020). The presentation points to a potential nuclear response to non-nuclear aggression, which is significant given that the wide range of tactical nuclear weapons within Russia's overall nuclear forces matches that of the U.S. (Rogers 2023a).

Since Russia invaded Ukraine in 2022, nuclear rhetoric and escalation management has been a major challenge (Arndt and Horovitz 2022). By March 2023, it was growing increasingly difficult to parse Russia's logic, or lack thereof, relating to the potential use of its nuclear weapons. Numerous news stories over a short time frame show the back-and-forth rhetoric being articulated by Russia's leadership. For example, on December 7, 2022, Russian news agency TASS reported the headline, "Risk of nuclear war is increasing – Putin" (TASS 2022a). On February 22, 2023, TASS reported, "Russia ready to defend itself with any weapon, including nuclear – Medvedev" (TASS 2023c). Then, there was a seeming reversal of direction, when on March 1/2, 2023, TASS reported, "No plans to use nuclear weapons in Ukraine – Russia's UN mission" (TASS 2023a). This was followed by another TASS headline on March 21/22, 2023, apparently in the same direction, "Russia, China convinced that nuclear war must never be unleashed – joint statement" (TASS 2023b). But then there was another apparent reversal of direction on March 24, 2023, when a headline in TASS announced, "Ukraine's attempt to take Crimea reason enough for Russia to use any weapons – Medvedev" (TASS 2023d).

In response to an expected Ukrainian counteroffensive against Russia, Western leaders are bracing for Putin to use "whatever tools he's got left," including nuclear threats and cyberattacks. U.K. officials at the G7 (Group of Seven) foreign ministers' summit held in Japan from April 16-18, 2023 said they must be prepared for Putin to use extreme tactics to attempt to hold Ukrainian territory (Crerar 2023). As if underscoring the point, on April 18, 2023, the last day of the ministers' meeting in Japan, Putin sent two Tu-95MS nuclear bombers over the Bering Sea and the Sean of Okhotsk as part of Russia's war games on its Pacific flank. These surprise drills were billed as a sudden check on combat readiness, but they obviously came at a time of already deep tensions with the West over the war in Ukraine (Stewart 2023). Putin also deployed nuclear submarines to the Pacific

Ocean as part of these surprise drills, with at least one submarine leaving a port in Kamchatka. The naval drills coincided with Chinese Defense Minister Li Shangfu's visit to Moscow, where he met with Putin on April 16, 2023. Japanese Chief Cabinet Secretary Hirokazu Matsuno stated that Tokyo had lodged a protest with Moscow over its military exercises around disputed islands near Japan's main northern island of Hokkaido. In fact, Japan has a territorial dispute with Russia dating back to the end of World War II over islands in the north Pacific. Kremlin spokesperson Dmitry Peskov responded that the drills were occurring "in strict accordance with international law," adding, "We are all well aware of the geography of these regional conflicts" (Nanu 2023).

If the real goal is to de-escalate rivalry between the U.S. and Russia—which it may not be for some neoconservatives in the U.S.—then one must take time to understand how and why Russian leadership feels vulnerable. We all must stop provoking one another if we want *Homo sapiens sapiens* to survive as a species, which means that all countries, including the U.S., need to stop engaging in provocative behavior involving WMDs, other advanced weapons systems, and military tactics and strategies. The species will not survive if we continue down the road of posturing and brinkmanship—euphemistically referred to during a recent interview as the type of "classical diplomacy" prevalent during the Cold War (DW 2022). As the global arms race intensifies, human beings do not seem to be waking up to the very real, extinction-level danger they face.

The U.S. must take a stronger leadership role to de-escalate geopolitical tensions by committing to a NFU policy with respect to nuclear weapons. As discussed above, a NFU policy requires the U.S. to commit never to be the first to use nuclear weapons in a conflict. Scientists actively are urging Biden to take this step, and, also, to cut the size of the U.S. nuclear arsenal (Sanger 2021). The Council for a Livable World (CLW), also mentioned above, supports a NFU policy, in keeping with its mission to support progressive national security policies and to help elect congressional candidates who support such policies. Its research arm, the Center for Arms Control and Non-Proliferation, provides research to members of Congress and their staffs. In addition to supporting NFU, the CLW also supports other "smart policies that limit nuclear risks." As CLW's website states:

> If the United States uses nuclear weapons first, a nuclear-armed adversary is likely to retaliate against us or our allies with nuclear weapons. Keeping the threat of nuclear first use on the table is an outdated and dangerous policy. A No First Use [NFU] policy just makes sense (CLW 2022).

Since any use of nuclear weapons, tactical or otherwise, would be insane, we must do everything possible to preserve the tradition that has prevailed since 1945 of nuclear nonuse in all instances of conflict (Tannenwald 2022). While a NFU policy is a useful element of a broader approach to de-escalate tensions that could lead to any use of a nuclear weapons in a situation of conflict, however, if we are truly serious about achieving peace on Earth— which I am—then the U.S. must take an even stronger leadership role beyond NFU and, also, beyond cutting the size of its nuclear arsenal. It goes without saying that we should stop 'modernizing' our nuclear arsenal, which makes no actual sense—as discussed in detail above in the section on the new fuzes.

To achieve a truly peaceful society, the same logic that applies to nuclear weapons and other WMDs also applies to weapons more generally. We need to learn how not to use them, rather than continue to develop more. To be truly useful to the world, the U.S. needs to teach others how to act with restraint, until the point that all advanced weapons systems, assault rifles, and handguns have been abolished.

As a species, *Homo sapiens sapiens* needs to choose sanity over insanity. This is the only path to a world free of violence in which human creativity and compassion can flourish. Earth is our beautiful and precious home, and we must learn how to preserve and protect it from the insanity of violent people. A good first step is to abolish all WMDs and other advanced weapons systems, and then to move on to abolishing assault rifles and handguns.

Arms Control "Non-solutions"

In the section above, I describe the major shortcomings of deterrence theory and strategy. But arms control—sometimes also referred to as arms control diplomacy, the diplomacy of arms control, or, simply, diplomacy—also has fallen short. Before describing how arms control has failed to resolve the nuclear crisis, I briefly mention some of the successes that arms control has achieved over the years and some of the aspirations of arms control advocates.

Arms control diplomacy has reduced superpower holdings of nuclear weapons by 85%, produced treaties limiting and then prohibiting nuclear testing, and produced guardrails, codes of conduct, and rules of the road relating to nuclear weapons. When the former Soviet Union dissolved, the diplomacy of arms control also provided safeguards against 'loose nukes' and 'dirty bombs.' Arms control also has resulted in treaties curbing nuclear proliferation and prohibiting chemical and biological weapons. Finally, arms control diplomacy has established protective norms. Outliers and norm breakers still exist, but without norms, there are no norm breakers. Arms control advocates hope that arms control diplomacy can keep the number of norm breakers small while recognizing that no hard problem is ever solved in perpetuity either by arms control or by deterrence (Krepon 2021a; 2021c).

During the Cold War, strategic arms control was built on treaties and treaties were built on numbers (Krepon 2021a). Historically, then, we are accustomed to arms control treaties and other forms of arms reduction agreements between the U.S. and Russia that measure the relative military force of both nations by the numbers of nuclear weapons each country holds—not by how destructive the weapons are. For example, the most recent of these agreements, New START, limits the U.S. and Russia to 1,550 warheads each. The numbers are higher, however, since bombers carrying many warheads are counted as carrying one. But the fact of the matter is that the size of a nation's weapons stockpile only crudely measures its overall strategic capabilities. Other factors include capability, range, and the accuracy of the associated delivery systems. So even though China's nuclear arsenal may have fewer warheads than that of Russia or the U.S., it still poses a significant strategic nuclear threat (Smith 2021).

Treaties that rely on numbers of nuclear weapons are very deceptive. Even though U.S. President Joe Biden and Russian President Vladimir Putin recently agreed to extend verifiable limits in New START until 2026, negotiating next steps has proven challenging. Already in 2021, Russian negotiators raised agenda items that seemed beyond reach even then, which was prior to Russia's recent invasion of Ukraine in 2022. At present, the greatest threats to nuclear peace relate to ground, air, and naval forces operating in proximity, and, also, to dangerous cyber and space practices. The U.S. therefore may try to prioritize codes of conduct rather than numerical arms control in any current and future conversations with Russia and China. Be that as it may, a new U.S.-Russian treaty mandating further reductions in the size of both countries' nuclear arsenals becomes harder to envision as China ramps up its force structure. China must be included in nuclear weapons discussions—and not merely in terms of a trilateral warhead counting exercise but instead by means of effective controls and reductions. Counting tactical nuclear weapons is a digression from more urgent items such as strengthening norms and buttressing codes of conduct (Krepon 2021a).

As an advocate of arms control, Krepon (2021b) argues that since deterrence without arms control invites apocalypse, we must enhance arms control initiatives and other diplomatic measures in order to diffuse tension between geopolitical rivals. Krepon (2021a) also suggests that to avoid nuclear war, we should pair arms control and deterrence together. Krepon's view is that national leaders will return to arms control after they realize that strengthening deterrence measures increases nuclear dangers and when they acknowledge that diplomacy is required to reduce these dangers. This will resonate with the past, he claims, as he recalls how former U.S. President John F. Kennedy and former Premier of the Soviet Union Nikita Khrushchev turned to arms control after the Cuban missile crisis, and how former U.S. President Ronald Reagan and former President of the Soviet Union Mikhail Gorbachev came together following Reagan's realization after 1983—a year in which multiple shocks to U.S.-Soviet relations occurred—that a paranoid Kremlin leadership believed that Armageddon was approaching. Krepon claims that at some point, nuclear-armed rivals will arrive at the same conclusions that Kennedy, Khrushchev, Reagan, Gorbachev, and other leaders before them reached, namely that nuclear war must be avoided, and deterrence by itself is, at best, half the solution. Krepon thinks that diplomacy and arms control can provide the help that deterrence needs, though I am skeptical that even deterrence and arms control together can take us where we need to go to eliminate the risk of a nuclear exchange. For decades, these two approaches in fact have been employed together, and their use in combination has not succeeded in eliminating the risk of nuclear war. To the contrary, the risk of nuclear war is probably the highest it has been since the U.S. detonated atomic bombs in 1945 over Hiroshima and Nagasaki, Japan, respectively.

Arms control proponents nevertheless continue to think that arms control treaties—such as treaties to reduce and limit the sizes of nuclear arsenals—are 'essential guardrails' around WMDs, including nuclear weapons. Even though arms control has resulted in an 85% reduction in the nuclear arsenals of Russia and the U.S. since the peak of the Cold War in the mid-1980s, everyone can see that arms control treaties and diplomacy more generally have broken down in recent decades—even more so in recent years. As a species, we are far from where we were in 1985 when Reagan and Gorbachev jointly declared that using nuclear weapons is not a rational option and that a nuclear war cannot be won and must never be fought. Now, even proponents of arms control realize that we must find new ways to ensure that nuclear strikes and cyberattacks against nuclear systems do not occur, either by mistake (Moniz 2020) or by design. All countries with nuclear weapons must realize that military attacks that target nuclear systems and installations and, also, cyberattacks, are steps along the road to an actual rather than a theoretical mutually assured destruction (MAD).

Krepon (2021a) suggests that because the geometry of nuclear competition is more complicated now, new arms control measures will borrow from the past but be different than they were during the Cold War. Furthermore, he argues that the existence of key norms—particularly the three norms of no battlefield use, no testing of nuclear weapons, and no new proliferation—

means that the options of deterrence strategists and national leaders are constrained. This may be true, but it does not constitute a solution to the current nuclear crisis on Earth.

Given the series of destabilizing nuclear tests that North Korea has conducted since 2006, people may be confused by Krepon's claim (2021a) that 'no testing' of nuclear weapons is a norm that has been upheld for almost five decades. Krepon's caveat with respect to this norm is that it has been followed by major and regional powers, which means that he leaves North Korea out of the discussion. This makes Krepon's analysis misleading, however, since the radioactive fallout from nuclear weapons tests, regardless of which country is conducting them, has a horrific, long-term negative impact on people, plants, animals, and the environment (EPA 2023; ICAN n.d.-b). One therefore should not exclude the actions of North Korea when discussing if the norm of 'no testing' has been upheld. At a broad level, it has not.

I also think Krepon's (2021a) norm of 'no proliferation' is misleading, especially given the steps that countries are taking to create fissile materials and to attempt to become nuclear powers. A case in point is Saudi Arabia working with China to try to acquire nuclear weapons capability. Krepon asks readers 'to imagine' that North Korea remains the last nuclear-armed state, but as the Saudi example shows, 'imagining' clearly is not enough if countries are taking actions to acquire nuclear weapons capability. One should not give Saudi Arabia a 'pass' here and 'imagine' that everything is going smoothly when it is not.

I do agree with Krepon (2021a; 2021c) that we must do everything possible to continue the decades-long norm of no battlefield use of nuclear weapons. While the other two norms—no testing and no proliferation—also are important, I think Krepon (2021a) overestimates our success as a species in having maintained them, given the considerations mentioned above.

At a broad level, however, the norm of no battlefield use of nuclear weapons—which is a key norm that has prevailed since 1945—may be one of the only things keeping humankind from catastrophic, global nuclear war right now. But Putin—at least rhetorically—is calling this key norm into question, and if it is broken in the future, by him or by anyone else, the nuclear floodgates could open in a devastating way. In other words, since the only thing holding the catastrophe of nuclear apocalypse at bay is *a norm*, breaking the norm most likely would be catastrophic. It is very uncertain if the fallout from breaking the norm of no battlefield use of nuclear weapons could be contained. What is abundantly clear is that nobody would win and that everyone would lose.

By the end of the Cold War, deep cuts were envisioned in nuclear arsenals. Dangerous military practices were relatively absent between major powers, which respected the territorial integrity and national sovereignty of others. But this at least relatively sane geopolitical inheritance was unnecessary and inconvenient in the eyes of Vladimir Putin, George W. Bush, and Donald Trump. Putin initiated the demise of arms control by disregarding provisions of the Conventional Forces in Europe Treaty. Bush withdrew from the Anti-Ballistic Missile Treaty, prompting, as forewarned, Putin's withdrawal from the second Strategic Arms Reduction Treaty (START), which prohibited land-based missiles carrying multiple warheads. As NATO expanded, Putin became more blatant in violating treaties, most notably the Intermediate-Range Nuclear Forces Treaty. Bush announced plans to deploy missile defenses in new NATO countries and to include Georgia and Ukraine in the queue for future NATO membership. Then the Russian army marched on Tbilisi, Georgia, after which Putin carried out hybrid warfare in eastern Ukraine and annexed Crimea on March 18, 2014. Putin then complained that Russia was the injured party when Trump withdrew from the Intermediate-Range Nuclear Forces and Open Skies treaties (Krepon 2021a).

As world leaders discard one treaty after another, Krepon (2021a) proposes that lengthening and strengthening the three norms of no battlefield use, no testing, and no new proliferation will be reassuring and stabilizing. Despite significant competition and some severe crises between the U.S. and the former Soviet Union, and, now, between the U.S. and Russia and the U.S. and China, arms control diplomacy has been at least partially successful by upholding the norm of no battlefield use of nuclear weapons. Krepon argues that success in norm strengthening happens one day at a time and one crisis at a time. Reinvention depends on diplomatic adeptness, creativity, and wisdom, he argues, though he recognizes that it also depends on the quality of relations between major powers. If competition between nuclear actors sharpens and if national leaders intensify that competition, then no proposals to reverse course will succeed.

Krepon (2021a) asks how high the barriers would be against use and testing of nuclear weapons in 2045 if humankind were able to achieve a century of non-battlefield use. He uses 2045 in his example because he is aiming for a century of non-battlefield use, and the two, battlefield atomic bomb detonations in recent human history both occurred in 1945. A century of non-battlefield use, a half-century of no testing of nuclear weapons by major and regional powers, and another quarter-century of successful nonproliferation might seem too ambitious, Krepon acknowledges. Perhaps, he says, but in 1945, it seemed ambitious to envision a world in which nuclear weapons would not be used in warfare for three-quarters of a century—and we have in fact achieved that, though barely. When conversations began about limiting nuclear testing in the Eisenhower administration, it similarly seemed ambitious to envision a world in which major and regional powers would not conduct tests for a quarter-century. Yet, we also have achieved that—albeit, again, just barely.

In 1962, not long after the Cuban Missile Crisis, forty-six countries collectively pledged to reduce the threat posed by nuclear weapons by signing the Nuclear Non-Proliferation Treaty (NPT). Some people argue that the NPT is the best foundation we possess at present to prevent the spread of nuclear weapons (Moniz 2020). Those who conceived of a global nonproliferation compact decades ago were rewarded on July 1, 1968 when sixty-two countries signed the NPT—though unfortunately absent were China, France, West Germany, other U.S. allies, Brazil, Argentina, and leading non-aligned states (Krepon 2021a).

Today, 188 UN member states and two observers (the Holy See and the State of Palestine) are parties to the NPT, which, given its almost universal membership, has the widest adherence of any international arms control agreement. The only current non-parties to the NPT are India, Israel, North Korea, Pakistan, and South Sudan (Moniz 2020). North Korea originally acceded to the NPT but announced it was withdrawing from the Treaty in 2003, having never come into compliance with it. Krepon (2021a) also states that the NPT has one severe test—Iran. In contrast, I am persuaded by Woolsey et al.'s (2016; 2020) arguments that Iran almost certainly already possesses nuclear weapons capability, meaning it has gone beyond 'test' status in the treaty sense of that term.

The signatories to the NPT are classified into two categories: nuclear-weapon states (NWS), and non-nuclear-weapon states (NNWS). Under the NPT, all countries with nuclear weapons, often referred to as states-parties, commit to pursue general and complete disarmament, while countries without nuclear weapons, i.e., the NNWS, pledge never to develop or acquire them (Kimball 2022a). The NPT defines NWS as those that have built and tested a nuclear explosive device before January 1, 1967—i.e., the U.S. (1945), Russia (1949), the U.K. (1952), France (1960) and China (1964). Four other states are known or believed to possess nuclear weapons—India, Pakistan, and North Korea have openly tested and declared that they possess nuclear weapons. Also, even though Israel almost certainly already possesses nuclear weapons, it may not have become a signatory to the NPT because it did not want to relinquish the country's policy of nuclear ambiguity,' or 'opacity' (Hebrew *amimut*) (Kristensen and Korda 2022b). Furthermore, even though Iran almost certainly possesses nuclear weapons (Woolsey et al. 2016; 2021), it is not considered a NWS under the NPT because it is not explicit about its actual nuclear status.

The first two pillars of the NPT are disarmament and forgoing development and acquisition of nuclear weapons. The third pillar ensures that states-parties can access, develop, and share nuclear technology for peaceful uses and applications (Kimball 2022b). Although the NPT has facilitated some progress on nuclear arms reductions, disarmament efforts currently are stalled. Accordingly, the entire approach to nuclear risks must be reconsidered, in part because so much has changed since the signing of the NPT, and since the U.S. and former Soviet Union agreed to bilateral arms control agreements. In addition to threats posed by Vladimir Putin and Kim Jong-un, new threats include hypersonic weapons, cybertechnology, and AI, all of which must be considered alongside longstanding threats such as technological advances with respect to nuclear weapons in North Korea (Moniz 2020), Iran (Woolsey et al. 2016; 2020), and, very alarmingly now, Saudi Arabia (Cohen 2021).

Krepon (2021a) believes that because the three norms of no battlefield use, no testing by major or regional powers, and no new proliferation are the most difficult norms for national leaders to break, extension of these norms is possible even during this period of heightened competition between countries. If a leader of a country were to authorize the first use a nuclear weapon since 1945, such an individual would live in infamy. The companion norm of no testing signifies recognition of the dangers associated with use. Nevertheless, in some instances, those who complain about troubling experiments de facto are blocking on-site inspections by opposing ratification of the Comprehensive Nuclear-Test-Ban Treaty (CTBT) (Krepon 2021a).

Five of the forty-four Annex 2 States (i.e., those that must ratify the CTBT and deposit their instruments of ratification with the Secretary-General of the UN for the Treaty to enter into force) have signed but not ratified the CTBT—China, Egypt, Iran, Israel, and the U.S. Indeed, the U.S. and China are the only remaining NPT Nuclear Weapon States that have not ratified the CTBT (NTI 2023a). Chinese officials repeatedly have said that they will not ratify the CTBT until the U.S. Senate consents to do so; India will not ratify until China does; and Pakistan will wait for India to ratify. Krepon's point is that a cascade of ratifications could begin with a super-majority vote in the U.S. Senate if Republican senators who claim to be concerned by China's nuclear buildup were to consent to the CTBT while demanding that all four instruments of ratification be deposited together. Meanwhile, test moratoria among major powers continue because the major power willing to resume testing would set off a very different cascade, with all four nuclear-armed rivalries likely following suit (Krepon 2021a). Failure to ratify the CTBT also demonstrates the limited usefulness of the United Nations Conference on Disarmament (CD), which has achieved very little since it was created in 1979. Even though the CD began substantive negotiations on a comprehensive nuclear-test-ban treaty in January 1994, it has not succeeded in bringing the actual CTBT—adopted by the UN General Assembly on September 10, 1996—into force (UNODA n.d.-a).

Despite stagnation at an international level, adherents of arms control diplomacy nevertheless want to move forward in creating a new forum for norm building. This approach follows the general recognition that communication is a central element in any arms control diplomacy. For example, Kevin Rudd (2021), a former prime minister of Australia and current President of the Asia Society and Chair of the International Peace Institute, emphasizes that because the current, tense geopolitical climate undermines strategic nuclear stability, diplomatic talks must occur to keep the nuclear weapons situation from careening completely out of control. Rudd emphasizes that "strategic transparency" is achievable even when relationships between nation-states are marked by mutual suspicion.

Krepon (2021a) suggests that harnessing deterrence with arms control could be achieved with a new forum for norm building. Krepon rules out the 65-member CD in Geneva as too unwieldy to succeed. As Krepon notes, the CD's last hurrahs in treaty making occurred during the Clinton administration. Since then, its procedures have empowered blocking action. In addition, little if any reason exists to expect that the permanent five members of the UN Security Council can create an effective forum to advance a relevant arms control agenda. Instead, Krepon argues that the geometry of nuclear competition is such that a new forum to focus on norm building and codes of conduct should be created in which all four pair of nuclear-armed rivals—the U.S. and China, the U.S. and Russia, China and India, and India and Pakistan (since Krepon does not consider Iran a nuclear weapons state)—are represented. Krepon also thinks that the U.K. and France should participate in such a forum, since they are two countries that he thinks have significant expertise and practical experience to offer. Krepon excludes Israel and North Korea from such a forum because, on his view, the addition of these countries would pose more potential problems than it would create potential benefits. I disagree with this point. Israel needs to be included in this type of discussion, were it to occur, otherwise leaders there might feel improperly sidelined. Furthermore, as difficult as it may seem, it also is important to include North Korea in nuclear weapons dialogue. After all, beneath all his bluster, that is what Kim Jong-un really wants—namely to be recognized and included as a member of the international community.

Krepon (2021a) outlines possible discussion items for arms control diplomacy. He suggests reconsidering the Johnson administration's negotiations with the Kremlin, which included plans to discuss limitations on medium-range, intermediate-range, and ocean-spanning missiles. Given Putin's deployment of these missiles in violation of the Intermediate-Range Nuclear Forces (INF) Treaty, U.S. rejoinders, and China's heavy investments in missiles of less-than-intercontinental-range, Krepon thinks this should be considered. Krepon also suggests addressing the contentious matter of including interceptors for national missile defenses. Furthermore, depending upon what means of delivery are included and excluded, Krepon argues that it may be possible to devise an effective arms control regime with equal aggregates of nuclear capable delivery vehicles and missile defense interceptors. Although China and Russia might protest equal aggregates depending upon units of account, counting rules, and range limits, equal aggregates might serve the interests of the U.S. and its allies, he says. Furthermore, Krepon acknowledges that difficult issues would need to be negotiated and that China would need to accept monitoring arrangements. He also suggests that the U.S. should consider the build-down concept of reducing while modernizing. Overall, I find Krepon's approach to dialogue too cumbersome to succeed. I also strongly oppose the idea of any country making engineering modifications to existing nuclear arsenals and missile defense interceptors—especially space-based ones—since both actions are escalatory in nature.

Krepon's (2021a) specific plan for norm building involves a seven-nation forum consisting of the U.S., Russia, China, India, Pakistan, the U.K., and France. Such a forum would be difficult to steer, he concedes, but he counters by stating that the nuclear dangers we face now are interconnected and unwieldy. Krepon suggests that when the nature of a problem seems intractably complex, the wisest course could be to expand the scope of the problem. Even as the four pair of nuclear-armed rivals compete, they have the most to lose if key norms are broken and the most to gain if they are extended. Krepon suggests that existing bilateral conversations on nuclear risk reduction should continue while noting that no effective channels of communication and substantive exchanges currently exist between India and China and between India and Pakistan. In both instances, border clashes between each pair of countries are becoming more intense.

One reason that Krepon (2021a) thinks that a non-hierarchical, seven-nation approach to norm building could succeed is because all seven countries are concerned about the intentions and capabilities of states with the most dynamic nuclear modernization programs. Since each state has its own reasons to engage and, also, its reasons for being wary, Krepon argues that if at least some of the states are willing to sit at the table, it becomes more difficult for any country to hold out. The ground rules for seven-nation talks seem most likely to avoid traps if the agreed focus of conversation is nuclear risk reduction and norm building, he argues. Sidebar conversations would be encouraged, since they could lay the groundwork for bilateral agreements, but Krepon suggests that tabling bilateral issues should be prohibited. He proposes that the first order of business could be to affirm the canonical Reagan-Gorbachev pledge that a nuclear war cannot be won and must not be fought. Krepon also suggests that both bilateral and multilateral thematic discussions on dangerous military practices might suggest common concerns and remedies while acknowledging that even if all seven states were to agree to attend such a forum, the intensification of rivalries could foreclose useful discussions. Furthermore, Krepon notes that numerous potential pitfalls would require multilateral diplomacy, and that the U.S. State Department would require reinforcements. Nevertheless, despite the difficulties involved, he contends that sufficient connective tissue exists to try. Potential benefits include new opportunities to engage China and to open obturated channels of conversation.

In Krepon's (2021a) vision for a norm building forum, lengthening and strengthening norms are important priorities given that dangerous military practices are on the rise. Over time, however, if a seven-nation, norm building forum were to prove worthwhile, he suggests that topics could evolve from norm building to consideration of guardrails, limits, and reductions for nuclear modernization programs. Because Krepon thinks that none of the states with the most dynamic modernization programs will be willing to relax requirements unless others do, he argues for a multilateral build-down concept according to which all seven states would agree to reduce the size of their arsenals as they modernize them. I disagree with Krepon here. Because modernization efforts are escalatory and destabilizing, I think that ceasing modernization programs is even more important than a build-down per se. Krepon nevertheless thinks that a build-down approach has the advantage of becoming all encompassing. He also argues that it is important to avoid a ratio-based, hierarchical, multilateral system such as the one tried before for naval arms control, which he believes has no practical chance of success. However, Krepon also argues that numerically based arms control cannot be forsworn, since Russia is adding new means of delivery to its strategic forces and China is increasing its deployments of land-based missiles.

In addition to creating a forum for seven-nation talks, Krepon (2021a) suggests that a new negotiating forum needs to be created to address trilateral nuclear arms control and reductions, given that the current landscape makes bilateral controls insufficient. China's continued effort in building up its own nuclear arsenal likely reflects mixed motives, just as the former Soviet Union built up its nuclear arsenal as it prepared for strategic arms limitation talks during the Johnson and Nixon administrations. Although some deterrence strategists may view Beijing's activities as early evidence of nuclear war-winning ambitions, it also is plausible that these activities reflect an attempt to seek increased leverage and to avoid disadvantage in any future negotiations. Krepon's argument is that Beijing certainly recognizes that it cannot "just say no" to strategic arms limitations indefinitely, but because China is increasing the size of its nuclear arsenal quickly, the Biden administration must speed up preparations for trilateral negotiations on numerical limitations that serve the national security interests of the U.S. and its friends and allies. As with bilateral U.S.-Soviet strategic arms limitation talks, trilateral discussions are likely to encounter stalls and unexpected delays, which provides time to consider scope and limitations.

According to Krepon (2021a), if norm-strengthening negotiations were to evolve into numerical accords, the fluidity of trilateral relations and opposition in the U.S. Congress will preclude treaty making. Instead, he suggests that if trilateral accords can be reached, they likely would occur in the form of executive agreements and be term limited, since arguments in favor of formality and agreements of indefinite duration are not persuasive when treaties, like executive agreements, can be discarded after U.S. national elections. Krepon also notes that one need not wait for resolution of discussions with respect to numbers before reaffirming norms, and that such reaffirmation is necessary for trilateral talks to succeed over time. Even if agreements are not reachable or are not as inclusive as one would like, discussions with Russia and China still could have utility, at a minimum by ascertaining useful assessments of different limitation parameters and by offering a venue to determine how best to proceed. Treaties and numbers matter greatly, but they are much harder to negotiate in a trilateral context that includes China together with the U.S. and Russia in contrast to a bilateral context that only includes the U.S. and Russia. Plus, in the future, even more nations may have seats at the table.

My own view is that if a multiple-nation, norm-strengthening approach is tried, at the outset, it should include as many nuclear-armed countries as possible. Because of the complex geometry of nuclear competition and the intransigence of domestic politics in the U.S., less formal constraints may be unavoidable. Since this will take time, Krepon argues that norms matter more than formalities, since norms are easier to extend than new strategic arms reduction treaties are to negotiate and nuclear numbers are to reduce. Krepon also argues that over time, with or without treaties, a hard focus on extending and reaffirming crucial norms can establish conditions for far fewer numbers of nuclear weapons. Krepon nevertheless acknowledges that if China and Russia choose to engage in dangerous military practices, arms control—whether by means of norm strengthening and/or reductions in numbers—will not succeed. In such a scenario, the U.S. also is likely to engage in dangerous military practices, he says, and the dynamics of such competition would invite crises, or worse. Perhaps then, says Krepon, competitors would become more inclined towards the measures of reassurance brought by the pairing of arms control and deterrence rather than relying on deterrence alone.

Even if one were to give Krepon the benefit of the doubt by recognizing that he wrote his analysis prior to Russia's 2022 invasion of Ukraine, I still find his vision regarding the potential of arms control far too optimistic. If arms control had the potential to work, it would have worked by now. Furthermore, pairing arms control with deterrence is nothing new—human beings have been doing that since 1945. By now, it should be clear to everyone that we do not have the luxury of time for such a step-by-very-slow-step approach.

In addition, the aspirations of arms control advocates with respect to what can be achieved tend to recede in time—i.e., what once seemed achievable by means of arms control almost invariably no longer does after one or two years. This pattern of creating arms control goals that later recede is one of the reason certain people use the phrase "[a]rms control non-solutions" (Woolsey et al. 2021) to describe arms control in general, and why arms control has lost much of its luster. Arms control often suffers from receding goals because leaders in both Russia and the U.S. have a tendency to cast aside treaties as inconvenient to their pursuit of freedom of action. Even though Republican presidents in the past produced great achievements in arms control, most current Republication senators and aspirants for higher office denigrate arms control and treaty-making as failed and unnecessary pursuits. Even though arms control provided guardrails in the past, the escalation of dangerous military practices makes these guardrails look exceedingly ineffective today. In Asia, for example, four nuclear-armed states—China, Pakistan, India, and North Korea—all are increasing their nuclear arsenals. This trend points towards tragedy unless something changes (Krepon 2021a).

Deterrence advocate and arms control critic Mike Pompeo attacks arms control along two lines. One avenue of critique is the idea that emphasizing arms control signals weakness. Here, Pompeo states that substituting "imagined arms control accords for nuclear capabilities is neither acceptable nor wise. Such pablum will be viewed by our adversaries as American fragility and thus as an inducement for caustic adventurism." Pompeo also criticizes arms control because he states that the entire process often becomes bogged down (Pompeo 2022). I think Pompeo is incorrect in his first criticism of arms control but correct in his second. At all times and in all things, one can communicate from a position of strength rather than weakness that one is willing to talk and negotiate. This is especially the case when one holds a moral position, since then one has deontological strength on one's side.

But Pompeo's second criticism of arms control is accurate. The process has become terribly bogged down over decades. Furthermore, results of arms control diplomacy often are unclear. For example, treaties between the U.S. and the former Soviet Union to limit nuclear weapons have yielded very uncertain results. SALT I and SALT II both were difficult to verify and were ineffective in limiting the former Soviet Union from dramatically increasing its nuclear forces. The stated reason that the U.S. provided for withdrawing from the Intermediate-Range Nuclear Forces Treaty (INF) in 2019 was Russian non-compliance. Although START I, SORT (Strategic Offensive Reductions Treaty), and the New START Treaty have led to historic reductions in U.S. and Russian nuclear arsenals, the fact that these treaties are bilateral and never included China has been problematic from the outset. Furthermore, New START also has limited utility since it does not address in a meaningful way destabilization from HGVs (Pompeo 2022). It also does not address new directed energy weapons (DEWs) such as high-energy lasers, high-power radio frequency and microwave devices, and charged and neutral particle beam weapons (CRS 2022a).

To summarize this section's discussion, those who favor arms control over deterrence think that the most immediate challenge for human beings is to forge international agreements to eliminate existing nuclear stockpiles, stop the production of fissile materials, cease uranium enrichment, stop processing plutonium, and take other steps to eliminate the nuclear risk to humankind. It sounds good on paper, and many of the observations and suggestions made above obviously are well-intentioned. Logical though many of these suggestions sound, however, they are overly optimistic for the times. Furthermore, Pompeo's one apposite criticism of arms control—that it has a history of becoming bogged down—is indubitably correct.

I was more optimistic about the potential of arms control diplomacy before Russia's recent invasion of Ukraine in February 2022. But as the war in Ukraine has unfolded, Russia has shown no compunction about threatening to use nuclear weapons. The Russian army also has disregarded basic security at both Chernobyl and Zaporizhzhia nuclear power plants also does not bode well for genuine arms control dialogue. We need something much more creative and radical to avert an actual nuclear apocalypse.

Acknowledging that arms control has not worked is not an argument to return to deterrence, however—since deterrence also does not work. What we need is a far more creative and radical approach to the entire nuclear situation before it is too late. This requires moving beyond both deterrence and arms control diplomacy to embrace a new approach to making Earth safe from the risks posed by nuclear weapons and, also, from nuclear power.

We also need to be more pragmatic, which means we must acknowledge just how deep the quagmire created by military competition in a sea of mutual, global distrust between great powers, other nation-states, nonstate actors, transnational, and supranational alliances really is. Any proposed solution must have the ethical and spiritual force necessary to pull our species through this very narrow impasse without *Homo sapiens sapiens* becoming extinct from nuclear war, the deployment of other WMDs and/or advanced weapons, climate change, and/or a combination of systemic and dynamic challenges such as those described in chapter 1. We are past the time when any package of deterrence plus arms control will prove adequate to calm people given current geopolitical competition between great powers. What we need now is a deontological solution that is 'diplomatic' in general, not merely in the limited sense of arms control diplomacy. In the next chapter, I argue that unilateral restraint is our best option, because it is a moral option aligned with the greater good that is based on optimizing the wellbeing of the whole.

Nuclear Risks from Terrorist Groups and Other Hostile Actors

In addition to the risk of nuclear exchange between nation-states, there also is the potential for terrorist groups, Violent Extremist Organizations (VEOs), and/or armed ethic groups to acquire and/or construct—and potentially to detonate—one or more atomic and/or thermonuclear bombs. A crude explosive made of HEU could be constructed relatively easily as compared to one using plutonium. It also is relatively easy to divert, smuggle, and hide HEU (Schaper 2013, I). In addition to these risks, nuclear power plants also are vulnerable to terrorist acts during armed conflicts.

Well-funded terrorist organizations with global reach continue to pose problems, especially groups that have access to nuclear know-how and a determination to develop and use WMDs. At least twenty-two countries—and likely some number more—possess weapons-usable nuclear materials. Much of this material is poorly secured. In addition, cyberattacks, which have long been a threat to financial institutions, corporations, and critical infrastructure, also threaten nuclear infrastructure. Nuclear power plants and weapons command-and-control and early warning systems potentially are very vulnerable to cyberattacks, with potentially catastrophic consequences (Moniz 2020). In addition, peer/near-peer/VEO adversaries and/or armed ethnic groups regularly engage in the strategies and tactics of Irregular Warfare (IW) (Cole 2019), which potentially could include a surprise nuclear attack.

The correlation between extremism and a tendency towards apocalyptic narratives makes the risk of nuclear weapons in the hands of violent extremists particularly problematic. Leaders of countries also are susceptible to delusional thinking. Any person regardless of station in life who gravitates towards apocalyptic scenarios, eschatological doctrines, and cataclysmic narratives is pushing human society towards the real possibility of an apocalypse. A particularly deep-seated delusion is that a 'final battle' will usher in a savior, Second Coming, 'New Age,' or some other major, societal change. While widespread nuclear war would usher in a major change for humankind, it would not be a 'New Age'—it would be extinction.

Nuclear Weapons Pointed at Population Centers and Infrastructure

It is not merely that human beings 'possess' nuclear weapons. We live in a world in which thousands of nuclear weapons are armed and pointed at numerous, major population centers and much of the critical infrastructure required to keep human society functioning. This includes power grids, freshwater pipelines, wastewater systems, highways, bridges, tunnels, railways, airports, utilities, food distribution hubs, grocery stores, financial centers, IT networks, communications systems, etc. This is not a theoretical problem—it is our actual reality, right now, here on Earth.

Nuclear Risks Amplified by Planetary Dynamics

Planetary dynamics make the precariousness associated with nuclear technologies very serious. Tectonic shifts, earthquakes volcanic eruptions, asteroid and/or comet impacts, and other devastating events all could trigger a nuclear accident.

Many nuclear power plants are positioned very poorly. The World Nuclear Association estimates that 20% of the nuclear reactors in the world operate in earthquake danger zones with significant seismic activity (Randall 2014). Climate change, including coastal flooding, further intensifies risks associated with nuclear installations (Polansky 2018; USAWC 2019). In addition, potential flooding from upstream dam failures also poses a serious risk to nuclear power plants (Perkins et al. 2011).

Decades-old Nuclear Power Infrastructure Susceptible to Accidents

Since 1984, the average age from grid connection of operating nuclear power plants has been increasing. As operational nuclear power plants become older, they are likely to be more prone to accidents. As of mid-2022, the average age from grid connection was 31 years, up slightly from 30.9

years as of mid-2021. A total of 270 reactors, which is approximately two-thirds of the world's operating fleet, have operated for 31 or more years. Of these, 105—or more than one in five—have operated for at least 41 years. In 2020, the average age of reactors in the U.S. passed 40 years, reaching 41.2 years by the end of 2021. The average age of reactors in France's fleet currently exceeds 36 years. As of the end of 2021, Russia's average reactor age was 28.4 years. South Korea's reactors have an average age of 22.4 years. China's average is 8.8 years (Himmelstein 2022).

Because of the possibility of accidents associated with nuclear power plants, the most prudent course of action is to stop building new nuclear power plants and to take offline those that already exist. Prior to Russia's February 2022 invasion of Ukraine, there was some indication that the world would move in this direction, since global demand for nuclear power as a percentage of global electricity peak share was decreasing. As compared to 1996, when 17.5% peak share of global electricity generation came from nuclear power, the percentage had fallen to 9.8% by 2021. This reduction occurred as more countries put on hold or abandoned their nuclear power strategies as compared to the number of countries expanding them (Himmelstein 2022). However, according to *The World Nuclear Industry Status Report 2022* (Schneider 2022), global nuclear power generation *rose* by 3.9% in 2021, the same rate by which it had dropped in 2020. In 2022, 411 reactors were operational globally, which was 26 fewer than in 2011. Another 29 nuclear reactors currently are in long-term storage. Meanwhile, more than half of the 53 nuclear reactors in China and India are under construction (Himmelstein 2022).

Today, thirty-three countries currently operate nuclear power reactors, with fifteen listed as actively pursuing nuclear technology (Himmelstein 2022). As counted by the number of operational reactor units in 2022, the countries with the most nuclear reactors are the U.S. (92), France (56), China (55), Russia (37), South Korea (24), India (19), Canada (17), Ukraine (15), the U.K. (11), and Japan (10) (Himmelstein 2022, citing to Schneider 2022). In 2020, nuclear energy newcomers Belarus and the United Arab Emirates (UAE) were added to the global list of countries with nuclear reactors (Himmelstein 2022).

The world's fastest growing nuclear energy program also is one of the youngest: China has used nuclear energy since the early 1990s and currently operates 55 nuclear reactors, the majority of which joined the grid in the past decade. The U.S. continues to have the largest number of nuclear reactor units, with 92 operational reactor units as of July 2022, down 12 since 2011. In comparison, Japan has reduced its operational reactors by 38 since 2011. Japan currently has 10 nuclear reactors, and many people think Japan will stop constructing new nuclear reactors. Around the world, however, only three countries that previously had nuclear energy programs shut off all their reactors: Italy in 1987; Kazakhstan in 1998; and Lithuania in 2009. During the current energy crisis, Germany has extended its nuclear power program, though it still plans to end nuclear energy production in 2023 (Himmelstein 2022).

Blackmail and Theft Involving Nuclear Weapons and Nuclear Power

Even though the odds of an accident involving a nuclear power plant in Ukraine increases the longer the war goes on, the international community has not found a way to protect nuclear facilities in that country from attacks. Both Ukraine and Russia have accused one another of shelling the Zaporizhzhia Nuclear Power Station (Diaz-Maurin 2022a; see also UN 2022), while Ukrainian President Volodymyr Zelensky has warned of the threat of a radiation leak and has accused Russia of "nuclear blackmail" (Foltynova 2022). In early May 2023, new concerns with respect to the safety of the power station were raised after Moscow ordered residents from Russian-occupied areas close to the facility to evacuate (Rising 2023; Regan et al. 2023).

In fact, the risks associated with blackmail involving both nuclear weapons and nuclear power have been significantly underestimated. During the war in Ukraine, Russia has engaged in implicit and explicit nuclear coercion, with Russia's leadership and propagandists clearly trying to exploit the perception that a deterrence gap exists such that Russia can use one or more short-range, tactical nuclear weapons if it so chooses. As mentioned above, the notion of a deterrence gap is a threshold below which individuals often mistakenly believe they can employ nuclear weapons without risk of significant retaliation (Copp 2022).

While the media have paid considerable attention to Russia's rhetoric regarding nuclear weapons, media discussion of Russia's dominance in the nuclear power cycle is quite tepid. But Russia's nuclear power cycle dominance is another arena in which potential nuclear blackmail can arise given various countries' reliance on Russia at various stages of this cycle. Since Russia's recent invasion of Ukraine in February 2022, the U.S., the European Union (EU), and other Western nations have implemented numerous packages of sanctions targeting various elements of Russia's lucrative energy industry (Foltynova 2022). Russia is a major energy supplier to the world of oil, gas, coal, and nuclear fuel and reactors (Bowen and Dabbar 2022). Although the U.S., the EU, and other Western nations have been trying to decrease their overall dependency on Russian energy, this effort has focused primarily on oil, gas, and coal. After shelling occurred near Ukraine's Zaporizhzhia nuclear plant, Ukrainian President Zelensky called on the international community to mount a stronger response that also included banning Russian imports from the nuclear power sector. But this has not occurred (Foltynova 2022), probably for the same reason mainstream media outlets all but ignore the topic. This is because the U.S. itself would be hurt economically by a ban on imports from Russia's nuclear power sector, since the U.S. itself relies on Russia with respect to certain elements of the U.S.' own nuclear power industry.

Not only is Russia's central role in the world's nuclear power cycle an issue for the U.S., but it also is a concern for many other countries. Starting with mining, Russia is among the five countries with the world's largest uranium resources. Russia is estimated to have around 486,000 tons of uranium, which is the equivalent of approximately 8% of global supply. According to 2019 estimates, the countries with the largest uranium resources in the world are Kazakhstan, Namibia, Canada, Australia, Uzbekistan, Russia, Niger, China, India, Ukraine, South Africa, Iran, Pakistan, Brazil, and the U.S. Among this lineup, Russia produced about 5% of the world's uranium from mines in 2021 (Foltynova 2022). Of note, it is estimated that Europe and the U.S. buy approximately $1 billion of civilian nuclear goods and services directly from Rosatom (Sokolski and Stricker 2023; see also Kessler 2022).

But uranium mining is just one piece of the nuclear process. Raw uranium is not suitable as fuel for nuclear plants. First it needs to be refined into uranium concentrate, converted into gas, and then enriched. As compared to its competitors, this is where Russia excels. In 2020, only four conversion plants were in commercial operation. These were in Canada, China, France, and Russia. Of these, Russia was the largest player. It had almost 40% of the total uranium conversion infrastructure in the world, and it therefore produced the largest share of uranium in gaseous form—i.e., uranium hexafluoride (Foltynova 2022).

Uranium enrichment is the next step in the nuclear cycle. According to data from 2018, that capacity is spread among a handful of key players. With respect to global uranium enrichment capacity, Russia had the largest share—approximately 46% in 2018—with similar results likely for immediately subsequent years. By 2030, however, Russia's share might decrease as a result of a significant increase in capacity in China. But in the immediate term, Russia is a significant supplier of both uranium and uranium enrichment services. The EU purchased about 20% of its natural uranium and 26% of its enrichment services from Russia in 2020. In 2021, the U.S. purchased about 14% of its natural uranium and 28% of its enrichment services from Russia (Foltynova 2022). Since Russia's recent invasion of Ukraine in February 2022, some members Congress in the U.S. called for bans on imports of enriched uranium from Russia during the 2022 to 2026 time frame (Bowen and Dabbar 2022).

Russia also is considered the world leader when it comes to the export of nuclear plant development. Russian state-owned Rosatom State Nuclear Energy Corporation is a major participant in this arena. Founded in 2007 by Vladimir Putin, Rosatom specializes in nuclear energy, nuclear non-energy goods, and high-tech products. Its initiation of nuclear reactor construction is far greater than that of the next most prolific providers of nuclear plant development, namely China, France, and South Korea. Between 2012 and 2021, Rosatom initiated construction of 19 nuclear reactors, such that by the end of 2021, 15 Russian-designed reactors were under construction in nations other than Russia. Although China started building 29 reactors during the same period, only two of them were initiated abroad. During this time frame, France began building two reactors abroad, and South Korea began building four abroad (Foltynova 2022; see also Bowen and Dabbar 2022).

Nuclear reactors made in Russia are known as VVER, which is an acronym from the hybrid Russian-English term 'vodo-vodyanoi enyergeticheskiy reactor' (translated fully into English as 'water-water energetic reactor'). VVERs originally were developed in the former Soviet Union. They use water both as a coolant and as a moderator. There are several versions of VVERs, such as the VVER-440 and VVER-1000, which are differentiated primarily based on the volume of power. Various types of VVERs currently operate in eleven countries, including Bulgaria, the Czech Republic, Hungary, and Finland. In addition, other countries such as Egypt, Turkey, and Argentina currently have VVERs under construction or plan to build them (Foltynova 2022).

Russia has exported more reactors in recent decades than any other major supplier. In 2021, there were 439 total nuclear power reactors in operation, 38 of which were operating in Russia, all of which were VVERs. Another 42 operating in other countries, including 15 in Ukraine, were of the Russian VVER type. Rosatom is a major potential provider of new nuclear power plants globally. Many plants around the world have been conducting multiyear design and site preparation efforts for the construction of VVERs, with many already under construction. Stopping work on or abandoning these projects would cause significant financial and energy supply impacts for those countries (Bowen and Dabbar 2022).

To keep nuclear reactors operating, nuclear plants require a regular supply of nuclear fuel, usually of a specific type. Countries also depend on Russia at this level of the nuclear cycle. Although there are several suppliers on the market, the Russian TVEL Fuel Company currently is the only authorized supplier of fuel needed for VVER-440s. Certain countries therefore rely on deliveries of nuclear fuel from Russia and could risk halting operations at their facilities temporarily if this fuel were not received. Slovakia has four nuclear reactors and the Czech Republic has two with this vulnerability. Westinghouse, an U.S. nuclear power company, is seeking ways to offer alternative fuel in Europe, but it will take time to obtain the necessary licenses and approvals. There also are concerns that fuel from the U.S. might be more expensive, and it is unclear how Westinghouse would handle the waste-management system. In contrast, Russia can supply high-assay low-enriched uranium, or HALEU, which is the type of fuel required for the more advanced reactors now under development by many companies across the world (Foltynova 2022), including by many nuclear startups in the U.S. Even for countries such as the U.S. that do not host VVERs, Russia remains a major supplier of several services involved in manufacturing nuclear fuel (Bowen and Dabbar 2022).

In 2020, the U.S. Department of Energy (DOE) announced a series of large, cost-share awards with certain private reactor developers. For the largest demonstrations, DOE would contribute a share of the demonstration costs if private entities more than matched that investment (Bowen and Dabbar 2022). Putting this in simple terms, the USG is subsidizing private industry's development of nuclear startups in the U.S.

Dozens of advanced nuclear startups such as the nuclear startup TerraPower founded by Bill Gates require HALEU, which is a concentrated form of fuel (Wesoff 2022). Currently, the only supplier with the capacity to provide HALEU on a commercial scale is Tenex, and Tenex is owned by Rosatom (Foltynova 2022; Wesoff 2022; see also Bowen and Dabbar 2022).

Because Russia has been the only commercial source of HALEU, some advanced reactor developers either were planning to obtain their first fuel load of HALEU from Russia, or at least considered doing so. In addition, prior to Russia's recent invasion of Ukraine in 2022, enriched uranium from Russia could have been considered for usage in some of the future advanced reactors under development in the U.S. In recent decades, various private companies such as TerraPower have been founded to pursue commercialization of different advanced reactor designs. Some of these newer reactor designs use uranium with significantly higher enrichments as compared to what light water reactors use (15-19.75%

uranium-235 enrichment, instead of 3-5%) (Bowen and Dabbar 2022; see also Foltynova 2022).

According to the American Office of Nuclear Energy, HALEU availability is crucial for the development and deployment of reactors. Since HALEU is available in the U.S. in limited quantities, the USG is seeking options to fund research regarding and development of new fuel sources (Foltynova 2022). Based on inputs from companies, the Nuclear Energy Institute reported in 2018 that estimated HALEU needs potentially could increase from tens of metric tons per year in the mid-2020s to over a 100 metric tons per year in the late 2020s (Bowen and Dabbar 2022).

Looking for new markets for its nuclear technology is one way Russia gains influence and reaps profits in countries that are new to nuclear energy. One reason countries want to cooperate with Russia is because it offers a 'whole package' solution. Not only does Russia build a nuclear plant and supply fuel, it also trains local specialists, purportedly helps with safety questions (a highly suspect claim given the behavior of the Russian army at both Chernobyl and Zaporizhzhia), runs scholarship programs, and disposes of radioactive waste. When seeking new business, Russia also offers loans backed by government subsidies that cover at least 80% of the construction costs of new nuclear power plants. For example, Russia has loaned $10 billion to Hungary, $11 billion to Bangladesh, and $25 billion to Egypt (Foltynova 2022).

Because Russian involvement in the Western nuclear power supply chain is so pervasive, it is difficult to create policy options to reduce or end that involvement (Bowen and Dabbar 2022). At least fifty countries have some level of nuclear cooperation with Russia. On August 27, 2022—over six months after Russia's February 2022 invasion of Ukraine—officials in Budapest confirmed that Russia's Rosatom will start building two new nuclear reactors in Hungary. As mentioned above, Russia currently operates nuclear reactors in eleven countries, with more under construction or being planned. With at least thirty countries around the world, Russia either has signed memoranda of understanding, which serve as a declaration of interest in nuclear technology, or intergovernmental agreements, which set an intention to cooperate on the building of nuclear plants. Most of these thirty countries are in Africa, even though experts warn that certain countries there may not be ready for nuclear power. My own view is that nuclear power is a bad idea anywhere. But Russian officials argue that nuclear technology represents an answer to the African continent's increasing demand for electricity. The fact that African countries represent the largest voting bloc in the UN is an additional reason that Russia is taking multiple steps to strengthen its ties in the region (Foltynova 2022).

Any country utilizing nuclear power depends on Russia at one level or another. Russia's dominance in the nuclear power cycle therefore is another reason that nuclear power is not a good option for human beings. Furthermore, if China were to gain even more influence in the world's nuclear power cycle, it, too, could hold the rest of the world vulnerable with respect to energy.

A subset of the G7 made up of five nations—the U.K., the U.S., Canada, Japan, and France—is attempting to address at least some of the world's overreliance on Russia with respect to nuclear power. On April 17, 2023, the five-nation alliance was announced at the G7's Nuclear Energy Forum held in Japan. The five nations agreed to leverage the capabilities and resources of each country's civil nuclear power sector to undermine Russia's dominance in nuclear power supply chains by supporting a stable supply of nuclear fuel and securing the development and deployment of fuels for more advanced nuclear reactors in the future (NIA 2023). The alliance intends to pool nuclear fuel capacity in order to challenge nuclear giant Rosatom Corp., which, as discussed above, currently dominates global atomic-fuel markets and nuclear supply chains. As the war in Ukraine continues, the alliance intends to cut off an important source of geopolitical currency for Russia, since even with Western sanctions over the war, Rosatom has remained the world's largest exporter of nuclear reactors and fuel (Tirone 2023).

Not only does the five-country alliance hope to displace Russia from the international nuclear energy market, it also wishes to do so as quickly as possible and over the long run. Geopolitically, the alliance wants to reduce reliance on Russia in the nuclear fuel supply chain while increasing the availability of commercial, free-market alternatives in the supply of civil nuclear technologies to third countries. The government in the U.K. is of the opinion that the agreement will strengthen its own nuclear energy sector, along with those of the other four countries in the alliance. Some people think this will increase domestic energy security while reducing the cost of electricity for families living in the U.K. Finally, the parties to the agreement are seeking ways to collaborate in multilateral spheres in order to advance energy security and economic resilience for partners around the globe (GOV.UK 2023).

What is unfortunate about the new plans described above is that the U.K. has decided to pursue an aggressive plan to increase its own reliance on nuclear energy. The country intends to shift the percentage of electricity it derives from nuclear power from 15% to 25% by 2050. Furthermore, the U.K.'s Energy Security Secretary Grant Shapps thinks that the world increasingly should turn to nuclear as a source of low-carbon and secure energy. My view is that the U.K. would be much better served by intensifying its focus on renewables. The U.K. has a solid track record in this area, with almost 40% of the U.K.'s power coming from renewables over the last year. One stellar example is U.K. development of four large wind farms, the technology for which it offers to other countries as a way of creating green jobs and boosting energy security. This is a much saner direction than the U.K.'s current plan for "a nuclear renaissance" and its turn toward nuclear technologies such as Sizewell and Small Modular Reactors from Korea and Japan. Instead, offshore wind and, also, solar can help make everyone less reliant on volatile international fossil fuel markets and particularly Russian oil and natural gas (GOV.UK 2023) without compromising public safety by pursuing dangerous nuclear technologies.

Beyond the specter of nuclear blackmail involving nuclear power plants, one also must be alert to the possibility of nuclear theft. One example of this is the ongoing vulnerability of advanced U.S. nuclear technology at the Zaporizhzhia nuclear plant in Ukraine. As initially reported by RBC Ukraine, and, later, by CNN on April 19, 2023, Andrea Ferkile, the Director of the U.S. Department of Energy's Office of Nonproliferation Policy, wrote a letter dated March 17, 2023 to Rosatom's Director General. The letter stated that the Zaporizhzhia plant contains U.S.-origin nuclear technical data that is export-controlled by the USG. The U.S. warned Russia not to touch this sensitive nuclear technology—a warning that is difficult if not impossible to monitor since Rosatom currently manages the planet, even though Ukrainian staff continues to operate it (Bertrand and Lister 2023).

Nuclear power is not the answer—nor, obviously, are fossil fuels. The only immediate route to energy self-sufficiency requires cultivating renewables more broadly. People say it cannot be done—but it can. Later, when human beings are more ethically and spiritually mature, we also can benefit from the technologies that various ETIs utilize. For now, however, we need to scale up solar, wind, geothermal, hydropower, ocean energy, and bioenergy as quickly as possible.

Chapter 7:
Abolishing Nuclear Weapons and
Other Nuclear Technologies

Efforts to Get the Nuclear Situation Under Control

Historically, the nuclear paradigm relates to the arising of fascism in Europe, which was prevalent from 1919 to 1945, and the attempt to counter it—since both Nazi Germany and the U.S. were racing against one another to develop a weapon to end the war decisively. Some people argue that development of the atomic bomb was 'necessary' to stop fascism, since Nazi Germany was trying to develop the atomic bomb and other advanced weapons systems. The argument is not robust, however, since developing the atomic bomb clearly did not stop fascism. Fascism continues today, often masked as orthodoxy, libertarianism, national security, and other fundamentalist ideologies.

Indeed, hostility and competition between U.S., China, Russia, and many other countries coincides now with the revival of fascist ideologies in many parts of the world, including in the U.S. (Stanley 2021). Unlike in the 1930s and the 1940s, however, current ideological conflict and the resurgence of fascism and in some cases even Nazism is underpinned by very large arsenals of nuclear and thermonuclear weapons. The obvious problem is this. Over the long run, ideologies are not overcome by means of weapons, regardless of how powerful the weapons may be. If we want to change human thinking so that society can become peaceful and just, that requires a process that is not militaristic in nature. In the meantime, while we work on transforming human consciousness, we must address how to abolish large arsenals of nuclear weapons spread across the globe.

To succeed in abolishing nuclear weapons, examining what actions have been tried in the past yet have failed provides insight regarding what steps may be useful now. Efforts to curb the spread of nuclear material and technology began only a short time after the world was introduced to the destructive potential of atomic weaponry in 1945. In 1946 the Acheson-

Lilienthal Report, authored in part by Manhattan Project physicist J. Robert Oppenheimer, advocated that an Atomic Development Agency be created to regulate fissile material and to ensure that state rivalries over the technology did not occur. Ultimately, neither Dean Acheson or David Lilienthal presented the U.S. plan to the United Nations Atomic Energy Commission (UNAEC). Instead, Bernard Baruch presented the Baruch Plan, which also would have established an Atomic Development Authority that answered to the UN Security Council. The plan called for the U.S. to disassemble its nuclear arsenal if an agreement could be reached first assuring the U.S. that the Soviets would not be able to acquire a bomb. The plan failed to achieve consensus within the UNAEC (Kimball 2018).

In 1993, UN resolution 78/57 L called for a "non-discriminatory, multilateral and internationally and effectively verifiable treaty banning the production of fissile material for nuclear weapons or other nuclear explosive devices." The resolution passed unanimously. In March 1995, the UN Conference on Disarmament (CD) took up a mandate presented by Canadian Ambassador Gerald Shannon, often referred to as the Shannon Mandate (CD/1299). This Mandate established an ad hoc committee directed to negotiate a Fissile Material Cut-off Treaty (FMCT) by the end of the 1995 session. A lack of consensus over verification provisions, together with interest by China and Russia in also holding space arms control discussions specifically relating to Preventing an Arms Race in Outer Space (PAROS), stalled efforts to begin FMCT negotiations (Kimball 2018).

ICAN on Who Perpetuates Nuclear Weapons

The International Campaign to Abolish Nuclear Weapons (ICAN) is one of the most effective organizations currently working to abolish nuclear weapons, voicing its concerns over nuclear weapons is appropriately strong, moral language (ICAN n.d.-c). Launched in 2007, its efforts focus on stigmatizing, prohibiting, and eliminating all nuclear weapons. As of 2022, this global civil society coalition had 652 partner organizations in 107 countries, with staff located in Geneva.

ICAN works to promote adherence to and full implementation of the Treaty on the Prohibition of Nuclear Weapons (TPNW) (UN 2017). On December 23, 2016, the UN General Assembly passed resolution 71/258 to convene a conference in 2017 to negotiate a legally binding instrument to prohibit nuclear weapons, leading towards their total elimination. The General Assembly encouraged all Member States to participate in the Conference, along with participation and contributions by international organizations and civil society representatives. ICAN drafted the TPNW and formally introduced it at the UN. The Conference adopted the TPNW on July 7, 2017. On September 20, 2017, the Secretary-General of the UN opened the TPNW for signature. Following the deposit with the Secretary-General of the 50th instrument of ratification or accession of the Treaty on October 24, 2020, the TPNW entered into force on January 22, 2021 (UNODA n.d.-c). The TPNW is the first legally binding international agreement that explicitly prohibits the manufacture, production, and development of nuclear weapons and actions that assist those prohibited acts (Snyder 2021, 7; see ICAN n.d.-a for the full text of the treaty in multiple languages).

ICAN has authored numerous, informative reports on various aspects of the nuclear weapons industry. Reading them in detail helps one understand the devastation and immoral profiteering resulting from nuclear weapons:

- September 2011 *Nuclear weapons spending: a theft of public resources: Challenging government investments in nuclear arms* (ICAN 2011);
- February 2013: *Unspeakable suffering – the humanitarian impact of nuclear weapons* (Fihn 2013);
- February 2015: *A Pledge to Fill The Legal Gap: Vienna Conference 2014* (ICAN 2015);
- March 2018: *Don't Bank on the Bomb: A Global Report on the Financing of Nuclear Weapons Producers* (Beenes and Snyder 2018);
- October 2018: *Nuclear Weapons Ban Monitor 2018* (ICAN 2018);
- October 2019: *Beyond the bomb: Global exclusion of nuclear weapons producers* (Beenes 2019);
- October 2019: *Nuclear Weapons Ban Monitor 2019: Tracking Progress towards a World Free of Nuclear Weapons* (ICAN 2019a);

- November 2019: *Schools of Mass Destruction: American Universities in the U.S. Nuclear Weapons Complex* (ICAN 2019b);
- May 2020: *Enough is Enough: 2019 Global Nuclear Weapons Spending* (ICAN 2020);
- June 2021: *Complicit: 2020 Global Nuclear Weapons Spending* (ICAN 2021);
- June 2021: *A Non-Nuclear Alliance: Why NATO Members Should Join the UN Ban on Nuclear Weapons* (Lennane and Wright 2021);
- September 2021: *Implications of Germany's accession to the Treaty on the Prohibition of Nuclear Weapons* (Hajnoczi 2021);
- November 2021: *Perilous Profiteering: The companies building nuclear arsenals and their financial backers* (Snyder 2021);
- January 2022: *Rejecting Risk: 101 Policies against nuclear weapons* (Snyder 2022);
- February 2022a: *No Place to Hide: Nuclear Weapons and the Collapse of the Health Care Systems* (ICAN 2022);
- April 2022: *Nuclear Weapons Ban Monitor 2021: Tracking progress towards a world without nuclear weapons* (NPAIP 2022); and
- June 2022: *Squandered: 2021 Global Nuclear Weapons Spending* (ICAN 2022b).

ICAN's website also lists additional fact sheets and informative materials.

Perhaps unsurprisingly given the global uptick in nuclear rhetoric, for example in the U.K. under Boris Johnson, not all countries support the TPNW. The U.K.'s Integrated Review of 2021 states, "We are strongly committed to full implementation of the NPT in all its aspects, including nuclear disarmament, non-proliferation, and the peaceful uses of nuclear energy; there is no credible alternative route to nuclear disarmament" (GOV.UK 2021). One certainly can interpret this as "a thinly-veiled reference to the [U.K.] government's hostility to the UN's new Treaty on the Prohibition of Nuclear Weapons [TPNW]" (Hudson 2021b). Part of this hostility likely derives from the fact that the TPNW makes the very existence of nuclear weapons illegal, while countries such as the U.S. and the U.K. would like to retain at least some nuclear weapons without abolishing their arsenals *in toto*. On the surface, however, governments in the U.K. have opposed the TPNW by arguing that it will damage the NPT instead of arguing for nuclear weapons per se. But given that the NPT came into force decades ago and we still face the threat of nuclear weapons, we clearly need a new approach (Hudson 2021a).

 A small number of companies—approximately two dozen in any given year—produce nuclear weapons inasmuch as they are open to private investment and produce the goods or services that specifically contribute to nuclear weapons development, testing, production, manufacture, possession, stockpiling, and/or use. The TPNW designates all these activities as illegal. While nuclear armed countries have facilities dedicated to the production, manufacture, stockpiling, and development of nuclear weapons, the management and operations of those facilities often are outsourced to the private sector. It is challenging to track movement related to this rapidly changing dynamic, since private companies often change names and/or change hands. In addition, a tangled web of joint ventures and consortia are involved in producing nuclear arsenals, often receiving multiyear, multi-billion-dollar contracts that are enormously profitable to specific companies. Furthermore, since most existing nuclear weapons are designed to be delivered by missiles launched from the ground, from airplanes, or from submarines, one also must research companies that produce the missiles designed specifically for nuclear weapons use, and, also, companies that produce key components for nuclear warheads. Also of note, the nuclear weapons industry is consolidating by means of mergers and acquisitions (M&A), such as the merger between Raytheon and United Technologies, or Northrop Grumman's acquisition of Orbital ATK. M&A activity gives companies competitive advantages when they compete for contracts. At the same time, however, reduction in the number of nuclear weapons industry contractors makes it easier for financial institutions and investors to divest from such companies by excluding them from investment (Snyder 2021, 4-11).

Who Profits from Nuclear Weapons Production

In-depth reading of the reports cited above makes it clear that greed—not a genuine pursuit of national security—fuels the nuclear weapons industry.

Nuclear weapons production is enormously profitable for many different companies in the industry, including the companies that produce nuclear weapons and the financial institutions that finance the production of these weapons.

Since the first atomic bomb was exploded in recent human history in 1945 in the desert outside Alamogordo, New Mexico, human beings have wired the entire planet for reciprocal obliteration. This complete insanity demonstrates the depth of human aggressivity, divisiveness, and what disturbingly appears to be a 'taste' for largescale violence. But it is not insanity per se that is driving the nuclear weapons industry today—it is greed on the part of people working for nuclear weapons producers, the financial institutions that support companies producing nuclear weapons, and lobbyists who work on behalf of the nuclear weapons industry.

In 2021, nine nuclear-armed states spent $82.4 billion on their nuclear weapons, which is more than $156,000 per minute. (ICAN 2022b, 4). The U.S. leads the way, followed by other major nation-states—the U.S. ($44.2 billion); China ($11.7 billion); Russia ($8.6 billion); the U.K. ($6.8 billion); France ($5.9 billion); India ($2.3 billion); Israel ($1.2 billion); Pakistan ($1.1 billion); and North Korea ($652 million) (ICAN 2022b, 10). As ICAN (2022b, 4) writes:

> The exchange of money and influence, from countries to companies to lobbyists and think tanks, sustains and maintains a global arsenal of catastrophically destructive weapons. Each person and organization in this cycle is complicit in threatening life as we know it and wasting resources desperately needed to address real threats to human health and safety.

ICAN (2022b, 4) also notes that collectively, at least twelve major think tanks that research and write about nuclear weapons in India, France, the U.K., and the U.S. received between $5.5 million and $10 million from companies that produce nuclear weapons. It adds, "The CEOs and board members of companies that produce nuclear weapons sit on some of their advisory boards, serve as trustees and are listed as 'partners' on their websites" (ICAN 2022b, 4). In 2021, this 'think tank' income from nuclear weapons producers went specifically to the following organizations: Atlantic Council; Brookings Institution; Carnegie Endowment for International Peace; Center for New American Security; Center for Strategic and International Studies; La Fondation pour la recherche stratégique (FRS): French Institute of International Relations; Hudson Institute; International Institute of Strategic Studies; Observer Research Foundation; Royal United Services Institute; and Stimson Center (ICAN 2022b, 7).

Providing financing for nuclear weapons production generates tremendous fees for financial service firms. When companies are hired by governments to build the key components for nuclear weapons and to manage and operate the facilities in which nuclear weapons are produced, many raise capital by means of loans and/or by issuing bonds. Many of the companies involved in nuclear weapons production also are publicly traded (Snyder 2021, 9, inc. fn. 5 on 113 citing to Brady 2021).

In 2021, fifteen new nuclear weapons related contracts valued at more than $30 billion were awarded, compared to $14.8 billion in new nuclear weapons contract awards in 2020. Most contracts awarded in 2021 are related to the Trident or Minuteman III systems and are multi-year contracts lasting an average of five years. However, Northrop Grumman's three new contracts for the Minuteman III system "were the highest, valued at more than $23 billion and extending through 2040" (ICAN 2022b, 18). Unsurprisingly, Northrop Grumman spent almost $12 million on lobbying efforts related to nuclear weapons in 2021 (ICAN 2022b, 7).

U.S. companies are particularly to blame when it comes to making money from nuclear weapons production. Between January 2019 and July 2021, combined, financial institution investments in the twenty-five nuclear weapons producers were around $340 billion, which is around half of the total outstanding investments of $685 billion. According to financial information from financial institutions participating in loans and underwriting deals since January 1, 2019, together with those financial institutions involved in loans that closed before that date but had not yet matured before October 1, 2020, all top ten investors in the nuclear weapons industry are U.S. companies. Listed from highest to lowest investment in nuclear weapons producers for 2021 are Vanguard, State Street, Capital Group, BlackRock, Bank of America, Citigroup, JPMorgan Chase, Wells Fargo, Morgan Stanley, and Wellington Management. Other financial services companies, including various firms in the U.S., are involved in bondholding, loans, shareholding, and underwriting for the nuclear weapons industry (Snyder 2021, 74-77). *In other words, there is an exceedingly intertwined connections between some of the largest financial institutions in the U.S. and the nuclear weapons industry.*

If financial services companies refused to help nuclear weapons producers raise capital by means of loans and/or bond issuance, and if investment companies and sovereign wealth funds refused to invest in companies associated with the nuclear weapons industry, the entire house of cards would cease to be profitable and, like all things that are not profitable, it quickly would come tumbling down. Fortunately, some pushback against financial firms that assist the nuclear weapons industry does occur. Divestment also has occurred in opposition to nuclear weapons production, with public divestment being particularly effective. For example, Ireland's sovereign wealth fund dropped €6 million shares of nuclear weapons related stock because of its general opposition to the nuclear weapons industry (Snyder 2021, 7, 9, inc. fn. 5 on 113 citing to Brady 2021).

The takeaway from this discussion is that money—not national security—drives the nuclear weapons industry. In fact, this is true of all weapons industries—they persist because of the profit margins and the financial fees they generate, not because weapons offer a route to security. But because of *how profitable* the nuclear weapons industry is, for nuclear weapons producers and financial services and investment firms alike, it has persisted intransigently. So, to make meaningful progress in abolishing nuclear weapons, greed is the problem we really must tackle.

Emerging from Denial regarding Nuclear Risks

Humankind is failing to assess accurately the enormous risks posed by nuclear weapons and other nuclear technologies. The same is true with respect to the risks posed by many weapons systems. Meanwhile, as the U.S., China, and Russia, and their allies compete in a dangerous and escalatory way, the world becomes less stable by the day.

The cycle of escalatory competition and conflict in which people are engaging reflects humankind's incorrect ontological view regarding the nature of reality. Instead of recognizing the wholeness of reality, most people are driven by a mechanistic and coarse interpretation of reality according to which they try to maximize individual and corporate gain. In contrast, Bohm emphasizes that the entire universe is meaning. He states, "The outward and the inward are one part of one total meaning." He adds, "The being of matter is its meaning; the being of ourselves is meaning; the being of society is its meaning. The mechanistic view has created a rather crude and gross meaning which has created a crude and gross and confused society." It is coarse and confused human society that remains mired in violence and must awaken ethically and spiritually in order to move past militarism. Weber observes, "This view, your view, would make human beings feel rooted and have their dynamic place in the whole scheme of things." Bohm replies, "At least they would have a chance to find it there. It's a view within which it makes sense to observe to find out where your place is" (Weber 1987).

One of the reasons that human society has become so coarse and confused is because people pursue science and technology without regard to ethics and morality. Western culture has produced the tremendous power of science and technology without reflecting sufficiently on how this same power can be used in dangerous and destructive ways (Bohm 1989, 16). If one regards the pursuit of scientific knowledge as separate from its applications and implications, however, there is no necessary connection between being a scientist who produces the knowledge and being a human who is responsible for its applications. Instead of compartmentalizing science and ethics, human beings must view scientific and technological knowledge and how that knowledge is applied as inseparable. In other words, the social responsibility for science and technology must be understood as part of scientific processes themselves (Bohm 1989, 12, comment by Maraca Wilkins in dialogue with Bohm). If technologies and weapons continue to be developed in the absence of a coherent perspective that also considers the risks associated with their destructive potential and uses, *Homo sapiens sapiens* has very little chance of surviving as a species.

The long-term, downside risks of permitting nuclear weapons to continue to exist on Earth are enormous. From a deontological perspective, nuclear weapons never benefited humankind. Nothing can justify what happened to tens of thousands of Japanese civilians, including thousands of children, when the United States dropped atomic bombs on Hiroshima and Nagasaki. Those actions were and are completely indefensible deontologically.

Writing about the risks posed by nuclear weapons, Bohm states the following:

> As [Carl] von Clausewitz said, war is the continuation of politics by other means. If politics is non-negotiable, then we try to settle it by war. But because of nuclear weapons war is not a feasible way of doing that any more. You cannot possibly use it to solve a crisis (Bohm 1989, 19).

ETI too is trying to make this point with respect to nuclear weapons, WMDs, and weapons more generally. We cannot rely on weapons to settle our differences. Humankind simply must transform beyond coarse militarism and rise to its true creative potential as a species. If we do not make this transition, we will descend into extinction because of the sheer power of the WMDs and other new weapons systems currently in existence together with those being developed.

Even though the seriousness of the nuclear situation is acute, many people respond with denial rather than facing the situation directly. Denial plays out in many forms, including in the inordinate degree of narcissism and escapism we see in human society today. I often wonder if there is a direct correlation between the horror of the reality facing an intelligent species and the extent of the denial and escapism to which that species succumbs. If we could resolve the nuclear crisis on Earth, perhaps it would encourage people to start addressing some of the other urgent, global issues currently facing humankind, instead of retreating into the nonstop pursuit of so-called entertainment.

It also is the main reason we cannot permit another explosion of a nuclear weapon to occur during a situation of conflict—because the 'form' of the explosion will build up force in implicate, etc., levels of order, making it more likely to occur again. It is the same reason that one episode of gun violence—e.g., shootings in schools, stores, and workplaces—quickly is followed by other, similar episodes.

I also often wonder how savagely cruel nuclear blasts must seem to a species such as ETI, which is aware of subtle aspects of existence and is not prone to projection, denial, and escapism. We can conclude that ETI in our midst now does not suffer from such human cognitive errors, all of which depend upon the incorrect views that mind and matter are separate and that the 'parts' of reality are independent and do not comprise a whole. ETI is not tripping over these mistaken views, which we can see by the seamless way ETI's UAP move through spacetime and the gentleness that ETI shows towards humankind.

One reason that nuclear blasts are particularly jarring to the holomovement is because the nuclear threshold is the same as the implicate/explicate threshold. Nuclear blasts span the threshold between something manifesting at the level of explicate order and something unmanifesting back to implicate, etc., levels of order. When this is done en masse—with multiple trajectories of hundreds of thousands of human and other beings' lives, together with the stunning grandeur of naturally existing trees, plants, etc., 'unmanifested' together in a nuclear blast—the entire holomovement is disrupted—jarred by an awful cacophony of violence. This is simply wrong at every possible level.

It is imperative we abolish nuclear weapons and decommission and dismantle infrastructure related to nuclear power before an existential catastrophe ensues. An estimated twenty-four to one hundred and fifty species go extinct *each day* on Earth (Pearce 2015). Human beings are arrogant to think that the risk of extinction does not apply to them. It does. To walk back from the existential precipice in front of us, people need to move out of denial and to realize that weapons never solve issues over the long run—they only make things worse. We need to stop trying to rationalize the production of weapons, to learn to think deontologically, and to replace the exchange of artillery with the exchange of creativity, empathy, and genuine dialogue. Until we learn to communicate and to care properly for ourselves and for others, *Homo sapiens sapiens* will remain a profound threat to itself, to other species around us, and to the Earth we call home.

Correlation Between UAP Activity and Nuclear Sites and Vessels

In the last two decades, reports indicate that UAP activity either has increased around U.S. military sites, and/or that it is being detected more often now because the military is using new sensor systems.

A correlation between UAP activity and nuclear sites and vessels is implied by the language of the National Defense Authorization Act for Fiscal Year 2022 (NDAA FY22), specifically Title XVI, Space Activities, Subtitle E— Other Matters, Section 1683(h)(2)(M)-(O). These subsections imply that UAP activity is correlated with nuclear weapons sites; nuclear-powered ships and submarines; facilities associated with the production, transport, and storage of nuclear weapons; nuclear power generating stations; nuclear fuel storage sites; and other sites or facilities regulated by the Nuclear Regulatory Commission (NDAA FY22, 581-82). In fact, it is well known that UAP has occurred for decades at nuclear weapons storage and launch sites and near underground nuclear launchers and launch control facilities (TT 2010). In addition to the correlation between UAP activity and nuclear weapons sites, UAP activity also often occurs in proximity to nuclear power plants (EE n.d.) and in conjunction with atomic testing (Basset 2003).

A correlation between UAP activity and nuclear weapons repositories occurs both in the U.S. and in Russia (Kaleka 2013; Mazzola 2017; Ridge 2016 <2003>; Fein 1989; Russian Roswell n.d.). Such a correlation almost undoubtedly also occurs in all countries with nuclear weapons and nuclear power installations. For example, it has been suggested that one of the reasons France has so many UAP sightings is because it has the most nuclear power plants in operation of any country in Europe—a total of 56 as of December 1, 2022 (ENS 2022).

Declassified USG documents going back at least to 1949 attest to the connection between facilities and vessels associated with nuclear weapons and UAP activity (Uda 2010). This correlation is attested to in an assortment of official documents including those from the FBI, CIA, U.S. Air Force, and U.S. Army (Ridge 2005b). The USG initiated the declassification of some of these documents, while others were declassified following FOIA requests. Many U.S. military personnel also have provided firsthand testimony verifying a correlation between UAP activity and nuclear sites and vessels, including retired USAF officers who witnessed UAP incidents at or near USAF ICBM and USAF Strategic Air Command (SAC) nuclear weapons facilities. Many individuals who had custody of WMDs have come forward to describe their personal experiences of UAP activity at nuclear sites.

After controlling for population and region of the country, UAP reports occur more frequently in the vicinity of nuclear sites and in restricted airspace over nuclear facilities. UAP have been reported at nuclear research facilities and at nuclear weapons storage bunkers at military bases on numerous occasions. Many of these reports, which often were made by highly trained government scientists and military personnel with top-secret military clearances, have occurred at highly restricted nuclear facilities and at other government research and production facilities. According to multiple, independent accounts, beams of light were directed down from UAP onto the nuclear storage bunkers and underground missile silos. In some instances, UAP also reportedly focused light rays or energy beams on nuclear materials (Johnson 2002, fn. 3 citing to Gestin 1973, 26).

Among U.S. states, New Mexico evidences a strong correlation between UAP activity and nuclear sites that extends back at least to the 1940s and includes many areas such as White Sands, Alamogordo, Socorro, Farmington, Roswell, Holloman AFB, and Kirtland AFB. Many different types of UAP have been reported in New Mexico, including fireballs often green in color, and, also, more structured objects. Official reports have been filed with respect to UAP events at locations associated with the development of the atomic bomb, such as Los Alamos and Sandia Base; and at locations where nuclear weapons have been stored, such as the Manzano Mountains south of Kirtland AFB. Similarly, in Texas, UAP activity has been reported at Killeen Base at Camp Hood (Ridge 2005b; see also Uda 2010 for declassified USG documents discussing early UAP events in New Mexico; Ridge 2005a for discussion of UAP sightings at Camp Hood).

The following, chronology details UAP incidents at specific nuclear sites. It is a representative chronology only and does not include all such events. Along with descriptions of the events, I intersperse discussion from three, mainstream news articles for context.

1945: On July 16, 1945, the first atomic bomb in recorded history was detonated in the desert near Alamogordo, New Mexico. After this event, it becomes easy to discern an ongoing pattern of UAP activity at multiple nuclear weapons sites. It has been stated that approximately five weeks later. In August, reports were made of an avocado-shaped object coming down in the area near the White Sands Proving Ground. Two young men, Jose Padilla and Remigio Baca, are said to have been doing chores and running errands at the Rancho Padilla in San Antonito, near San Antonio, New Mexico (not Texas). Witnesses are said to have heard a sound and to have seen a grounded, avocado-shaped craft, bulging at one end, with occupants. It also was reported that the U.S. Army later removed the craft (Good 2007 <2006>, 53-66). Of note, however, the veracity of this account has been questioned (Johnson 2023).

1947: In July, a UFO is said to have come down near Roswell, NM. Of note, Roswell Army Airfield was home to the 509th Bomb Group, the world's only atomic bomber squadron at the time (Hastings 2017<2008>).

1948: On or around March 25, a UFO is said to have come down on a mesa in Aztec, New Mexico. It is alleged that the Los Alamos Radar Station in nearby El Vado, New Mexico tracked the landing (Explore Aztec n.d.).

A high security recovery operation lead by the Air Force and 5th Army Division was responsible for the removal of this craft. The recovery operation took approximately two weeks with all the remains being taken to Los Alamos Laboratory for scientific study and evaluation by some of the worlds' leading scientists. Later it was rumored to have been taken to Wright-Patterson Air Force Base in Dayton, Ohio. The recovery of this craft by the USG and military was one of the most secretive recoveries of a spacecraft with origins unknown since the similar recoveries in Roswell, New Mexico eight months earlier. The space craft was approximately 100 feet in diameter and eighteen feet tall. It was one of the most intact crafts that the USG had recovered at that time.

Formerly classified USG documents released in accordance with FOIA confirm that as early as December 1948, UAP described as discs and saucers began to be sighted in proximity to the Atomic Energy Commission (AEC) installation in Los Alamos, New Mexico. A little more than three years earlier, this laboratory tested the first atomic bomb in the desert near Alamogordo, New Mexico (Hastings 2016; see also Johnson 2002). In addition, numerous, amorphous green fireballs were sighted near military installations in Los Alamos and near the Sandia National Laboratories atomic weapons installation close by in Albuquerque, New Mexico (King 2020 <2018>). At least until February 1949, sightings of green fireballs in New Mexico continued to be documented (Carpenter n.d.-a; n.d.-b).

1950s: Documents state that throughout the 1950s, as many as 150 observations were made of discs and saucers in Los Alamos and in proximity to the Sandia National Laboratories. Other declassified documents discuss UAP sightings at the Oak Ridge National Laboratory in Tennessee [currently managed by UT-Battelle for the U.S. DOE], the

Hanford AEC Site plutonium processing plant in Washington State, and the Savannah River AEC Site, a plutonium manufacturing facility located in South Carolina (Hastings 2016; see also Ridge 2005b for more on the Hanford and Savannah River events).

1950: At 3:30 p.m. on October 15, the security staff at the AEC facility in Oak Ridge, Tennessee witnessed the approach of two shiny silver objects. One object vanished and the other stopped 5-6 feet above the ground and hovered some 50 feet from the observers (Wright 2019, 21).

In December, another UAP event occurred at Oak Ridge (Johnson 2002; Ridge 2005b).

1952: In March, a UAP event associated with uranium mines occurred in the southern part of the Belgian Congo (Ridge 2005b).

In July, UAP events occurred at Hanford AEC, Savannah River AEC, and Los Alamos AEC (Johnson 2002).

1963: On January 10, the *New York Times* published an article entitled, "Extraterrestrial Life." The article discusses how a large group of scientists submitted a report to the USG urging that the search for extraterrestrial life be made the principal scientific objective of the U.S. space program (NYT 1963).

1963-64: During the six-to-twelve-month period after the Cuban Missile Crisis of October 1962, when the 579th Strategic Missile Squadron had been placed on high alert at Walker Air Force Base (AFB) in New Mexico, a former Atlas ICBM launch officer at the base reported that on several occasions while he was on alert in the underground launch capsule at the Atlas Site 9, missile guards at ground level reported a silent UFO hovering over the site. Within a month or so, between three and ten incidents occurred of this type, during which UFOs hovered over the silo and shone lights down on the silos without making any noise. Another Atlas ICBM launch officer from the same squadron reported that in the fall of 1964, while on alert in the launch capsule at Atlas Site 7, his missile commander received a call from a sister site, either Site 6 or 8. The commander there reported an enlisted man at the site, possibility a security guard, had reported to him that a UFO was hovering directly over the site and shining a very bright light. It sped away, returned hovered, etc. The UFO 'zoomed' from the direction of Site 6 in the direction of Site 8 and hovered awhile, and it also hovered directly over both sites. A former Atlas ICBM Launch Facilities Specialist LFS from the same squadron said that he had been at one of the Atlas ICBM launch sites one evening in 1964 or 1965 when the missile commander directed him topside to view two UFOs moving in unison. The commander said the base was tracking the objects on radar and had scrambled two jet fighters to intercept them. This same Launch Facilities Specialist stated that UFO sightings at the base were common. He also said that on one occasion, he observed a UFO display at Walker AFB itself in contrast to at one of the remote sites associated with the base. He stated that at least half the people in his barracks at the base saw the UFO display, during which two to five lights moved in unison. In addition, a former Atlas Missile Facilities Specialist from the same squadron wrote to NICAP on December 20, 1964 to report an ongoing series of UFO events at his squadron's Atlas ICBM sites. Missile security told the Missile Facilities Specialist that the events had been classified top secret. The Missile Facilities Specialist also stated that UFO incidents were so frequent that some of the missile guards were balking at reporting for duty (Hastings 2006).

1964: A UFO is reported to have landed at Holloman AFB (Rivas 1974; Pasetta 1988).

1965: At 1:30 a.m. on August 1 at Francis E. ("F.E.") Warren AFB in Wyoming, personnel at the base made a telephone report to USAF Project Blue Book at Wright-Patterson, AFB in Ohio. Around this date, other UAP events also apparently occurred at Warren AFB. Also in August, several UAP sightings specifically were reported at Warren's Minuteman missile sites. One individual on duty at the Quebec Flight Launch Control Facility (LCF) said he had been called from the underground launch capsule and told to take his partner on duty outside and look up. Eight stationary lights grouped together in four pair were directly overhead. Another sighting of nine UAP occurred at 4:05 a.m. On another night, someone at one of Q-

Flight's Launch Facilities reported that he and his duty partner were sitting in a Security Alert Team (SAT) camper parked next to the missile silo when their vehicle began to shake. Both men saw a large, unidentified object hovering directly above the camper. When they were debriefed by an OSI agent later, they were told not to discuss the incident. A Non-Commissioned Officer In Charge (NCOIC) who had been on duty at Q-Flight's Launch Facilities the night of the incident verified that he also had seen the object as it hovered over the launch facilities. Of note, an almost identical account was made by a former USAF missile guard stationed at Minot AFB, North Dakota in 1968. Returning to the summer of 1965 at F.E. Warren AFB, a former Air Policeman said that departing night shift guards informed him about a sighting of an unidentified, anomalous object moving at incredible speed at Oscar Flight LCF. A few days later, this Air Policeman sighted UAP himself in front of the LCF (Hastings 2006; see also Johnson 2002).

1966: In March, a USAF document, "SUBJECT: UFO Report," describing two UAP at Minot AFB in North Dakota was sent to Wright-Patterson AFB. The UAP were witnessed for 3.5 hours and tracked on radar as they hovered and maneuvered near three Minuteman nuclear missile sites, "traveling very slowly back and forth and up and down." Radio transmission was interrupted by static when one of the objects came close to one of the missiles sites, but the static stopped when the object climbed in altitude. Part of the incident description follows:

> The UFO then began to sloop and dive. It then appeared to land 10 to 15 miles South of MIKE 6. "Mike 6" Missile Site Control sent a strike team to check. When the team was about 10 miles from the landing sight [sic], static disrupted [sic] radio contact with them. Five (5) to eight (8) minutes later, the glow diminished and the UFO took off. Another UFO was visually sighted and confirmed by radar. The one that was first sighted passed beneath the second. Radar also confirmed this. The first, made for altitude toward the North and the second seemed to disappear with the glow of red. A3C SEDOVIC at the South Radar base confirmed this also. At 0619Z[ulu], two and

one half (2½) hours after first sighting, an F-106 interceptor was sent up. No contact or sighting was established (Uda 2010).

Other UAP events occurred in August, also at ICBM sites at Minot AFB (UFOs & Nukes n.d.; see also Huyghe 1979).

In late June or early July at Ellsworth AFB in South Dakota, a former Electro-Mechanical Team technician for the 44th Missile Maintenance Squadron said he and another technician were dispatched to Launch Facility silo Juliet-3 to correct an electrical malfunction. Both the commercial power supply to the site and the emergency power systems had failed simultaneously. This rendered the Minuteman I missile temporarily inoperable, meaning that the ICBM had dropped off alert status. Meanwhile, an armed Air Force Security Alert Team had been ordered to investigate a triggered security alarm at nearby Launch Facility Juliet-5, where the missile there abruptly had dropped off alert status. Like the missile at silo Juliet-3, it also had lost commercial electrical power. In addition, its diesel-powered generator, which was designed to charge backup batteries, had failed to start. When the Security Alert Team arrived at Juliet-5, they reported that a UFO was sitting on the ground inside the security fence that surrounded the missile silo. The object was round, appeared metallic, and was resting on a tripod landing gear. In addition, an intense glow enveloped an entire, adjacent missile silo about four miles away. The Launch Commander called the Missile Command Post at Ellsworth AFB and described what was occurring. He then was told that a helicopter was being sent to the first site to investigate. When the helicopter was about five minutes away, the object ascended vertically at enormous speed. The next morning, someone wearing civilian clothing debriefed the two technicians in the commander's office, but the technicians denied having seen anything (Hastings 2006; see also Hastings 2016).

1967: On March 24/25 at Malmstrom AFB, an entire flight of ten Minuteman ICBMs shut down simultaneously immediately after an unidentified anomalous object was sighted in the vicinity by Air Force Security Police. On the same night, an anomalous object also was seen in Belt, Montana, not too far to the southeast of Malmstrom AFB (Hastings 2006; see also Johnson 2002). Retired Captain Robert Salas, who was a USAF SAC Launch

Officer/Engineer, has spoken about this event extensively. He states that a glowing red object hovered outside the front gate of the Malmstrom AFB nuclear installation at which he was stationed, after which multiple nuclear missiles started shutting down in rapid succession before the object departed at high speed (MWL 2021; TT 2010; see also Malmstrom n.d. for a brief history of this Air Force Base).

In early 2023, Salas and Dr. Robert Jacobs provided testimony to AARO (Boswell 2023), having been recommended to AARO by Robert L. Hastings (2023). As mentioned below, in 1964, an unidentified anomalous object fired multiple beams of light and shot down a nuclear missile with dummy warheads after it was launched at Vandenberg Air Force Base [now, Vandenberg Space Force Base] in California (Kaleka 2013; Hastings 2016; Mazzola 2017; see also Mazzola 2021; seniorsam 2015).

A now-declassified USAF teletype message, classified "SECRET," discusses the March 1967 unexplained, simultaneous failure of ten Minuteman nuclear missiles at Malmstrom AFB. At the time of the incident, a large, it was reported that a round UFO hovered near one of the ICBMs (Uda 2010). Portions of this message include:

SUBJECT: LOSS OF STRATEGIC ALERT, ECHO FLIGHT, MALMSTROM AFB

REF: MY SECRET MESSAGE …, 17 MAR 67, SAME SUBJECT.

ALL TEN MISSILES IN ECHO FLIGHT AT MALMSTROM LOST STRAT[EGIC] ALERT WITHIN TEN SECONDS OF EACH OTHER. THIS INCIDENT OCCURRED AT 0845Z[ULU] ON 16 MARCH 67. … INVESTIGATION AS TO THE CAUSE OF THE INCIDENT IS BEING CONDUCTED BY MALMSTROM TEAT. TWO FITTS HAVE BEEN RUN THROUGH TWO MISSILES THUS FAR. NO CONCLUSIONS HAVE BEEN DRAWN. THERE ARE INDICATIONS THAT BOTH COMPUTERS IN BOTH G&C'S [sic] [GUIDANCE & CONTROLS] WERE UPSET MOMENTARILY.

CAUSE OF THE UPSET IS NOT KNOWN AT THIS TIME (Uda 2010).

In addition to the March 1967 UAP events at Malmstrom AFB, UAP events also were reported at Minot AFB and at Los Alamos in 1967 (Johnson 2002).

1968: In August, UFO events occurred at Ellsworth AFB in South Dakota (Johnson 2002).

1973-74: In the fall of 1973, when over half of the Launch Control Centers (LCCs) at F.E. Warren AFB in Wyoming had been converted to Minuteman III missiles, First Lieutenant Walter F. Billings, a former Minuteman ICBM launch officer when he was on alert at Golf LCC, the crew at India LCC noted an Outer Security Zone alarm on one of their missiles. When the Security Alert Team (SAT) was sent to investigate, they observed a bright, unidentified anomalous object hovering over the silo. A rumor circulated that that SAC headquarters at Offut AFB had sent members of the USAF OSI (Office of Special Investigations) by helicopter to investigate the incident. Another rumor circulated that sometime that night, the object erased the target tapes of three warheads. Target tapes are used to guide H-bombs (hydrogen bombs) to their targets. This claim regarding the target tapes initially was cited as a reason to disbelieve the account, since although Minuteman I missiles used target tapes, the Minuteman III did not. Later it was explained that individuals who originally were trained on the old system often still used the term 'tapes' when referring to the newer, plug-in unit used by the Minuteman III missiles (Hastings 2006).

In late 1973 at F.E. Warren AFB, another incident occurred involving the entire missile maintenance crew. All seven members of the crew saw an anomalous object when they were conducting maintenance on a Minuteman III missile. Yet another incident occurred in the early spring of 1974, also at F.E. Warren AFB, when an anomalous object is reported to have landed near the Charlie LCC (Hastings 2006).

1975: People in four U.S. states that border Canada—Montana, North Dakota, Minnesota, and Wisconsin—witnessed numerous UAP. Montana borders Saskatchewan, Alberta, and British Columbia; North Dakota borders Manitoba and Saskatchewan; Minnesota borders Manitoba and Ontario; and Wisconsin's land area is south of Ontario, with Wisconsin's Lake Superior located between Wisconsin and Ontario. Also in 1975, numerous Canadian residents saw UAP, particularly in Manitoba and Ontario. Many UAP sightings also occurred at the Canadian Forces Station in Falconbridge, Ontario. Falconbridge is a radar station located in Canada, very close to Michigan's Upper Peninsula. At the request of the Canadians, U.S. jets were scrambled but were unsuccessful in intercepting the UAP (Falconbridge n.d.).

A USAF document describes many UAP events that appeared in the NORAD Command Director's Log in 1975 relating to the Weapons Storage Areas (WSAs) at Loring AFB in Maine and Wurtsmith AFB in Michigan, and that appeared in the 24th NORAD Region Senior Director's Log relating to ICBM sites at Malmstrom AFB in Montana. This document is dated October 1977 (Uda 2010).

Between October 27 and November 10, a series of incidents occurred in which nocturnal lights and unidentified "mystery helicopters" visited several different military bases and missile sites across the northern tier of the U.S. Numerous UAP were reported over Weapons Storage Areas (WSAs) for nuclear weapons at Loring AFB in Maine, Wurtsmith AFB in Michigan, Grand Forks AFB and at Minot Air Force Base in North Dakota, and Malmstrom AFB in Montana. For example, multiple UAP reports were made in conjunction with Malmstrom AFB near Great Falls, Montana. F-106 interceptors were scrambled out of Malmstrom AFB in response to multiple reports of UAP visits to nearby missile sites near Moore, Harlowton, and Lewistown, and to several missile sites around Malmstrom AFB itself (Johnson 2002, inc. fn. 2 citing to Maccabee 1985; see also Huyghe 1979).

Also in November, a two-way radio discussion occurred at Malmstrom AFB between Air Force Security Police personnel who stated that an anomalous object was hovering over the base's Weapons Storage Area (WSA). The 24th NORAD Region Senior Director's Log for the base reported that on November 7, a call from the 341st Strategic Air Command (SAC) stated that a UFO had been seen at various missile locations. Two to seven UFOs were reported on November 8. Two F-106s were scrambled to intercept the UFOs, which extinguished their lights when the jets approached their position and re-illuminated themselves after the fighters returned to base. The NORAD Combat Operations Center (NORAD-COC) in Colorado Springs was immediately informed (Hastings 2006). Over the years, too, NORAD-COC has received other UAP reports at nuclear-armed SAC bases.

UAP events of 1975 are discussed in an article, "What Were Those Mysterious Craft?" which appeared in the *Washington Post* in 1979. The article states:

> During two weeks in 1975, a string of the nation's supersensitive nuclear missile launch sites and bomber bases were visited by unidentified, low-flying and elusive objects, according to Defense Department reports.
>
> The sightings, made visually and on radar by air and ground crews and sabotage-alert forces, occurred at installations in Montana, Michigan and Maine, and led to extensive but unsuccessful Air Force attempts to track and detain the objects.
>
> Air Force and Defense Department records variously describe the objects as helicopters, aircraft, unknown entities and brightly lighted, fast-moving vehicles that hovered over nuclear weapons storage areas and evaded all pursuit efforts.
>
> In several instances, after base security had been penetrated, the Air Force sent fighter planes and airborne command planes aloft to carry on the unsuccessful pursuit. The records do not indicate if the fighters fired on the intruders.
>
> The documents also give no indication that the airspace incursions provoked much more than local command concern.

But a Nov. 11, 1975, directive from the office of the secretary of the Air Force instructed public information staffers to avoid linking the scattered sightings unless specifically asked.

An Air Force press officer who deals with UFO inquiries said he could have no comment yesterday on questions about general security and military responses related to the rash of sightings at strategic installations in 1975.

The Defense Department position, cited in that memo and reiterated yesterday by a departmental spokesman, is that formal investigation of unidentified flying objects (UFOs) ended in 1969 and that there were no plans for renewed Air Force investigation.

[This portion of the article is reproduced below under '1976.']

[Todd] Zechel, [a Ground Saucer Watch (GSW) investigator and] a former NSA employe who now lives in Wisconsin, said that the 1975 incidents around the missile and bomber facilities would not have been revealed had it not been for a "leak" from a Pentagon source.

That tip, he said, led to the information request that produced the reports on the "flap," as a rash of UFO incidents is called, in the last days of October and the first two weeks of November 1975.

The Air Force and NORAD data provided detailed accounts of sightings of unexplained objects from Loring Air Force Base in Maine, Wurtsmith AFB in Michigan and Malmstrom AFB in Montana, all with a two-week period.

At those and other missile-launching sites in the northern tier of states, military personnel reported that the objects hovered over nuclear weapons storage areas, in some cases as low as 10 feet from the ground, and missile silos, before they departed.

The reports referred to the objects in some cases as "helicopters," although no witness made a positive identification. The sounds the objects emitted were described as being similar to helicopter noise.

In one such instance, on Nov. 7, 1975, at Malmstrom AFB, Capt. Thomas W. O'Brien, who had just left duty as a missile launch officer, said an aircraft resembling a helicopter approached the silo area.

He and his deputy heard what they thought was a helicopter rotor over the building where they were resting. The unidentified deputy looked out the window and saw "the silhouette of a large aircraft hovering about 10 to 15 feet above the ground" and about 25 feet from the launch-area fence.

He reportedly saw two red and white lights on the front, a white light on the bottom and another on the rear. Darkness prevented him from seeing markings or personnel on the object. The object left after a minute or so of hovering, the report said.

Military crews at two other nearby launch facilities reported moving lights in the air on the same evening, but said they heard no sounds.

NORAD commanders' activity logs during that period of time reported another sighting at another unidentified launch facility in which witnesses said they saw the object "issuing a black object from it, tubular in shape." Standard radar surveillance provided no clues as to the presence of anything other than known craft in the area.

More detail appeared in reports of sightings on Oct. 30 and 31 over Wurtsmith AFB, where an "unidentified helicopter" flew around the base and hovered over weapons-storage bunkers.

Investigators subsequently determined that no military, commercial or private helicopters known to be based in the area could have been around Wurtsmith at those times. The crew of a KC135 tanker plane, already airborne, spotted the object near the base and attempted to give chase, but couldn't keep up with it.

Several sightings occurred at the Maine air base as well, where objects hovered over the weapons area. Radar and visual sightings were made, and another KC135 was sent aloft to oversee pursuit efforts by a helicopter borrowed from the Maine National Guard — Loring had none of its own.

The object eventually disappeared toward the Canadian border, where Canadian air force jets were on alert. There was no indication whether the Canadian planes spotted the object (Sinclair and Harris 1979, typographical corrections made to the original).

The varied descriptions of the objects suggests that two things could have been occurring, namely that both extraterrestrial UAP and human-made, conventional helicopters could have been present at the AFBs. In other words, the USAF may have 'taken advantage' of a situation in which genuine, extraterrestrial UAP were present to mount its own, helicopter-based 'testing' of certain bases' personnel.

1976: An unidentified, anomalous object was sighted over Iran. This 1976 event is described as follows in the 1979 *Washington Post* article cited above:

Yet another Air Force intelligence report indicated extensive interest in a 1976 incident over Iran, when two Iranian Air Force F4 Phantom fighter planes were scrambled to encounter a brightly lighted object in the skies near Tehran.

That object was tracked by Iranian ground radar, seen independently by the crew of a commercial airliner and pursued by the F4[s], which, according to the report, experienced a breakdown of their electronic communications devices when they neared the object.

The report, compiled by American officials, said that the electronic weapons system of one of the planes went dead when its pilot prepared to fire an AIM9 missile at a smaller object that appeared to roar out from the larger vehicle.

The planes' electronic equipment reportedly became operative after they veered away from the smaller object, which had returned to the larger light, the report said. Iranians described the larger object, with colored, fast-flashing lights, as the size of a Boeing 707 jetliner (Sinclair and Harris 1979).

The 1976 UFO event in Iran also was discussed in another prominent newspaper article, "C.I.A. Papers Detail U.F.O. Surveillance," which appeared in the *New York Times* in 1979:

Among the documents [released by the CIA] are several detailed reports of Air Force attempts to either intercept or destroy U.F.O.'s [*sic*].

In a 1976 incident in Iran, one report says, two F-4 Phantom jet fighter-bombers pursued a large U.F.O. that seemed to send out smaller craft. One of the smaller craft "headed straight toward the F-4 at a very fast rate of speed," the report said. "The pilot attempted to fire an AIM-9 missile at the object but at that instant his weapons control panel went off and he lost all communications." The pilot eluded the craft, then watched as [it] "returned to the primary object for a perfect rejoin," the report continued (NYT 1979, typographical corrections made to the original).

FOIA/Ground Saucer Watch (GSW): Documents relating to the 1975 and 1976 sightings were obtained by means of a FOIA request by Ground Saucer Watch (GSW), as described in the 1979 *Washington Post* article cited above:

The information on the 1975 and 1976 sightings—records from the Air Force and the North American Air Defense Command (NORAD)—was turned over to Ground Saucer Watch (GSW), a Phoenix-based organization that monitors UFO reports.

GSW obtained the information through a Freedom-of-Information request to the Air Force, one of a number it has made to government agencies involved in UFO investigations.

A similar request to the CIA, made both by GSW and the *Washington Post*, resulted in the CIA's turning over almost 900 pages of documents related to its monitoring of UFO reports since the 1950s.

The CIA was directed by a U.S. District Court judge here last year to turn over to Ground Saucer Watch UFO data unrelated to national security.

The agency, according to GSW officials and attorneys, apparently has withheld some UFO records, and GSW says it intends to seek further court action in the case.

The Air Force and other federal military and intelligence agencies have maintained consistently that sightings of unidentified flying objects have logical explanations—that the UFOs are not visitors from another world.

The CIA documents are largely a collection of worldwide intelligence reports, newspaper articles and agency memoranda relating to UFO sightings and theories of extraterrestrial life.

The CIA's position, reiterated yesterday by a spokesman, is that it has had no involvement with UFOs since 1953, when a special study panel [i.e., the Robertson Panel] concluded that they presented no threat to national security.

While memos from as recently as 1977 are included in the 879 pages, the CIA spokesman said the agency continues to be "a passive recipient" of UFO data, even though none of the material is analyzed.

Todd Zechel, a GSW investigator[, a former NSA employee,] and director of another organization, Citizens Against UFO Secrecy, said, "We've had to pry loose every item of information we have. I am inclined to believe the government doesn't know any more about

UFOs than we do, but if UFOs are what they say—nothing—why don't they open their files totally?"

Zechel and William Spaulding, a Phoenix engineer and director of GSW, said that Defense, the Air Force and the National Security Agency (NSA) have refused to turn over certain other information that would shed more light on military encounters with unidentified flying objects (Sinclair and Harris 1979, typographical corrections made to the original).

The 1979 *New York Times* article, cited to above, says this about the GSW topic:

The C.I.A. has repeatedly said that [it] investigated and closed its books on U.F.O.'s [*sic*] during 1952, according to Ground Saucer Watch [GSW], a nation-wide research organization of about 500 scientists, engineers and others who seek to scientifically prove or disprove the existence of U.F.O.'s [*sic*], but 1,000 pages of documents obtained under a freedom of information suit, show "the Government has been lying to us all these years," it [i.e., GSW] said.

"After reviewing the documents, Ground Saucer Watch believes that U.F.O.'s [*sic*] do exist, they are real, the U.S. Government has been totally untruthful and the cover-up is massive," William Spaulding, head of the group, said.

Mr. Spaulding, an aerospace engineer with AiResearch, one of the largest producers of specialized aerospace components, said the documents show that United States embassies are used to help gather information on U.F.O. sightings and that the information "seems to be directed to the C.I.A., the White House and the National Security Agency."

A C.I.A. memo of Aug. 1, 1952, recommends continued agency surveillance of "flying saucers," saying, "It is strongly urged, however, that no indication of C.I.A. interest or concern reach the press or public, in view of their probably alarmist tendencies to accept such interest as 'confirmatory' of the soundness of 'unpublished facts' in the hands of the U.S. Government," the document said. [Note, this memo is quoted above in chapter 4.]

[This section of the article is reproduced above under '1976.']

A major point of concern, a C.I.A. document of Oct. 2, 1952, shows, is that U.F.O. sightings could mask Russian air attacks or "psychological warfare." The report—to the director of Central Intelligence from the assistant director for the Office of Scientific Intelligence—recommends that the National Security Council be advised of the "implications of the 'flying saucer' problem"; that the matter be discussed with the Psychological Strategy Board, and that the C.I.A. help "develop . . . a policy of public information which will minimize concern and possible panic resulting from the numerous sightings of unidentified objects."

[Note, this memo is quoted above in chapter 4.]

A document dated November 1975, directs against acknowledging any pattern in sightings. "Unless there is evidence which links sightings, or unless media queries link sightings, queries can best be handled individually at the source and as questions arise," it said. "Response should be direct, forthright and emphasize that the action taken was in response to an isolated or specific incident."

Mr. Spaulding says the documents show that there are links and patterns in the sightings. From that evidence, he says, he believes U.F.O.'s [sic] are here on surveillance missions.

"We find a concentration of sightings around our military installations, research and development areas," he said. "The U.F.O. phenomenon is following what our own astronauts are doing on other planets—we send a scoutship, we take soil samples and then we land."

Mr. Spaulding said he has sworn statements from retired Air Force colonels that at least two U.F.O.'s [*sic*] have crashlanded and been recovered by the Air Force.

One crash, he said, was in Mexico in 1948 and the other was near Kingman, Ariz., in 1953. He said the retired officers claimed they got a glimpse of dead aliens who were in both cases about four feet tall with silverish complexions and wearing silver outfits that "seemed fused to the body from the heat."

Mr. Spaulding said his group is waiting now for a Federal judge to rule on the last phase of its C.I.A. suit, which seeks access to 57 items that would provide "hard evidence" of U.F.O.'s [*sic*] or "retrievals of the third kind." That evidence includes motion pictures, gun camera film and residue from landings, he said.

Among the films they want is 40 to 48 frames taken in 1952 by Ralph Mayher, then a cameraman for KYW-TV in Cleveland and now a member of Ground Saucer Watch [GSW]. The Air Force borrowed the film in 1957 and has never returned it. The official finding was that the object had been a meteor, Mr. Spaulding said.

"We're past the story-telling stage," Mr. Spaulding said. "We have to have in black and white to satisfy the scientific community. We have to establish the existence of the object to all the people in Missouri and then figure out who's driving it" (NYT 1979).

1980: In August, UAP events occurred at F.E. Warren AFB in Wyoming, at Kirtland AFB in New Mexico, and at Sandia Labs, also in New Mexico (Johnson 2002).

At F.E. Warren AFB in Wyoming, one of the Law Enforcement (LE) personnel said that on multiple occasions when he was on duty as Desk Sergeant, the security personnel at the Weapons Storage Area (WSA) would report lights overhead. These incidents were clustered temporally and are reported to have occurred two or three nights in a row before a break of some number of nights. In certain instances, personnel were called in and told to report to the armory to obtain weapons. Of note, declassified documents from the USAF Office of Special Investigations (OSI) confirm similar reports of UAP occurring in August 1980 near the Manzano WSA outside Kirtland AFB in New Mexico (Hastings 2006).

In December, a well-known series of UAP sightings occurred near the joint U.K.-U.S. Bentwaters RAFB (Royal Air Force Base) in Suffolk, England. During at least one event, an anomalous object was observed directing 'laser-like' beams of light near and/or into the Weapons Storage Area (WSA) (Hastings 2006; see also Hall 2001).

<u>1982:</u> Some people have stated that during at least one incident in the former Soviet Union, at least one ICBM was armed remotely very briefly before it returned to regular functioning (Hastings 2016; PRPS 2021)— though I disagree with description of the event as an "attack" (PRPS 2021).

<u>1991:</u> In October, UAP events occurred at Chernobyl in Ukraine and at Arkhangel'sk Missile Base in Russia (Johnson 2002).

<u>1992:</u> On October 27, UAP were witnessed at Ellsworth AFB in South Dakota. Just before midnight, two Minuteman missile maintenance personnel reported that a vehicle controller and a Minuteman Electro-Mechanical Team technician were approaching the operations hangar of the 44th Field Missile Maintenance Squadron (FMMS). During their approach, they saw a group of UAP moving in formation, or what may have been lights arranged across the surface of a very large craft. The lights/craft moved towards the Minuteman missile maintenance hangar, hovered about 300-500 feet above the ground, then moved away. One witness drew lights that appeared like a string of pearls delineating the presumed boundary of an unseen object, which was shaped like a kidney bean. A second witness drew something similar, but with some lights positioned away from object's edge and with the object shaped more like a boomerang. The second witness stated that the object was approximately 300 feet long and was much larger than an airplane (Hastings 2006).

At Malmstrom AFB one spring morning, an anomalous, bright, white light maneuvering in unusual ways with sudden changes in direction was witnessed at Alpha Flight missile silo A-3 and at silo A-10, which was approximately ten miles away (Hastings 2006).

1995: In the early hours of January 20 at Malmstrom AFB, two security police officers who comprised an Alert Response Team were driving to the India Flight Launch Control Facility when they noticed a strange light in the southern sky. As they passed Minuteman silo I-4, one of the men radioed a missile maintenance team working there and asked its leader whether he also could see the light. The team leader responded affirmatively. Earlier that month, on the evening of January 5, an Air Force officer at Malmstrom called NUFORC (National UFO Reporting Center) and stated that an individual in Shelby, Montana had reported a UFO. Shelby is located near the northern boundary of the Quebec Flight Minuteman missile field, with silo Q-18 situated less than two miles east of town. Another UFO sighting was reported to Malmstrom AFB on January 7. On January 18, an anonymous caller to NUFORC reported multiple UFO sightings between Fairfield and Deer Lodge, Montana. Nearly half of the countryside between Fairfield and Deer Lodge lies within the boundaries of what were at the time Malmstrom's Minuteman missile fields. Fairfield lies near the geographic center of the Hotel Flight Minuteman missile field, with silo H-9 situated at the western edge of town, slightly to the north of Route 408. If one travels southwest from Fairfield in a straight line towards Deer Lodge, one eventually exits the Hotel missile field and crosses directly into the Golf Flight missile field. According to the January 18 report, although no evidence placed the UAP close to specific silos within either Hotel or Golf Flight missile fields, the caller from Deer Lodge nevertheless relayed sighting reports made by persons calling from the middle of so-called Rocket Ranch country. The three sighting reports—from January 5, 7, and 18—were published by NUFORC (Hastings 2006).

1996: According to Tech. Sgt. Jeff Goodrich (USAF Ret.), former Team Chief of Missile Handling, 341st Maintenance Squadron, Malmstrom AFB, on February 2, he and another officer observed a loose formation of five triangular-shaped objects flying above Great Falls, Montana, which is located just west of Malmstrom. At the time of the sighting, both men had been working at the Missile Roll Transfer Building, a remote site located some miles from the main base. To be sighted from that location, the UAP would have passed over the India Flight missile field (Hastings 2006).

1997-2018: Many UAP sightings continued to occur in proximity to nuclear sites, with the descriptions here providing only a representative sample of such events.

2019: In late March and early April, UAP were detected near the U.S. Army's Terminal High Altitude Area Defense (THAAD) battery near the North West Field at Andersen Air Force Base on the Island of Guam. Guam is strategically important in the context of geopolitical dynamics involving China and North Korea, and the THAAD battery is tasked with defending the island from ballistic missile attacks (Rogoway and Trevithick 2020a).

Also in 2019, hundreds of witnesses in Colorado saw up to thirty UAP simultaneously in proximity to an arsenal of nuclear strategic ballistic missile forces (Andresen 2022b, 312).

2018-2021: Three recent reports discuss UAP sightings at Lawrence Livermore National Laboratory in Livermore, California. One recent report discusses a sighting of an unidentified aerial object at Sandia National Laboratories in New Mexico. All four of these reports are sourced from the Protective Force Division at Lawrence Livermore National Laboratory (Rogan 2022).

UAP and Nuclear-powered Vessels

In addition to UAP activity at sites associated with nuclear weapons, much UAP activity also occurs in proximity to nuclear-powered vessels (Andresen 2022b, 308, 313-314, 316).

In November 2004, well-documented UAP activity occurred around the USS *Nimitz* (Beaty 2019 <2018>), an aircraft carrier with two nuclear reactors (Pike 2000). In 2015, UAP activity occurred around the USS *Theodore*

Roosevelt (History n.d.), which also is an aircraft carrier with two nuclear reactors.

UAP activity around nuclear-powered carrier strike groups indicates that ETI is concerned that accidents could occur involving the nuclear technologies propelling these vessels, and, also, that the vessels themselves could be involved in conflict, making their nuclear propulsion systems vulnerable to attack. In fact, nuclear-powered submarines have a long history of accidents. Two U.S. nuclear submarines, USS *Thresher* and USS *Scorpion*, are sitting at the bottom of the Atlantic Ocean where they sank in the 1960s (Keane 2021). More recently, in 2019, many senior Russian naval officers were killed in an accident involving one of Russia's deep-submergence vehicles (DSV). Russia's DSVs may have been conducting covert sea trials of a weapons system comprised of a large caliber nuclear torpedo (Muraviev 2019).

Beginning on July 14, 2019 and continuing over several days, groups of UAP followed USN vessels in the area around California's Channel Islands. The events centered on the *Arleigh Burke* class destroyer USS *Kidd*, the nearby USS *Rafael Peralta*, and the USS *Russell*. As many as six UAP are reported to have swarmed the ships at one time. In the case of the USS *Rafael Peralta*, a white light hovered over the ship's flight deck for more than ninety minutes, which is much longer than the twenty-eight-minute maximum flight time of a Phantom IV drone, for example. One object matched the ship's speed of sixteen knots as it continued to hover over the vessel's helicopter landing pad. For almost three hours on July 15, 2019, significant UAP activity also occurred around the USS *Russell* between San Clemente Island and San Diego, but closer to shore in comparison to the UAP activity that had occurred on July 14 (Kehoe and Cecotti 2021).

UAP and Nuclear Power Plants

In addition to UAP events associated with nuclear weapons sites and nuclear-powered vessels, many UAP sightings also occurs at nuclear power generating stations (Andresen 2022b, 308, 314).

On numerous occasions, UAP have been reported over nuclear power installations (Johnson 2002). In the late 1980s, for example, a security officer at Cooper Nuclear Station reported an unidentified object flying down the Missouri River about 150 feet in the air and hovering in front of the intake. After observing the object for a few minutes, he contacted a second security officer, who also observed it. The next evening, the first security guard observed the object return when he again was posted at the intake. It moved into and hovered over the protected area north of the reactor building. The object was silent, triangular, and had a circle of rotating lights on the bottom. It was approximately one-third the size of the reactor building. On the second night, the security guard called the security break room, so most of the officers on shift also observed the object. After hovering for a few minutes, the object exited the protected area and went back up the river to the north, as it had done the previous night (The Black Vault 2017, 6, 21).

Two recent series of UAP events at nuclear power generating stations occurred in 2019 and 2022, in the U.S. and Sweden, respectively. Whereas the media describe unidentified anomalous phenomena near nuclear weapons facilities and nuclear-powered vessels as "UAP," it uses the term "drones" to describe the objects involved in these two series of events — even though the accounts strongly suggest that human-made drones were not involved. The first series of events began on September 29, 2019 and continued successive nights. During this period, a swarm of 'drones' appeared at the Palo Verde Nuclear Generating Station near Tonopah, Arizona. Palo Verde is the most powerful nuclear plant in the U.S. (Rogoway and Trevithick 2020b). This is not an isolated series of events, either. Nuclear Regulatory Commission (NRC) documents report that between 2015 and 2019, approximately sixty 'drone' sightings occurred at twenty-four different nuclear power facilities in the U.S. (Tingley 2022a; Trevithick 2022).

In Sweden, so-called drone sightings were reported over active nuclear power facilities in Forsmark, Oskarshamn, and Ringhals (Mossige-Norheim and Strömberg 2022). According to the Swedish Police Authority, on January 14, 2022, 'drones' were seen over Forsmark, which is the largest single energy-producing facility in Sweden and, also, over Oskarshamn (Tingley 2022a). 'Drones' also were reported at Ringhals and over the Barsebäck Nuclear Power Plant. The Barsebäck site was decommissioned in 2015 and is scheduled to be demolished by 2028. In addition, 'drones' were reported over government buildings in the Swedish capital of Stockholm, including Stockholm Palace, the official residence of Sweden's King Carl XVI Gustaf. One unmanned aerial craft also was reported over the Swedish Riksdag, or legislature. 'Drones' also were reported near Kiruna Airport, which is north of the polar circle and near Luleå Airport, near the Gulf of Bothnia separating Sweden from Finland (Trevithick 2022).

Not only have UAP been detected in the vicinity of nuclear power plants, but evidence also exists that UAP have taken action to stabilize dangerous situations after at least two nuclear accidents. During the disaster at Chernobyl on April 26, 1986, about three hours after the initial explosion and during the height of the fire, technicians reported that they observed a fiery sphere, similar in color to brass, within one thousand feet of the damaged Unit 4 reactor. Two bright red rays are reported to have shot out from the object in the direction of the reactor. The object hovered in the area for about three minutes, after which the rays vanished and the object moved slowly away to the northwest. Radiation levels taken immediately before the object appeared were 3,000 milliroentgens/hour, but they declined to 800 milliroentgens/hour after the UFO directed its rays on the damaged reactor. Many people therefore conclude that the object reduced the radiation level (Stonehill 1998, 68-69; MWL 2020). Numerous incident reports collated from the Ukrainian National UFO Database suggest that UAP continue to monitor the Chernobyl site.

Witness accounts also suggest that an unidentified anomalous object took actions to stabilize the situation after the Fukushima Daiichi Nuclear Power Plant melted down in the aftermath of the March 2011 earthquake in Japan (Ancient Aliens 2014). During the earthquake, many people saw orb-type UAP coming out of the water near the Fukushima plant. UAP also may continue to monitor the site (Narayanan 2019).

UAP and Volcanos

In addition to UAP activity at nuclear sites, UAP activity also has been reported for years in the vicinity of volcanoes (Andresen 2022b, 315-16). For example, a recent video shows a UAP at the La Palma volcano at Cumbre Vieja (Old Summit) on the island of Tenerife in Spain's Canary Islands (iceage2012 2021). As is well known, ash from volcanic eruptions impacts Earth in devastating ways.

The possibility that a volcanic eruption could damage or even destroy a nuclear reactor is a horrific potential risk. In fact, a scenario involving a volcanic eruption destroying a nuclear reactor cannot be ruled out anywhere. For example, certain areas of the world such as Japan have numerous active volcanoes. Around 110 of the approximately 1,500 active volcanoes in the world are in the so-called Ring of Fire, which includes Japan within it (Jozuka 2015).

As of March 2021, ten years after the Fukushima disaster—and despite the number of active volcanoes in the region—residents of the area have agreed that five nuclear power plants with a total of nine reactors can resume operations. This is unfortunate, since in addition to the risks associated with potential accidents, security also may be lax at some of Japan's nuclear power plants. For example, in September 2020, an employee identification card was used in an unauthorized manner to gain access to the central control room at the Kashiwazaki-Kariwa nuclear power plant in Japan (nippon 2021).

ETI's Concern about the Nuclear Situation

The nexus between nuclear technologies and ETI/UAP activity is clearly documented. The idea that ETI may take an active role in mitigating disasters associated with nuclear power plants, including disasters caused by earthquakes, and potentially by volcanic eruptions, is imminently plausible given observations of UAP at Chernobyl and Fukushima. Although it makes sense that an intelligent, extraterrestrial species would help humankind in this way, it is wrong to count on ETI/UAP here. It is our responsibility as human beings to address and resolve the precariousness of the situation involving nuclear technologies on Earth.

ETI/UAP activity occurs near military installations and/or vessels, especially those involving nuclear technologies, and around nuclear power plants because ETI is trying to bring our attention to the existential risk we are posing to ourselves, to other species, and to the environment by means of nuclear technologies. Even though humankind is a Type 0 civilization according to the Kardeshev Scale, humankind's understanding of nuclear fission means that it has at its disposal enough energy to self-annihilate. Earth is rigged and ready to explode, with nuclear weapons pointed by nation-states at a vast array of 'strategic targets' across the planet. Five factors make the situation even worse: the high degree of fragmentation in human society; the persistence of dominance hierarchies in which only a few individuals have ultimate control over whether to launch nuclear weapons or not; geopolitical competition; dangerous military practices; and incessant media manipulation involving disingenuous rhetoric that is intended to help an elite few consolidate wealth, power, and control in society.

ETI's UAP routinely appear at many different nuclear sites in a way that demonstrates that ETI is assessing the risk that nuclear weapons and other nuclear technologies pose to humankind more accurately than human beings are (Andresen 2022b, 304-16). Furthermore, ETI seems much more concerned about the wellbeing of *Homo sapiens sapiens* than many human beings are themselves. ETI's concern for humankind is clear from the actions of its UAP in monitoring human nuclear capabilities and taking store of humankind's nuclear arsenals without taking hostile actions. In addition, many people who witness UAP around nuclear installations come away with the impression that ETI is sending a clear message to humankind—'What are you humans doing with nuclear weapons?' Many witnesses feel that ETI is communicating to human beings how dangerous nuclear weapons really are and is sending a message to people that they should not possess let alone use nuclear weapons and other nuclear technologies (Ridge 2016 <2003>; Hastings 2016; Mazzola 2021; Ancient Aliens 2019).

We know that UAP can disappear both from the visual and infrared spectrums at will, and that they also can disappear from radar. Yet, ETI goes out of its way to make sure that many of its UAP appearances around nuclear sites and nuclear-powered vessels are seen and/or otherwise detected. ETI clearly is drawing human attention to these sites and to the precariousness of the situation associated with them. Literally and metaphorically, ETI is shining a light on the danger associated with nuclear technologies. Even more, by switching nuclear missiles on and off, ETI clearly is indicating that nuclear weapons launch protocols *are hackable from a distance*. The only sane response to that reality is to abolish all nuclear weapons before a nefarious human actor and/or nation-state learns to hack these weapons systems. ETI's message is clear—humankind must abolish nuclear weapons and other nuclear technologies for its own wellbeing and for the wellbeing of the whole (Andresen 2022b, 312, 316, 319).

Considerable evidence also supports the conclusion that ETI does not want any weapons—particularly nuclear weapons—in space. In 1964, an unidentified anomalous object fired multiple beams of light and shot down a nuclear missile with dummy warheads after it was launched at Vandenberg Air Force Base [now, Vandenberg Space Force Base] in California (Kaleka 2013; Hastings 2016; Mazzola 2017; see also Mazzola 2021; seniorsam 2015). Dr. Robert Jacobs has described this event in detail, including to AARO (Boswell 2023). A seemingly surreal resonance exists between the event in 1964 and an event at Vandenberg Air Force Base in 2022, when a test rocket carrying a component for a future U.S. nuclear-armed ICBM blew up eleven seconds after takeoff (Starr 2022).

Observations of UAP traveling slowing very close to the Earth's surface and then disappearing almost instantaneously also may relate to the nuclear issue. UAP may be conducting detailed mapping of specific terrain and/or the subsurface, especially in proximity to installations associated with nuclear technologies. ETI may want to ensure that it could render nuclear weapons at these installations inoperable if human beings took steps to obliterate itself, destroy other species, and devastate the environment. UAP also could be sampling soil to assess and, possibly, to remediate radiation and other environmental damage that already has resulted from prior nuclear detonations and from uranium mining and plutonium processing—and, also, in conjunction with prior nuclear power plant disasters such as those at Chernobyl and Fukushima.

ETI appears sensitive to how quickly human beings can assimilate its presence and is treading as lightly as it can while simultaneously drawing attention to the existentially dangerous situation humankind has created for itself because of nuclear weapons and other nuclear technologies, and as a result of the existence of advanced weapons systems more generally. This requires very careful balancing, since increasing information regarding ETI/UAP too quickly could trigger precisely what ETI is seeking to avoid, namely overreaction in the form of the launching of a nuclear weapon. One need only consider how much of an overreaction the U.S. displayed in shooting down four aerial objects in February 2023 (Evans and Wright 2023) to realize how subtle this balance really is for ETI.

In a fragile, interconnected, global society such as our own, the smallest perturbation in one aspect of the system can have a massive impact on many other aspects of the system. Whereas open acknowledgement of the existence of an extraterrestrial presence can occur now, more transparency on UAP propulsion and/or materiel must wait for a calmer day. The current geopolitical situation on Earth simply is too tightly strung. Releasing information on systems that can utilize almost unfathomably high amounts of energy will remain unwise until nuclear weapons and other nuclear technologies are abolished and until geopolitical hostilities have transformed into geopolitical cooperation and genuine goodwill.

The frequent appearance of UAP around sites associated with nuclear weapons technologies cycles back to the initial development of these technologies. Oppenheimer used Bohm's mathematics to develop the atomic bomb, and now it is by means of Bohm's interpretation of the quantum theory that one gains real insight into ETI/UAP. This recalls Bohm's (1986a, 94) observation that similar forms resonate together in the implicate order. By understanding this dynamic more generally—not merely a single instance of it—I am confident that human beings can change all sorts of things for the better. But this will require understanding resonance between local and global aspects of existence, the cognitive-emotional content of metaphor, and the nature of time, among many other profound issues.

Nuclear Weapons and the War in Ukraine

Only four days into the war in Ukraine, Putin made his first threat of escalation by implying that he would use tactical nuclear weapons if NATO became heavily involved in supporting Ukraine. The most recent iteration of this threat is Moscow's announcement that it will forward base Russian nuclear weapons in Belarus, primarily nuclear-armed versions of the Iskander missile which plans have being placed close to Belarus's western border with NATO states. In addition, Russia will train Belarusian pilots to fly planes capable of carrying nuclear weapons. This development in Russia's war strategy is occurring as Ukrainian forces are set to launch a major offensive that could take them very close to the border with Crimea. Many experts are expressing concern that losing de facto control of Crimea would be a defeat too far for Putin, and that this risk could push him to decide to use one or more nuclear weapons in Ukraine (Rogers 2023b).

Without doubt, the current nuclear crisis with respect to Ukraine is the most dangerous nuclear-related situation humankind has experienced since the 1962 Cuban Missile Crisis. But even though the nuclear crisis could worsen in Ukraine suddenly at any time in the coming months, people in most Western countries do not seem to be registering the situation as a major issue of concern. This arguably could be considered the biggest mistake of the post-Cold War era. The harsh reality is that a nuclear weapon could be detonated in Ukraine unless something changes soon.

While Putin and others in Russia have made statements indicating that the use of tactical nuclear weapons is an option in the war in Ukraine, what many people do not realize is that for over five decades, NATO's own policy permits the alliance to use nuclear weapons first if the alliance is losing a conventional war (Rogers 2023b). Indeed, Putin's threats that he may use nuclear weapons if Russia is losing the war in Ukraine very much parallel NATO's policy of "flexible response," which dates to 1968. According to this policy, NATO states that it will consider using nuclear weapons if the alliance is losing a conventional war (Rogers 2023a). Indeed, in many ways, the statement made by Russian Security Council Deputy Chairman Dmitry Medvedev on February 22, 2023, that "Russia will defend itself by any means, including with nuclear weapons" (TASS 2023c), is little more than a restatement of NATO's own flexible response policy, which also permits the first use of nuclear weapons under certain circumstances. Even though it remains an element NATO nuclear strategy and applies to the alliance as a whole and, also, to its individual members, NATO's policy of flexible response is not widely understood. While many people think that relatively small-scale use of nuclear weapons is not a part of war planning, this is inaccurate. NATO continues to see the value of using nuclear weapons first if a conventional assault by an opponent threatens the alliance, in the same way that NATO's nuclear-armed members see that value when acting individually (Rogers 2023a).

NATO's policy of flexible response originally was codified in final form on January 16, 1968, in document MC14/3, *Overall Strategic Concept for the Defence of the NATO Area (NATO 1968). The policy states that if* NATO were losing a conventional war with the Soviet bloc, the alliance would be prepared to use nuclear weapons first to prevent defeat. Although the Soviets previously claimed that they had a No First Use (NFU) policy, people in NATO did not take this position seriously given that the former Soviet Union was in possession of an enormous range of tactical nuclear weapons. By the early 1970s, flexible response had become clearly established under NATO's Nuclear Operations Plan, which outlines two levels with respect to the use of tactical nuclear weapons against Soviet forces—selective options, and general response. Selective options include various plans, many of which assume first use of nuclear weapons against Warsaw Pact conventional forces. At the lowest level of deployment, for example, this potentially could include up to five, small air-burst nuclear detonations intended as warning shots to demonstrate NATO's intent (Rogers 2023a).

From a conceptual point of view, NATO views first use of nuclear weapons more from the perspective of winning a war as opposed merely to compensating for a perceived, conventional imbalance in force. For example, a U.K. government presentation of NATO nuclear policy demonstrates the political nature of the objective of maintaining the capability of selective, sub-strategic use of weapons of theater warfare. The goal is to demonstrate in advance that NATO has the capability and will to use nuclear weapons in a deliberate, politically controlled way in order to restore deterrence "by inducing the aggressor to make the decision to terminate his aggression and withdraw" (Rogers 2023a). One sees this same intent expressed in NATO's April 11, 2023 statement regarding nuclear deterrence policies and forces (NATO 2023).

To be sure, NATO's posture with respect to first use of nuclear weapons is backed by capability. As mentioned above, the U.S. bases tactical nuclear weapons at six bases in five NATO countries (Belgium, Germany, Italy, the Netherlands, and Turkey), and it maintains aircraft, training, and support facilities at these bases. Individual NATO members also can deploy nuclear forces. During the 1982 Falklands War (Guerra de las Malvinas in Spanish), for example, the U.K., which is a prominent NATO member, deployed task force ships with nuclear bombs. Credible reports also state that a strategic missile submarine was sent on patrol in the mid-Atlantic within missile range of Argentina. (Rogers 2023a; see also Rogers 1996). In fact, the U.K. apparently sent one of its nuclear-armed Polaris missile submarines carrying sixteen ballistic missiles with thermonuclear warheads to the Falkland Islands, the territorial sovereignty of which the U.K. and Argentina continue to dispute (Norton-Taylor 2023).

At a practical level, as the war in Ukraine continues, the U.S. has been upgrading the nuclear storage facilities at Lakenheath, which is its main air base in the U.K. Now, after a fifteen-year lapse (Rogers 2023b), the USAF's move to upgrade its old nuclear support facilities means that air-delivered tactical nuclear weapons may be deployed at Lakenheath once again (Rogers 2023a). Interestingly, as mentioned above in chapter 3, Lakenheath was the site of an important UFO report from August 1956 (Thayer 1971).

Upgrading Lakenheath indicates that the U.K. government's stance today is similar to the one it held in 1982, namely something more aggressive than the intent of achieving relatively stable deterrence in accordance with the principle of 'mutually assured destruction' (MAD) (Rogers 2023b). The MAD policy was declared fully in the early 1960, primarily by U.S. Secretary of Defense Robert McNamara. A decade earlier, by the early 1950s, when the former Soviet Union also had emerged as a nuclear power, the public thought that a balance of terror would bring stability and that nobody would be the first to attack for fear of destruction. However, the corollary was that if a nuclear-armed state did use nuclear weapons, then its opponent would do likewise, leading to a major nuclear exchange and global devastation. Governments emphasized the stability of MAD, which influenced public opinion sufficiently in nuclear-armed states such that limited nuclear wars have not been discussed often at a broad, societal level. But such discussions have been relatively common in military and security circles, where war-fighting has been embedded in nuclear planning. For example, throughout the 1950s, the U.S. and the Soviet Union expanded the sizes of their nuclear arsenals while they also developed a range of nuclear weapons, many of which has specific purposes within domains of war-fighting. In addition to massive H-bombs, these included nuclear-tipped torpedoes, surface-to-air missiles, anti-submarine nuclear depth bombs, land mines, short-range battlefield missiles and mortars, and nuclear-capable artillery. The public's assumption that MAD was working as a deterrence principle seemed nowhere in sight. Instead, within military circles, the extent of planning for limited nuclear use—which extended nuclear planning to NATO for nuclear war in Europe—was discussed in some open source security journals. For example, by 1982, the U.S. had a portfolio of 40,000 nuclear targets with multiple levels of operation, from limited wars to a full nuclear exchange (Rogers 2023a, citing to Ball 1983 regarding U.S. targeting in 1982).

By 'upgrading' Lakenheath, the U.S. is strengthening its links to the U.K.—which is the one NATO country where until recently some openness existed with respect to the country's nuclear posture of seventy years. One of the founding members of NATO and the third state to develop nuclear weapons after the U.S. and Russia, the U.K. recognized the perceived usability of nuclear weapons immediately after they were developed, with the U.K. successfully testing its own first atomic bomb in October 1952. By the 1950s, certainly, planners in the U.K. envisaged nuclear use against an opponent with massive conventional forces but which had yet to acquire nuclear weapons. Not long after U.K. armed forces had started to deploy tactical nuclear bombs, then Defence Minister and later Prime Minister, Harold Macmillan, stated in the House of Commons:

> [T]he power of interdiction upon invading columns by nuclear weapons gives a new aspect altogether to strategy, both in the Middle East and in the Far East. It affords a breathing space, an interval, a short but perhaps vital opportunity for the assembly, during the battle for air supremacy, of larger conventional forces than can normally be stationed in those areas (Rogers 2023a, citing to JD 2002).

As implied by the quote above, under certain circumstances, the U.K. would consider using nuclear weapons in response to a non-nuclear attack involving chemical or biological weapons—and it also would consider using nuclear weapons to pre-empt such an attack. While there was some willingness within the U.K. from the early 1990s until 2010 to decrease the size of its nuclear arsenal, by early 2021, former U.K. Prime Minister Boris Johnson's government announced an increase in the overall size of the U.K.'s nuclear arsenal. It also announced a revised U.K. nuclear posture that included the potential use of nuclear arms against non-nuclear weapons states judged to be "in material breach of [their] non-proliferation obligations" (Rogers 2023a).

As Kate Hudson, General Secretary of the Campaign for Nuclear Disarmament (CND), stated in 2021, the announcement of Johnson's government that the U.K. was increasing the size of its nuclear arsenal and was prepared to use nuclear weapons against a non-nuclear state was tantamount to tearing up the NPT. The illegal move would have contravened the U.K.'s legal obligations under the NPT, which the U.K. ratified in 1970. They obligations require that countries in possession of nuclear weapons must disarm and that those countries that do not possess nuclear weapons do not acquire them. Even earlier, when former Prime Minister Tony Blair's government first pursued the replacement of the Trident system in 2005, Matrix Chambers (barristers' chambers) proffered a legal opinion finding that replacing Trident would be a material breach of the NPT, since the NPT requires that each of the parties to it pursues negotiations in good faith on effective measures to end the nuclear arms race and on nuclear disarmament (Hudson 2021b).

Meanwhile, rhetorically, the U.K. asserts that it is committed to the NPT, which makes its actual nuclear posture at any one time all the more opaque (Hudson 2021b). For example, the U.K.'s Integrated Review of 2021 states, "We are strongly committed to full implementation of the NPT in all its aspects, including nuclear disarmament" (GOV.UK 2021). Furthermore, polling in the U.K. shows that a clear majority of people support nuclear disarmament. Although there have been attempts to build a narrative in the U.K. that some countries are 'responsible' and can be trusted with nuclear weapons while others are not (Hudson 2021a), the reality of high-level moves within the U.K. suggest otherwise. Despite the Review's discussion of rules-based order and the country's prominent role in global diplomacy, Johnson was willing to forego the U.K.'s legal obligations under the NPT and to ratchet up global tensions, presumably to reinforce the notion of a so-called global Britain as a notable force in the world. The U.K.'s decision under Johnson to increase its nuclear arsenal helps one understand why countries, especially those of the Global South, have lost trust in the NPT process, especially since any breach of the NPT could encourage other countries to pursue nuclear weapons. Decisions such as Johnson's also call into question the role of the U.K. and what it stands for as a nation. As Hudson states, "Rearming with weapons of mass destruction is not something that we can accept" (Hudson 2021b).

Today, the war in Ukraine has reached a violent stalemate. NATO perceives itself so heavily invested in the war that it cannot imagine allowing Ukraine to be defeated should Russia regain the initiative. Meanwhile, if Ukraine pushes Russian forces back significantly, especially in Crimea, the risk of Putin escalating the war to a new level almost certainly would increase. This is no ordinary state-on-state conflict, since the intensifying NATO involvement in the war means, de facto, that the actual conflict is between nuclear-armed Russia and the NATO nuclear-armed alliance. The fact that both have policies of nuclear first use should concern everyone, until the point that the situation is resolved.

Furthermore, while Russia has struggled to maintain its conventional forces since the end of the Cold War, it has invested heavily in maintaining and developing its nuclear forces. This means that nuclear weapons figure significantly in Russian military thinking, since Russia it is with respect to the nuclear category of power that Russia can claim to be a superpower. Recognizing the very real risk of a nuclear crisis over Ukraine, on January 24, 2023, those who run the *Bulletin of Atomic Scientists* moved the Doomsday Clock to 90 seconds to midnight, the closest it has been in its 75-year history. The world needs to take Putin's threat of tactical nuclear use very seriously instead of downplaying it or even dismissing it out of hand (Rogers 2023a).

Time for Unilateral Restraint

Cycles of violence underpinned by thousands of nuclear and other devastating weapons systems do not make human beings safer. They simply amplify the precariousness of life, placing humankind, all species on Earth, and Earth itself, at great risk. Weapons will never end societal dysfunction. Only a correct ontology and correspondingly better worldview can supplant maladaptive thinking.

In the case of nuclear weapons, deterrence, arms control, and often both together have proven insufficient to resolve our current crisis. Instead of acting sanely, humankind is going in precisely the opposite direction. The U.S. is 'modernizing' its nuclear arsenal in a destabilizing manner, China is increasing the size of its nuclear arsenal, and Russia is doing everything it can—including suspending its participation in New START (Chernova et al. 2023)—to terrorize the world by implying that it is considering the use of nuclear weapons in the war in Ukraine. Meanwhile, those who finance the nuclear weapons industry and the defense contractors who manufacture nuclear weapons continue to profit, even though their actions make global society unstable and threaten the longevity of our species. Life is precarious enough to begin with, and amplifying this precariousness with nuclear weapons and inherently unstable nuclear power plants is completely insane.

Time has run out for updating our deterrence theory and planning multi-nation forums for norm building. We need a much more immediate and much more deliberate solution. The path forward is opposite to the neoconservative path the U.S. embarked upon at the beginning of this century and which both parties have followed unblinkingly since. What we need now is unilateral restraint. The U.S. needs to stop modernizing its nuclear arsenal and it needs to start dismantling its nuclear weapons — regardless of what Russia, China, or North Korea say they will or will not do in response. Because unilateral restraint is deontological rather than strategic in nature, this is the correct, next step from a moral point of view. It also may be the only current option with sufficient moral power to cut through the defensive aggressivity of people suffering from emotional and spiritual impasses and maladaptive repetition compulsions who continue to insist on leading humankind towards extinction.

The United States should lead the process of unilateral restraint, since nuclear weapons were first developed in the U.S., and the U.S. is the only country that has used nuclear weapons on the battlefield. The leaders of other countries may take some time to adjust — and we should monitor this carefully to make sure they do. If they do not, we can reassess. But in the meantime, we must try.

Of course, the leaders of any country — including Russia — could break free of their own paranoia and take a strong, moral stand by demonstrating unilateral restraint with respect to nuclear weapons. World history twists and turns in unusual ways. Aśoka, the last major emperor of the Mauryan dynasty of ancient India, who brutally conquered the state of Kalinga, an independent feudal kingdom on the East Coast of the country, converted to Buddhism partly because of his remorse regarding his own ruthless actions. People do change.

The point is that people simply must come to their senses, and it does not matter who comes to his or her senses first. Someone — *anyone* — simply needs to stand up and lead from a place of moral correctness rather than from a position of militaristic strategizing. Otherwise, if we do not abolish nuclear weapons and find an alternative to nuclear energy, *then it is almost certain that a nuclear weapon will be used in conflict, or that a nuclear accident will occur at a nuclear power plant because of a natural disaster or because a nuclear power plant is targeted during conflict.* It is mutually assured insanity.

Human beings are smarter than this. Now we simply must show it.

Chapter 8:
Next Steps

Confirm the Existence of ETI

It is time for the U.S. Government (USG)—or any government with sufficient evidence to do so—to confirm that there is an ETI currently operating UAP in our midst on and around Earth. Such a confirmatory statement regarding the existence of an ETI can be brief yet official. For example, the Office of the Director of National Intelligence (ODNI) in the U.S. could issue such a statement in a document that it could title an *Interim Assessment*, using language in keeping with its *Preliminary Assessment* from June 25, 2021. Or ODNI simply could state on its website (dni.gov) that an extraterrestrial intelligence (ETI) is operating some of the UAP being witnessed and otherwise detected on and around Earth. With such an admission, the USG also can state straightforwardly that while the technology exhibited by certain UAP is beyond the range of current human capabilities, extraterrestrial UAP have not taken aggression actions towards humankind in the almost eighty years the USG has known about them.

ODNI's *2022 Annual Report* indicates—albeit very indirectly—that humankind should prepare itself for the extraordinary acknowledgement that an advanced ETI is operating in our midst. While such a statement may seem extraordinary from one vantage point, it could not be more natural when considering the enormity of the universe in which we live. ETI's presence in our midst on and around Earth promises significant changes in human society. New forms of knowledge will provide us with the opportunity we need to spread our wings as intelligent beings while remaining mindful of the importance of contributing constructively and peacefully to the whole.

A statement from ODNI that some UAP are extraterrestrial in origin should be made expeditiously, because people need time to assimilate this information. Again, if the U.S. is unwilling to take this step, other countries such as Canada, China, or Russia could be first to confirm the existence of an extraterrestrial presence in our midst. All three of these countries, and perhaps others, have sufficient evidence to support the claim, and any country with such evidence can issue a statement first.

The USG or any government with sufficient evidence to do so also should release a set of 50-100 still, high-resolution images of extraterrestrial UAP. Such a set of images should show an array of different UAP and should be large enough to demonstrate that certain UAP are extraterrestrial in origin but not so large that a major nation-state is unwilling to release it.

Issue Apology for Decades of Disinformation

The USG should take the lead in issuing a sincere apology to the public for propagating decades of disinformation with respect to UAP. Even a concise apology would go a long way in healing the rift that currently exists between the public and government on this topic. Such an apology could be issued online at an appropriate website, for example The White House website in the U.S. (whitehouse.gov).

Keep Information on UAP Propulsion and/or Materiel Classified

I do not support the U.S. declassifying information it may possess on UAP propulsion and/or materiel. If human beings mature ethically and spiritually—with spiritual maturity being interpreted as coming to a correct ontological view—and if great power competition wanes and the threats of plutocracy, kleptocracy, authoritarianism, and totalitarianism are eliminated, that would be the right time for classified information on UAP propulsion and/or materiel to be shared. Now, however, the downside risk of sharing such information is too great, because such information theoretically could be weaponized by an autocratic regime, an extremist group, or even by an apocalyptically minded individual. Humankind simply must experience an ethical and spiritual awakening before information specifically relating to UAP propulsion and/or materiel can be shared more widely in a manner that is safe.

Elsewhere (Andresen 2022b, 294-95), I distinguish the concept of *openness* from the notion of *transparency*. My argument is simple. Right now, we can and should be quite open about the existence of an ETI operating on and around Earth. But blanket transparency on UAP propulsion and/or material is premature. This is not because I like secrecy—far from it—or because I think that people who currently have access to such information are particularly enlightened—again, far from it. It is because I think that the amount of energy to which UAP have access is so immense that if the information ever made it into the hands of despots, tyrants, and/or terrorists, it could lead to the obliteration of humankind, other species, and potentially the entire planetary system of Earth.

The amount of energy to which UAP have access *is being significantly underestimated by many people.* Human beings simply are not ready yet to handle this much energy—many people cannot even handle their own impulses, guns, and nuclear weapons responsibly, let alone the amount of energy to which ETI/UAP have access.

Name It Like It Is

Although DoD's new office name, All-domain Anomaly Resolution Office (AARO), is better than the prior one, Airborne Object Identification & Management Synchronization Group (AOIMSG), it is important to separate study of human-made UAP that nevertheless may have usual features and/or represent breakthrough yet human-made technology from the study of UAP that are extraterrestrial in origin. Furthermore, the study of extraterrestrial UAP needs to be removed from the purview of DOD.

One idea that can be considered is creating an Extraterrestrial Research Office (ERO) within the Department of Education (ED). Such an office could focus on education regarding and constructive communication with ETI removed from a militarized context. Human beings must learn to acculturate with ETI in a peaceful, constructive, and creative manner, without viewing the human-ETI relationship through a national security lens. Regardless of species, we all participate together in this incredible universe, which is our common home.

Begin Simple, Consistent Communication Signaling to ETI/UAP

Communication is key to almost everything, including constructive and creative acculturation with ETI. If we can learn to communicate constructively with ETI/UAP for the betterment of the whole, we can learn to communicate constructively with one another for the same reason.

My view is that the ETI in our midst now is so sophisticated that any form of signalling will work as a method of communication, as long as it is done with kind, genuine, and non-violent intentions. Making simple, human-initiated attempts to communicate will help de-mystify the ETI/UAP phenomenon and will encourage all human beings to develop a constructive relationship with ETI.

Blinking lights and music may be a good place to start with human-initiated communication with ETI. One can establish a very simple and consistent set of phrases correlated with a pattern of blinking lights, for example: two blue blinking lights repeated at four-second intervals means 'Hello'; three white blinking lights repeated a four-second intervals means 'We have no hostile intent towards you, ETI/UAP'; etc. Engaging in a sincere yet simple way would help usher in more complex communication later. Fancy quantum signalling system is not required here—remember, this is *communication*, not a competition to demonstrate who has the most sophisticated technology. A few lights that blink according a consistent pattern that correlates with simple and kind words and phrases will do.

After some communicative reciprocity has been established, new patterns of blinking lights, meaning additional things, can be added to the mix— maybe even to the tune of some gentle, harmonious, instrumental (i.e., nonvocal) music. I suggest Debussy's Rêverie, one of the most elegant pieces of piano music that exists—at least among human composers. Anyone can try reaching out to ETI. What I am underscoring here with this very light approach is that it is time to stop thinking like political and military beings and to start thinking like kind and creative beings—

especially where extraterrestrials are involved. If we also can manage this with respect to other humans, that truly would be revolutionary.

Introduce Curricula on ETI/UAP into the Educational System

Curricula on ETI/UAP are needed at all levels of the educational system, starting with elementary school, and going all the way up to the university level. Adult education courses on this topic and classes in local houses of worship and community centers also will be very helpful. Education is a very important step in preparing people for widespread and creative acculturation with ETI.

In all classes at all levels, extraterrestrial life should be presented matter-of-factly, so individuals do not feel afraid and instead can come into a creative relationship with ETI themselves.

Cultivate Deontological Thinking and Ethical and Moral Action

Homo sapiens sapiens will not survive much longer unless we move from strategic and tactical thinking on existence to a deontological one. Unless we progress beyond an adversarial mindset, we will go extinct because of the combination of advanced weapons systems, AI, and ruthless actors who attempt to maximize conventional gains for themselves without demonstrating respect for the preciousness of the whole.

The issue in front of us now is that the war in Ukraine has the potential to lead to something even more serious, with repercussions for the very survival of humankind. Given the large number of UAP reports that have been made by many credible witnesses since the start of the conflict in Ukraine, there are strong indications that ETI is observing what is occurring there. We should take note of this and use it to see beyond ourselves as human beings so we stop eradicating one another in one violent conflict after another across the globe.

Any planning to 'unite the world against ETI' should cease, since it is an altogether irrelevant idea. ETI/UAP pose no threat to us, or to anyone. ETI understands the whole nature of reality and is working for the good. Instead, we must move beyond repetitive habits according to which many people spend their entire lives strategizing based on greed for territory, dehumanization of the 'other,' and will to power. We must find a way past all that.

By means of the same dynamic discussed earlier in the book according to which similar forms cohere and resonate together in implicate order (Bohm 1986a, 94), militarism creates patterns that perpetuate themselves. To break free of militarism and to move beyond the societal mess we are in, including by means of ending needless suffering and immoral killing, we must see beyond ourselves by breaking free of the focus on 'self' and purported 'self-interest.' This is the cause of people hurting one another, and, also, hurting themselves in the process.

Creating a vision that encourages people to think about existence in a manner that includes perceiving implicate, etc., levels of order, finding our way to intraspecies peace, and establishing peaceful relations with the ETI in our midst now are important actions to pursue/ Human beings are not separate islands unto themselves. We are all embedded in a coherent, whole, existence in which everything is deeply interpenetrative with everything else. We therefore must become more familiar with and cultivate perception and understanding of subtle aspects of reality, such as implicate, etc., levels of order. This will bring individuals and our species the peace that it so desperately needs, and it will help us address the myriads of challenges currently in front of us.

Lead with Compassion

We must lead with compassion and love, not on the basis of short-term, self-interested strategizing. Moving from a strategic and tactical mindset and utilitarian ethical framework to an ontological mindset and deontological ethical framework is feasible, and it will result in the mitigation of suffering, the promotion of social justice, and contribution to the wellbeing of the whole.

Now is never too late.

References

116th United States Congress. 2020. "Intelligence Authorization Act for Fiscal Year 2021." U.S. Senate Select Committee on Intelligence, June 17, 2020. https://www.intelligence.senate.gov/publications/intelligence-authorization-act-fiscal-year-2021.

Aborigines Protection Society. 1837. *Report of the Parliamentary Select Committee on Aboriginal Tribes, (British Settlements.)* Reprinted, with Comments, by the "Aborigines Protection Society." London: Published for the Society, by William Ball, Aldine Chambers, Paternoster Row, and Hatchard & Son, Piccadilly. https://apo.org.au/sites/default/files/resource-files/1837-02/apo-nid61306.pdf.

AcqNotes. 2022. "Technology Development. Independent Research & Development." AcqNotes. https://acqnotes.com/acqnote/tasks/indepen-dent research-development.

AFP (Agence France Presse). "Russia Will Use International Space Station 'Until 2028.'" *Barron's*, April 12, 2023. https://www.barrons.com/news/russia-will-use-international-space-station-until-2028-fdb0b535.

Aharonov, Yakir, and Lev Vaidman. 1996. "About Position Measurements Which Do Not Show the Bohmian Particle Position." In *Bohmian Mechanics and Quantum Theory: An Appraisal*, edited by James T. Cushing, Arthur Fine, and Sheldon Goldstein. Boston Studies in the Philosophy of Science 184: 141-54. Dordrecht: Kluwer Academic Publishers.

Air Force. n.d. "Brigadier General George Francis Schulgen." Air Force. https://www.af.mil/About-Us/Biographies/Display/Article/2995977/george-francis-schulgen/.

Air Force NWC (Air Force Nuclear Weapons Center). n.d. "Sentinel ICBM." Air Force Nuclear Weapons Center. https://www.afnwc.af.mil/Weapon-Systems/Sentinel-ICBM-LGM-35A/.

Alcubierre, Miguel. 1994. "The warp drive: hyper-fast travel within general relativity." *Classical and Quantum Gravity* 11: L73–77. https://doi.org/10.1088/0264-9381/11/5/001.

Aldrich, Jan. 1998. "Investigating the Ghose Rockets." *International UFO Reporter* 23, no. 4 (Winter): 9-14.

Almar, Ivan. 1995. "The Consequences of a Discovery: Different Scenarios." In *Progress in the Search for Extraterrestrial Life: 1993 Bioastronomy Symposium*, edited by G. Seth Shostak. ASP (Astronomical Society of the Pacific) Conference Series, Volume 74. Santa Cruz, California, August 16-20, 1993. San Francisco: Astronomical Society of the Pacific. http://www.aspbooks.org/ publications/74/499.pdf.

Ambarzumian, Viktor Amazaspovich [Ambartsumian]. 1972. "First Soviet-American Conference on Communication with Extraterrestrial Intelligence (CETI)." *Icarus* 16, no. 2 (April): 412-14.

Ambrose, Soren. 2015. "Opinion: The World Has Reached Peak Plutocracy." Our World: Brought to you by the United Nations University, April 28, 2015. https://ourworld.unu.edu/en/opinion-the-world-has-reached-peak-plutocracy.

An, Daniel, Krzysztof A. Meissner, Pawel Nurowski, and Roger Penrose. 2022. "Apparent evidence for Hawking points in the CMB Sky." arXiv, submitted August 6, 2018 (v1), last revised August 9, 2022 (v5). https://arxiv.org /abs/1808.01740.

Anand, Nupur, Anirban Sen, David French, and Isla Binnie. 2023. "How JPMorgam's Dimon won the First Republic deal." *Reuters*, May 2, 2023. https://www.reuters.com/business/finance/how-jpmorgans-dimon-won-first-republic-deal-2023-05-02/.

Ancient Aliens. 2014. "Alien Encounters." Season 8, Episode 4 of *Ancient Aliens*. History Channel, August 15, 2014. https://play.history.com/shows/ancient-aliens /season-8/episode-4.

———. 2019. "The Nuclear Agenda." Season 14, Episode 14 of *Ancient Aliens*. History Channel, September 6, 2019. https://play.history.com/shows/ancient-aliens /season-14/episode-14.

Andresen, Jensine. 2022a. "Cartographies of Knowledge and Academic Maps." In *Extraterrestrial Intelligence: Academic and Societal Implications*, edited by Jensine Andresen and Octavio A. Chon Torres, 2-6. Newcastle upon Tyne, U.K.: Cambridge Scholars Publishing.

———. 1999. "Crisis and Kuhn." In *Catching up with the Vision: Essays on the Occasion of the 75th Anniversary of the Founding of the History of Science Society*. Supplement, *Isis (75th Anniversary Issue)* 90 Supplement: S43-S67.

———. In preparation. *Extraterrestrial Mind*.

———. 2022b. "Mind of the Matter, Matter of the Mind." In *Extraterrestrial Intelligence: Academic and Societal Implications*, edited by Jensine Andresen and Octavio A. Chon Torres, 281-330. Newcastle upon Tyne, U.K.: Cambridge Scholars Publishing.

———. Forthcoming. *Safe Space*. Turin, Italy: Tuthi.

———. 2021. "Two Elephants in the Room of Astrobiology." In *Astrobiology: Science, Ethics, and Public Policy*, edited by Octavio A. Chon Torres, Ted Peters, Joseph Seckbach, and Richard Gordon, 193-231. Hoboken, NJ/Beverly MA, Wiley/Scrivener.

Andresen, Jensine, and Octavio A. Chon Torres, editors. 2022. *Extraterrestrial Intelligence: Academic and Societal Implications*. Newcastle upon Tyne, U.K.: Cambridge Scholars Publishing.

AP (The Associated Press). 2022. "Putin says Russia could adopt US preemptive strike concept." *ABC News*, December 9, 2022. https://abcnews.go.com/International/wireStory/putin-russia-adopt-us-preemptive-strike-concept-94876110.

———. 2023. "Putin says Russia will station tactical nuclear weapons in Belarus." *NPR (National Public Radio)*, March 25, 2023. https://www.npr.org/2023/03/25/1166089485/putin-russia-tactical-nuclear-weapons-belarus.

Arkin, William M., and Marc Ambinder. 2022. "As Russia-Ukraine Tensions Rise, U.S. 'Stress Tests' New Nuclear War Plan." *Newsweek*, January 29, 2022. https://www.newsweek.com/russia-ukraine-tensions-rise-us-stress-tests-new-nuclear-war-plan-1674324.

Arndt, Anna Clara, and Liviu Horovitz. 2022. *Nuclear rhetoric and escalation management in Russia's war against Ukraine: A Chronology*. Research Division International Security I WP NR. 03, September 2022. Stiftung Wissenschaft und Politik. German Institute for International and Security Affairs. https://www.swp-berlin.org/publications/products/arbeitspapiere/Arndt-Horovitz_Working-Paper_Nuclear_rhetoric_and_escalation_management_in_Russia_s_war_against_Ukraine.pdf.

Arquivo Nacional. 2018. "Conheça o fundo sobre OVNIs do Arquivo Nacional." Arquivo Nacional. https://www.gov.br/arquivonacional/pt-br/canais_atendimento/imprensa/noticias/conheca-o-fundo-sobre-ovnis-do-arquivo-nacional.

Arts, Steven A. 2003. *Mystery Airships in the Sky*. Frederick, MD: America Star Books.

Baker, David. 2017. *Nuclear Weapons: 1945 onwards (strategic and tactical delivery systems. Operations Manual.* Sparkford, Yeovil, Somerset, U.K.: Haynes Publishing.

Ball, Desmond. 1983. "Targeting for Strategic Deterrence." *The Adelphi Papers* [currently, *Adelphi Series*] 23, no. 185: 1-3. International Institute for Strategic Studies [Published online as "Targeting for strategic deterrence: Introduction," May 2, 2008. https://www.tandfonline.com/doi/abs/10.1080/05679328308457441 ?journalCode=tadl19.]

Ballester Olmos, Vincente-Juan. 1976. *A Catalogue of 200 Type-I UFO Events in Spain and Portugal.* Center for UFO Studies, April 1976. http://cufos.org/books/ Catalogue_of_200_Type_I_UFO_Events_in_Spain_and_Portugal.pdf.

Ballester Olmos, Vincente-Juan, and Ole Jonny Brænne 2007. "NORWAY FOTOCAT." Academia.edu. https://www.academia.edu/92870951/Norway_ FOTOCAT_spreadsheet.

Basset, Laurent, director. 2003. *UFOTV Presents: The Secret – Evidence That We Are Not Alone.* https://www.amazon.com/UFOTV-Presents-Secret-Evidence-Alone/dp/B00NYFSBD6. [A 2007 version of this video is available at https://www.microsoft.com/en-us/p/ufotv-presents-the-secret-evidence-we-are-not-alone/8d6kgwzl5wl4?activetab=pivot%3aoverviewtab.]

Basterfield, Keith. 2022. "The BAASS AAWSAP CAPELLA data warehouse resurfaces." Unidentified Aerial Phenomena – scientific research, May 8, 2022. https://ufos-scientificresearch.blogspot.com/2022/05/the-baass-aawsap-capella-data-warehouse.html.

———. 2018. "Bigelow Aerospace Advanced Space Studies (BAASS) and Vallee's 'Capella' Project." Unidentified Aerial Phenomena – scientific research, November 8, 2018. https://ufos-scientificresearch.blogspot.com/2018/11/bigelow-aerospace-advanced-space_8.html.

BBC (BBC News). 2010a. "Brazil air force to record UFO sightings." *BBC News*, August 12, 2010. https://www.bbc.com/news/world-latin-america-10947856.

———. 2002. "Boeing tries to defy gravity." *BBC News*, July 29, 2002. http://news.bbc.co.uk/2/hi/science/nature/2157975.stm.

———. 2010b. "New Zealand releases government files." *BBC News*, December 22, 2010. https://www.bbc.com/news/world-asia-pacific-12057314.

Beaty, David C., director. 2019 <2018>. "The Nimitz Encounters." May 26, 2019. https://www.youtube.com/watch?v=PRgoisHRmUE. [Also uploaded in 2018.]

Beaujon, Andrew. 2019. "A New Exhibit at the National Archives Explores Government UFO Documents." *Washingtonian*, November 21, 2019. https://www.washingtonian.com/2019/11/21/a-new-exhibit-at-the-national-archives-explores-government-ufo-documents/.

Beck, Lewis White. 1971-1972. "Extraterrestrial Intelligent Life." *Proceedings and Addresses of the American Philosophical Association* 45: 5-21. [Reprinted in: Regis, Jr., Edward, editor. 1985. *Extraterrestrials: Science and alien intelligence*, edited by Edward Regis, Jr., 3-18. Cambridge, U.K.: Cambridge University Press.]

Beenes, Maaike. 2019. *Beyond the bomb: Global exclusion of nuclear weapons producers.* October 1, 2019. Utrecht: ICAN and PAX. https://www.dontbankonthebomb .com/wp-content/uploads/2019/10/201910_Beyond-the-bomb_final.pdf.

Beenes, Maaike, and Susi Snyder. 2018. *Don't Bank on the Bomb: A Global Report on the Financing of Nuclear Weapons Producers.* March 1, 2018. Utrecht: PAX and ICAN. https://www.dontbankonthebomb.com/wp-content/uploads/2018/10/2018_Report_web.pdf.

Belfield, Haydn. 2023. "There's no libertarian approach to preventing the end of the world." *Vox*, March 7, 2023. https://www.vox.com/future-perfect/2023/3/7/23618766/peter-thiel-existential-risk-oxford-union-silicon-valley-technology-artficial-intelligence.

Belinfante, Frederik J. 1973. *A Survey of Hidden-Variables Theories.* International Series of Monographs in Natural Philosophy. Oxford: Pergamon Press.

Bell, Elizabeth A., Patrick Boehnke, T. Mark Harrison, et al. 2015. "Potentially biogenic carbon preserved in a 4.1 billion-year-old zircon." *Proceedings of the National Academy of Sciences (PNAS)* 112, no. 47: 14518-14521. https://www.pnas.org/doi/10.1073/pnas.1517557112.

Bell, John S. 1966. "On the Problem of Hidden Variables in Quantum Mechanics." *Reviews of Modern Physics* 38, no. 3: 447-52. https://journals.aps.org/rmp/abstract/10.1103/RevModPhys.38.447.

———. 1987. *Speakable and Unspeakable in Quantum Mechanics.* Cambridge, U.K.: Cambridge University Press.

Bendett, Samuel. 2023. "Rogozin Sending Russian Uncrewed Ground Vehicles to Ukraine—And Does It Matter?" Modern War Institute at West Point, February 10, 2023. https://mwi.usma.edu/bureaucrats-gambit-why-is-dmitry-rogozin-sending-russian-uncrewed-ground-vehicles-to-ukraine-and-does-it-matter/.

Bennett, Jay. 2020. "Here's How Much Deadlier Today's Nukes Are Compared to WWII A-Bombs." *Popular Mechanics*, December 13, 2020. https://www. popularmechanics.com/military/a23306/nuclear-bombs-powerful-today/.

Berendzen, Richard, editor. 1973. *Life beyond Earth and the Mind of Man*. NASA SP-328. Washington, D.C.: National Aeronautics and Space Administration (NASA).

Berger, Eric. 2022. "Dmitry Rogozin may be in some trouble in Russia." *Ars Technica*, December 5, 2022. https://arstechnica.com/science/2022/12/dmitry-rogozin-may -be-in-some-trouble-in-russia/.

Berliner, Don. 1976. "The Ghose Rockets of Sweden." *Official UFO* 1, no. 11 (October): 30-31, 60-64.

Bertrand, Natasha, and Tim Lister. 2023. "US warns Russia not to touch American nuclear technology at Ukrainian nuclear plant." *CNN*, April 19, 2023. https://www.cnn.com/2023/04/18/politics/us-warns-russia-zapori-zhzhia-nuclear-plant/index.html.

BFD (*Black Files Declassified*). 2020a. "American UFOs." Season 1, Episode 2 of *Black Files Declassified*. Science Channel, April 9, 2020. https://www. sciencechannel.com/tv-shows/black-files-declassified/.

———. 2020b. "Secrets of the Space Force." Season 1, Episode 1 of *Black Files Declassified*. Science Channel, April 2, 2020. https://www.science channel.com/tv-shows/black-files-declassified/.

———. 2020c. "To Catch an Alien." Season 1, Episode 3 of *Black Files Declassified*. Science Channel, April 16, 2020. https://www.sciencechannel.com/tv-shows/ black-files-declassified/.

Billingham, John. 2014. "SETI: The NASA Years." *In Archaeology, Anthropology, and Interstellar Communication*, edited by Douglas A. Vakoch, 1-21. The NASA History Series, NASA SP-2013-4413. Washington, D.C.: National Aeronautics and Space Administration (NASA), Office of Communications, Public Outreach Division, History Program Office. https://www.nasa.gov/sites/default/files/ Archaeology_Anthropology_and_Interstellar_Communication_TAGGED.pdf.

Billingham, John, Roger Heyns, David Milne, and G. Seth Shostak. 1994. *Social Implications of Detecting an Extraterrestrial Civilization: A Report of the Workshop on the Cultural Aspects of SETI*. Mountain View, CA: SETI Institute.

The Black Vault. 2017. "Response to Freedom of Information Act (FOIA) Request. FOIA 2017-0368." The Black Vault, March 23, 2017. https:// documents. theblackvault.com/documents/nrc/FOIA-2017-0368-NRC-UFO.pdf.

———. 2001. "Report of Meetings of Scientific Advisory Panel on Unidentified Flying Objects Convened by Office of Scientific Intelligence, CIA, January 14-18, 1953." The Black Vault, August 7, 2001. https://documents.theblackvault.com/ documents/ufos/robertsonpanelreport.pdf.

The Black Vault Originals. 2023a. "November 11, 2014, Chilean Navy Helicopter UFO Encounter Video Released." February 23, 2023. https://www. youtube.com/watch?v=fk00f3Q_2rE.

———. 2023b. "Highly Classified NRO System Detects Possible 'Tic-Tac' Object in 2021." March 9, 2023. https://www.youtube.com/watch?v=dFclfz837Dc&t=11s.

Blanchette, Jimmy. 2023. "Extraterrestrial Reveal The Keys of The Universe,Yielding A New Physics Capable of Transforming Human Civilization." February 3, 2023. https://www.youtube.com/watch?v=utyDfwTCepA&t=2120s.

Bloomberg. 2023. "Credit Suisse Chairman Lehmann: Entering a 'Multipolar World.'" *Bloomberg*, January 17, 2023. https://www.bloomberg.com/news/ videos/2023-01-17/credit-suisse-chairman-lehmann-entering-a-multipolar-world.

Bohm David. 1984 <1957>. *Causality and Chance in Modern Physics*. London: Routledge & Kegan Paul.

———. 1953. "Comments on a Letter Concerning the Causal Interpretation of Quantum Theory." *Physical Review* 89: 319-20.

———. 1987. "Hidden variables and the implicate order." In *Quantum Implications: Essays in Honour of David Bohm*, edited by Basil J. Hiley and F. David Peat, 33-45. London and New York: Routledge & Kegan Paul.

———. 1983. "Hidden Variables and the Implicate Order." Paper presented at a conference entitled 'David Bohm's Implicate Order: Physics and Theology,' sponsored by the Center for Theology and the Natural Sciences (CTNS), Berkeley, California, April 22-23, 1983. [First published June 1985 in *Zygon: Journal of Religion & Science* 20, no. 2 (June): 111-24; later reproduced in 1987, *Quantum Implications: Essays in Honour of David Bohm*, edited by B. J. Hiley and F. D. Peat, 33-45 (see 43 for quoted material). London: Routledge.]

———. 1986a. "Creativity: the signature of nature." In *Dialogues with Scientists and Sages: The Search for Unity*, edited by Renée Weber, 91-104. London and New York: Routledge & Kegan Paul.

———. 1986b. "The implicate order and the super-implicate order." In *Dialogues with Scientists and Sages: The Search for Unity*, edited by Renée Weber, 23-49. London and New York: Routledge & Kegan Paul.

———. 1989. "Meaning and Information." In *The Search for Meaning: The New Spirit in Science and Philosophy*, edited by Paavo Pylkkänen, 43-62. Wellingborough, UK: Crucible. [Page numbers refer to the PDF of Bohm's chapter available at https://www.implicity.org/Downloads/Bohm_meaning+information.pdf.]

———. 1990. "A new theory of the relationship of mind and matter." *Philosophical Psychology* 3, no. 2: 271-86. https://dl.icdst.org/pdfs/files/05aaa28b931575ca2 e0baeadc5ab0d2c.pdf.

———. 1952a. "A Suggested Interpretation of the Quantum Theory in Terms of 'Hidden' Variables. I." *Physical Review* 85, no. 2 (January 15): 166-79. https://quantum.country/assets4/Bohm1952.pdf.

———. 1952b. "A Suggested Interpretation of the Quantum Theory in Terms of 'Hidden' Variables. II." *Physical Review* 85, no. 2 (January 15): 180-93. https://cqi.inf.usi.ch/qic/bohm2.pdf.

———. 1951. *Quantum Theory*. Englewood Cliffs, NJ: Prentice-Hall, Inc.

———. 2005 <1980>. *Wholeness and the Implicate Order*. Taylor and Francis e-Library. [First published 1980 by London and New York: Routledge & Kegan Paul.]

Bohm, David, and Basil J. Hiley. 1982. "The de Broglie Pilot Wave Theory and the Further Development of New Insights Arising Out of It." *Foundations of Physics* 12, no. 10: 1001-16.

———. 1975. "On the intuitive understanding of nonlocality as implied by quantum theory." *Foundations of Physics* 5: 93-109. https://link.springer.com/article/ 10.1007/BF01100319.

———. 1987. "An ontological basis for the quantum theory." *Physics Reports* 144: 323-48.

———. 1993. *The Undivided Universe: An Ontological Interpretation of Quantum Theory*. London and New York: Routledge.

Bohm, David, Basil J. Hiley, and Panayiotis N. Kaloyerou. 1987. "An Ontological Basis for the Quantum Theory. 1. Nonrelativistic Particle Systems. 2. A Causal Interpretation of Quantum Fields." *Physics Reports* 144, no. 6: 323-75.

Bohm, David, and Rupert Sheldrake. 1986. "Matter as a meaning field." In *Dialogues with Scientists and Sages: The Search for Unity*, edited by Renée Weber, 105-26. London and New York: Routledge & Kegan Paul.

Bohr, Niels. 1934. *Atomic Theory and the Description of Nature*. Cambridge, U.K.: Cambridge University Press.

———. 1958. *Atomic Theory and Human Knowledge*. New York: Wiley.

———. 1935. "Can Quantum-Mechanical Description of Physical Reality be Considered Complete?" *Physical Review* 48: 696-702. https://journals.aps.org/pr/abstract/10.1103/PhysRev.48.696.

———. 1949. "Discussions with Einstein on Epistemological Problems in Atomic Physics." In *Albert Einstein: Philosopher-Scientist*, edited by Paul Arthur Schilpp, 199-241. Cambridge, U.K.: Cambridge University Press. https://www.marxists.org/reference/subject/philosophy/works/dk/bohr.htm. [This article is also found in: Schilpp, Paul Arthur, editor. *Albert Einstein: Philosopher-Scientist*. Volume 7 of The Library of Living Philosophers. Evanston, IL: Open Court.]

Boswell, Josh. 2023. "TWO Air Force vets have testified to Pentagon's UFO office about seeing mysterious objects TURN OFF ten nuclear warheads and blast test missiles out of the sky at US bases." *Daily Mail*, February 21, 2023. https://www.dailymail.co.uk/news/article-11776067/Air-Force-vets-testified-witnessing-UFOs-TURN-nuclear-warheads.html.

Bowen, Matt, and Paul Dabbar. 2022. "Reducing Russian Involvement in Western Nuclear Power Markets." Columbia | SIPA Center on Global Energy Policy, May 23, 2022. https://www.energypolicy.columbia.edu/research/commentary/reducing-russian-involvement-western-nuclear-power-markets.

Bowman, Emma, and James Doubek. 2023. "The U.S. military shot down an unidentified object over Canada's Yukon territory." *NPR (National Public Radio)*, February 11, 2023. https://www.npr.org/2023/02/11/1156347424/us-military-shot-down-unidentified-object-canada.

Boylan, Richard J. 1996. "Effects on Human Consciousness and Spirituality of the Upcoming Announcement of UFO Reality." drboylan.com. https://www.drboylan.com/effectsc2.html.

Boyle, Alan. 2016. "How aliens and Apollo astronaut Edgar Mitchell got tangled up in WikiLeaks emails." *GeekWire*, October 10, 2016. https://www.geekwire.com/2016/aliens-apollo-astronaut-edgar-mitchell-wikileaks/.

Bradsher, Keith, and Clifford Krauss. 2022. "China Is Burning More Coal, a Growing Climate Challenge." *The New York Times*, November 3, 2022. https://www.nytimes.com/2022/11/03/business/energy-environment/china-coal-natural-gas.html.

Brady, Niall. 2021. "Ireland's state fund drops nuclear holdings." *The Sunday Times*, June 27, 2021. https://www.thetimes.co.uk/article/state-fund-drops-nuclear-holdings-ireland-h50whb8mh.

Brennan, John O. 2021. "The Global Beacon – Special Guest Ex CIA Director John O. Brennan." Interview by Ret. General Wesley K. Clark. Global Beacon, June 24, 2021. https://podcasts.apple.com/us/podcast/the-global-beacon-special-guest-ex-cia-director-john/id1546100412?i=1000526882668.

———. 2020. "John O. Brennan on Life in the CIA: What working in intelligence has taught him about human nature." Episode 111 of *Conversations with Tyler Cowen*, December 16, 2020. https://conversationswithtyler.com/episodes/john-o-brennan/. [The video version of this interview is available at https://www.youtube.com/watch?v=LdQ7L0ugZJc.]

———. 1993. "Requested Information on UFOs." Memorandum for Richard Warshaw, Executive Assistant, DCI, September 30, 1993 (as cited in Haines 1997 <1995>, fn. 90 on 83).

Breslo, Jim. 2020. "SPACE: USS Princeton Radar Tech Kevin Day: Fighter Tr…" Season 5, Episode 13 of *SPACE RACE*. Hidden Truth Show with Jim Breslo, January 27, 2020. https://hubhopper.com/episode/s5e13-space-uss-princeton-radar-tech-kevin-day-fighter-training-exercises-were-cancelled-due-to-tic-tac-ufo-encounter-1580134592.

Brown, Melvin R., and Basil J. Hiley. 2000. "Schrödinger revisited: An algebraic approach." arXiv, submitted May 4, 2000 (v1), last revised July 19, 2004. https://arxiv.org/abs/quant-ph/0005026. [PDF from February 1, 2008 is available at https://arxiv.org/pdf/quant-ph/0005026.pdf.]

Browne, Ed. 2022. "Dmitry Rogozin: Aliens Could Have Visited Earth, Russia Investigating UFOs." *Newsweek*, June 13, 2022. https://www.newsweek .com/russia-space-agency-roscosmos-dmitry-rogozin-ufos-aliens-1715162.

Bruhn, Gerhard W. n.d. "Remarks on Burkhard Heim's IGW Successors J. Hauser and W. Droescher and their Theory." Website of Gerhard W. Bruhn, Department of Mathematics, Darmstadt University of Technology. https://www2. mathematik.tu-darmstadt.de/~bruhn/IGW.html.

Bryson, Steve, Michelle Kunimoto, Ravi K. Kopparapu, et al. 2020. "The Occurrence of Rocky Habitable Zone Planets Around Solar-Like Stars from Kepler Data." arXiv, submitted October 28, 2020, last revised November 3, 2020 (v2). https://arxiv.org/abs/2010.14812.

BUFOF (Brazilian UFO Files). 2015. "Brazilian UFO Files." Internet Archive, December 27, 2015. https://archive.org/details/BrazilianUFOFiles/1968%20 Envelope%2006%20Conteu%CC%81do%201968_ENVELOPE_06_CONTEUDO /page/n7/mode/2up.

Bugos, Shannon. 2022. "U.S. Nuclear Modernization Programs." Arms Control Association, January 2022. https://www.armscontrol.org/factsheets/USNuclear Modernization.

Burris, Sarah K. 2019. "Security officials told unknown object near Congress was 'hovering' – no one has any idea what happened." *Raw Story*, November 26, 2019. https://www.rawstory.com/2019/11/security-officials-told-unknown-object-near-congress-was-hovering-no-one-has-any-idea-what-happened/.

Business Wire. 2023. "Bank of America, Citigroup, JPMorgan Chase, Wells Fargo, Goldman Sachs, Morgan Stanley, BNY Mellon, PNC Bank, State Street, Truist and U.S. Bank to Make Uninsured Deposits Totaling $30 Billion Into First Republic Bank." *Business Wire*, March 16, 2023. https://www.businesswire .com/news/home/20230316005695/en/Bank-of-America-Citigroup-JPMorgan -Chase-Wells-Fargo-Goldman-Sachs-Morgan-Stanley-BNY-Mellon-PNC-Bank-State-Street-Truist-and-U.S.-Bank-to-Make-Uninsured-Deposits-Totaling-30-Billion-Into-First-Republic-Bank.

Carlson John B., and Peter A. Sturrock, 1975a. "Stanford Workshop on Extraterrestrial Civilization: Opening a New Scientific Dialog." *Aeronautics and Astronautics* 13, no. 6 (June): 63-64.

———. 1975b. "Stanford Workshop on Extraterrestrial Civilization: Opening a New Scientific Dialog." *Origins of Life* 6, no. 3: 459-70. https://link.springer.com/ article/10.1007/BF01130352.

Carlson, Peter. 2002. "50 Years Ago, Unidentified Flying Objects From Way Beyond the Beltway Seized the Capital's Imagination." *The Washington Post*, July 21, 2002. https://www.washingtonpost.com/archive/lifestyle/2002/07/21/50-years-ago-unidentified-flying-objects-from-way-beyond-the-beltway-seized-the-capitals-imagination/59f74156-51f4-4204-96df-e12be061d3f8/.

Carpenter, Joel. n.d.-a. "Green Fireball Chronology." PROJECT 1947. http://www.project1947.com/gfb/gfbchron.html.

———. n.d.-b. "Green Fireballs Over Los Alamos." PROJECT 1947. http://www.project1947.com/gfb/gfbintro.html#gr.

———. n.d.-c. "Guided Missiles and UFOs." PROJECT 1947.
    Part One: http://www.project1947.com/gr/grchron1.htm.
    Part Two: http://www.project1947.com/gr/grchron2.htm.
    Part Three: http://www.project1947.com/gr/grchron3.htm.
    Part Four: http://www.project1947.com/gr/grchron4.htm.

CDSE (Center for Development of Security Excellence). n.d. "Student Guide. Course: Special Access Program (SAP) Overview." Center for Development of Security Excellence: Defense Counterintelligence and Security Agency. https://www.cdse.edu/Portals/124/Documents/student -guides/SA001-guide.pdf.

———. 2021. *Special Access Program Security Annual Refresher*. Student Guide." Center for Development of Security Excellence. https://www. cdse.edu/Portals/124/Documents/student-guides/SA002-guide.pdf.

CEFAA (Comité de Estudios de Fenómenos Aéreos Anómalos). n.d. Comité de Estudios de Fenómenos Aéreos Anómalos. http://www.cefaa.gob.cl/.

CGTN (China Global Television Network). 2023. "Full text: China's Position on Political Settlement of Ukraine Crisis." *CGTN (China Global Television Network)*, February 24, 2023. https://news.cgtn.com/news/2023-02-24/Full-text-China-s-Position-on-Political-Settlement-of-Ukraine-Crisis-1hG2dcPYSNW/index.html.

Chan, Queenie Hoi Shan. 2018. "An Ingredient For Life In Our Solar System: Salt." Science Friday, January 12, 2018. https://www.sciencefriday.com/segments/a-dash-of-salt-to-go-with-your-solar-system/.

Chan, Queenie Hoi Shan, Michael E. Zolensky, Yoko Kebukawa, Marc Fries, Motoo Ito, Andrew Steele, Zia Rahman, Aiko Nakato, A. L. David Kilcoyne, Hiroki Suga, Yoshio Takahashi, Yasuo Takeichi, and Kazuhiko Mase. 2018. "Organic matter in extraterrestrial water-bearing salt crystals." *Science Advances* 4, no. 1 (January 10). https://advances.sciencemag.org/content/advances/4/1/eaao3521 .full.pdf.

Chapman, Lizette. 2023. "Thiel's Founders Fund Withdrew Millions From Silicon Valley Bank." *Bloomberg*, March 10, 2023. https://www.bloomberg.com/news/ articles/2023-03-11/thiel-s-founders-fund-withdrew-millions-from-silicon-valley-bank?leadSource=uverify%20wall.

Chernova, Anna, Nathan Hodge, and Lauren Kent. 2023. "Russia is suspending its participation in New START nuclear weapons treaty, Putin says." *CNN*, February 21, 2023. https://www.cnn.com/europe/live-news/russia-ukraine-war-news-2-21-23/h_363e991bf29355db991746ae8750ddb1.

Chester, Keith. 2007. *Strange Company: Military Encounters with UFOs in World II*. San Antonio, TX/New York, NY: Anomalist Books.

Chronicle Herald. 2020. "Russian cosmonaut captures possible UFO footage." *The Chronicle Herald*, August 24, 2020. https://www.pressreader.com/canada/the-chronicle-herald-metro/20200824/281625307680122.

CLW (Council for a Livable World). 2022. Council for a Livable World. https://livableworld.org/.

CNA (CNA [Center for Naval Analyses] Russia Studies Program). 2020. *Foundations of State Policy of the Russian Federation in the Area of Nuclear Deterrence*. CNA [Center for Naval Analyses] Russia Studies Program, June 2020. https://www.cna.org/reports/2020/06/Foundations%20of%20State%20Policy%2 0of%20the%20Russian%20Federation%20in%20the%20Area%20of%20Nuclear %20Deterrence.pdf.

cnes (Centre National d'Études Spatiales). 2006. "LE GEIPAN OUVRE SES DOSSIERS." Centre National d'Études Spatiales (cnes). https://cnes.fr/fr/web CNES-fr/5847-le-geipan-ouvre-ses-dossiers.php.

CNN (CNN Editorial Research). 2021. "CIA Directors Fast Facts." *CNN*, September 24, 2021. https://www.cnn.com/2013/11/12/us/cia-directors-fast-facts/index.html.

CNN Opinion. 2022. "Opinion: The most likely nuclear scenario." *CNN Opinion*, September 28, 2022. https://www.cnn.com/2022/09/28/opinions/how-close-putin-nuclear-war-de-bretton-gordon/index.html.

Cobb, Wendy Whitman. 2023. "What Is a UFO? The U.S. Shot down Three Mysterious Objects as Interest and Concern Increase Over Unidentified Craft." *Government Executive*, February 17, 2023. https://www.govexec.com/oversight

/2023/02/what-ufo-us-shot-down-three-mysterious-objects-interest-and-concern-increase-over-unidentified-craft/383036/.

COBEPS (Comité belge d'étude des phénomènes spatiaux). 2021. "Catalogue des Observations Belges: 979 - août 1989." Comité belge d'étude des phénomènes spatiaux. https://datastudio.google.com/u/0/reporting/7e5a715a-212b-473d-8491-992f94b87605/page/6kSMC?s=l6cZ8UAgv8U.

Cofield, Calla. 2017. "600-Year-Old Starlight Bolsters Einstein's 'Spooky Acton at a Distance.'" *Space.com*, February 13, 2017. https://www.space.com/35676-einstein-spooky-action-starlight-quantum-entanglement.html.

Cohen, Zachary. 2021. "US intel and satellite images show Saudi Arabia is now building its own ballistic missiles with help of China." *CNN*, December 23, 2021. https://www.cnn.com/2021/12/23/politics/saudi-ballistic-missiles-china/index.html.

Cohen, Zachary, Kristin Wilson, Noah Gray, Rene Marsh, and Barbara Starr. 2019. "'Slow-moving blob' that may have been a flock of birds caused White House lockdown." *CNN*, November 26, 2019. https://edition.cnn.com/2019/11/26/politics/white-house-lockdown-airspace/index.html.

Cole, Reyes. 2019. "The Myths of Traditional Warfare: How Our Peer and Near-Peer Adversaries Plan to Fight Using Irregular Warfare." *Small Wars Journal*, March 28, 2019. https://smallwarsjournal.com/jrnl/art/myths-traditional-warfare-how-our-peer-and-near-peer-adversaries-plan-fight-using.

Collyns, Dan. 2013. "Peru's UFO investigations office to be reopened." *The Guardian*, October 27, 2013. https://www.theguardian.com/world/2013/oct/27/peru-ufo-investigations-office-reopening.

Committee (Committee on Science and Astronautics). 1968. Hearings before the Committee on Science and Astronautics, U.S. House of Representatives, Ninetieth Congress, Second Session, July 29, 1968 [No. 7]. https://babel.hathitrust.org/cgi/pt?id=uc1.$b654608&view=1up&seq=5.

Conte, Michael. 2020. "Pentagon officially releases UFO videos." *CNN*, April 29, 2020. https://edition.cnn.com/2020/04/27/politics/pentagon-ufo-videos/index.html.

Cooper, L. Gordon. 1978. "Letter to Ambassador Griffith, Mission of Grenada to the United Nations, 866 Second Avenue, Suite 502, New York, NY 10017." https://home.ssl.berkeley.edu/forum_thread.php?id=61318.

Copp, Tara. 2022. "US Military 'Furiously' Rewriting Nuclear Deterrence to Address Russia and China, STRATCOM Chief Says." *Defense One*, August 11, 2022.   https://www.defenseone.com/threats/2022/08/us-military-furiously-rewriting-nuclear-deterrence-address-russia-and-china-stratcom-chief-says/375725/.

———. 2023. "A look at the uranium-based ammo the UK will send to Ukraine." *Associated Press (AP) News*, March 23, 2023. https://apnews.com/article/depleted-uranium-ukraine-russia-tanks-a92a4784dfcbd1ff221813154b7f3a8e.

Cordesman, Anthony H. 2005. *Proliferation of Weapons of Mass Destruction in the Middle East: The Impact on The Regional Military Balance*. Working Draft: Revised March 25, 2005. Washington, D.C.: https://csis-website-prod.s3.amazonaws.com/s3fs-public/legacy_files/files/media/csis/pubs/050325_proliferation%5B1%5D.pdf. [Page number refers to the PDF available at this link.]

COSETI (The Columbus Optical SETI Observatory). 2015 <1996>. "The Columbus Optical SETI Observatory." The Columbus Optical SETI Observatory, first uploaded April 7, 1996, last updated November 22, 20215. http://www.coseti.org/.

Cowen, Trace William. 2020. "Classified UFO Documents Could Be Released Within 180 Days Thanks to COVID-19 Relief Bill." *Complex,* December 30, 2020. https://www.complex.com/life/2020/12/classified-ufo-documents-released-180-guests-covid-19-bill.

Cox, Billy. 2009. "Memory-metal files are missing." *Herald-Tribune*, May 21, 2009. https://www.heraldtribune.com/story/news/2009/05/21/memory-metal-files-are-missing/28868099007/.

Credit Suisse. 2014. "Multipolar World to Blame for Geopolitical Tensions?" Credit Suisse, August 29, 2014. https://www.youtube.com/watch?v=vE6a1GbWy0o.

———. 2022. "Zoltan Pozsar." Credit Suisse. https://www.credit-suisse.com/microsites/americas/global-trading-forum/en/speaker-bios/zoltan-pozsar.html.

Creighton, Jolene. 2014. "The Kardashev Scale – Type I, II, III, IV & V Civilization." Futurism, July 19, 2014. https://futurism.com/.

Creitz, Charles. 2019. "UFO investigations expert claims group could have physical evidence to make 'some sort of definitive conclusion.'" *Fox News*, October 4, 2019. https://www.foxnews.com/media/suspected-ufo-material-pentagon-official.

Crerar, Pippa. 2023. "West prepares for Putin to use 'whatever tools he's got left' in Ukraine." *The Guardian*, April 18, 2023. https://www.theguardian

.com/world/2023/apr/18/west-prepares-for-putin-to-use-whatever-tools-hes-got-left-in-ukraine.

Crowe, Michael J. 1986. *The Extraterrestrial Life Debate, 1750-1900: The Idea of a Plurality of Worlds from Kant to Lowell*. Cambridge, U.K.: Cambridge University Press.

CRS (Congressional Research Service). 2022a. *Department of Defense Directed Energy Weapons: Background and Issues for Congress*. Updated September 13, 2022. Washington, D.C.: Congressional Research Service. https://sgp.fas.org/crs/weapons/R46925.pdf.

———. 2022b. *The National Security Council: Background and Issues for Congress*. Washington, D.C.: Congressional Research Service. Updated October 19, 2022. https://sgp.fas.org/crs/natsec/R44828.pdf.

———. 2021a. *Renewed Great Power Competition: Implication for Defense—Issues for Congress*. Washington, D.C.: Congressional Research Service. Updated August 3, 2021. https://crsreports.congress.gov/product/pdf/R/R43838/76.

———. 2021b. *U.S. Strategic Nuclear Forces: Background, Developments, and Issues*. Updated July 13, 2021. Washington, D.C.: Congressional Research Service. https://crsreports.congress.gov/product/pdf/RL/RL33640/67.

Crux. 2023. "Russia Reveals Nuke Sub Plan After Sending Zircon Hypersonic Missiles To The Atlantic | Ukraine War." *Crux*, January 5, 2023. https://www.youtube.com/watch?v=xCOnQOCRgs8.

———. 2022. "What Are Suitcase Nukes And Has Russia Already Deployed Them In US & Other Allied Nations?" *Crux*, December 31, 2022. https://www.youtube.com/watch?v=tS_7eZVt854.

CSRI (Credit Suisse Research Institute). 2017. "'Getting Over Globalization'" – Outlook for 2017." Press Release. Credit Suisse, January 19, 2017. https://www.credit-suisse.com/about-us-news/en/articles/media-releases/_getting-over-globalization---what-to-watch-for-in-2017--201701.html. [Report is available at https://www.credit-suisse.com/about-us/en/reports-research/studies-publications.html.]

CTV News. 2020. "Cosmonaut sees UFOs while filming the Aurora Borealis." August 20, 2020. https://www.youtube.com/watch?v=acMxxJsJFFs.

CUN (CUN Italia Network). 2022. Centro Ufologico Nazionale. https://www
.centroufologiconazionale.net/. [Database is available at https://www.
centroufologiconazionale.net/avvistamenti/Casistica1900-2021.pdf.]

Dalton, Jane. 2019. "Killing off animals and plants now threatens humanity itself,
UN experts warn in urgent call for actions." *Independent*, May 6, 2019.
https://www.independent.co.uk/climate-change/news/un-nature-biodiversity-
report-2019-humans-animals-earth-paris-a8899926.html.

David, Leonard. 2020. "Scientists call for serious study of 'unidentified aerial
phenomena.'" *Space.com*, October 12, 2020. https://www.space.com/
unidentified-aerial-phenomena-scientific-scrutiny.

Davies, Paul C. W. 1995. *Are We Alone? Philosophical Implications of the Discovery of
Extraterrestrial Life*. New York: Basic Books.

———. 2010. "The Nature of the Laws of Physics and Their Mysterious Bio-
Friendliness." *Science and Religion in Dialogue*, Volume 2, edited by Melville Y.
Stewart, 767-88. Hoboken, NY: Wiley-Blackwell.

Davis, Nicola. 2018. "Earthlings likely to welcome alien life rather than panicking,
study shows." *The Guardian*, February 16, 2018. https://www.theguardian
.com/science/2018/feb/16/earthlings-likely-to-welcome-alien-life-rather-than-
panicking-study-shows.

Deamer, D. 2012. *First Life: Discovering the Connections between Stars, Cells, and How
Life Began*. Berkeley/Los Angeles/London: University of California Press.

Deasy, Hugh. 2009. "The physics of UFOs." [Original is no longer available online;
revised document is available at https://www.anomalistik.de/images/
pdf/ufo/ufo-poster%20deasy%20mufon-ces.pdf.]

Declaration (Declaration of Principles Concerning Activities Following the
Detection of Extraterrestrial Intelligence). 1989. *Declaration of Principles
Concerning Activities Following the Detection of Extraterrestrial Intelligence*. Paris:
International Academy of Astronautics (on behalf of various international
organizations). https://iaaspace.org/wp-content/uploads/iaa/Scientific%20
Activity/setideclaration.pdf.

Denmark. 2016. "Danish UFO Files." Internet Archive, January 16, 2016.
https://archive.org/details/DanishUFOFiles/mode/2up.

Department of the Army. 2019. "Cooperative Research and Development
Agreement between To The Stars Academy of Arts and Science, Inc. and The
U.S. Army Combat Capabilities Development Command Ground Vehicle

Systems Center." The Black Vault, October 2019. https://documents2
.theblackvault.com/documents/army/TTSA-ARMY-CRADA.pdf.

Derham, William. 1714. *Astra-Theology: Or a Demonstration of the Being and Attributes
of God from a Survey of the Heavens.* London: W. and J. Innys.

Devonshire-Ellis, Chris. 2022. "The New Candidate Countries For BRICS
Expansion." Silk Road Briefing, November 9, 2022. https://www.silk
roadbriefing.com/news/2022/11/09/the-new-candidate-countries-for-brics-
expansion/.

Dewdney, Christopher, Peter R. Holland, and Antonios Kyprianidis. 1987. "A
Causal Account of Non-local Einstein-Podolsky-Rosen Spin Correlations."
Journal of Physics A: Mathematical and General 20: 4717-732. https://iopscience
.iop.org/article/10.1088/0305-4470/20/14/016.

Dews, Fred. 2014. "Communications, Technology, and Extraterrestrial Life: The
Advance Brookings Gave NASA about the Space Program in 1960." Brookings
Institution, May 12, 2014. https://www.brookings.edu/blog/brookings-now
/2014/05/12/communications-technology-and-extraterrestrial-life-the-advice-
brookings-gave-nasa-about-the-space-program-in-1960/.

Diaz-Maurin, François. 2022a. "The 2022 nuclear year in review: A global nuclear
order in shambles." *Bulletin of the Atomic Scientists*, December 26, 2022.
https://thebulletin.org/2022/12/the-2022-nuclear-year-in-review
-a-global-nuclear-order-in-shambles/#post-heading.

———. 2022b. "Nowhere to Hide: How a nuclear war would kill you—and almost
everyone else." *Bulletin of the Atomic Scientists*, October 20, 2022.
https://thebulletin.org/2022/10/nowhere-to-hide-how-a-nuclear-war-would-
kill-you-and-almost-everyone-else/#post-heading.

Discogs. 2007. "National Press Club: Pilots To Tell Their UFO Stories Monday,
November 12, 2007." Discogs, November 12, 2007. https://www.discogs
.com/group/thread/647103.

DOS ([U.S.] Department of State). n.d. "222. Memorandum From the Director of
Central Intelligence (Hillenkoetter) to the National Security Council."
Department of State, Office of the Historian. https://history.state.gov/
historicaldocuments/frus1945-50Intel/d222.

Doyle, Stephen E. 1993. "Social Implications of NASA's High Resolution Microwave
Survey." International Astronautical Federation (IAF) Congress, October 19,
1993. Graz, Austria.

Drake, Frank. 1976. "On hands and knees in search of Elysium." *Technology Review* 78: 22-29.

Dröscher, Walter, and Jochem Häuser. 2007. "Advanced Propulsion Systems from Artificial Gravitational Fields." Abbreviated Version. 43rd AIAA/ASME/SAE/ASEE Joint Propulsion Conference and Exhibit, 8-11 July 2007, Cincinnati, OH. http://www.hpcc-space.com/publications/documents/ AIAA5595JCP2007DarkAbbreviated.pdf. [Since the author's surname is Häuser, I use the spelling with diacritic in the in-text citations and bibliography. However, the actual cover page to this document lists "Hauser" without the diacritic.]

―――. 2008. "Gravity-Like Fields and Space Propulsion Concepts." ResearchGate, July 2008. https://www.researchgate.net/publication/242544190_Gravity-Like_Fields_and_Space_Propulsion_Concepts. [Since the author's surname is Häuser, I use the spelling with diacritic in the in-text citations and bibliography. However, the actual cover page to this document lists "Hauser" without the diacritic.]

―――. 2015. *Physics, Astrophysics, and Cosmology of Gravity-Like Fields: An Elementary PRIMER for Breakthrough Physics for Propulsion and Energy Generation Technologies.* HPCC-Space GmbH, Hamburg, Germany. ResearchGate, November 2015. https://www.researchgate.net/profile/J-Haeuser/publication /281347752_Introduction_to_the_Physics_Astrophysics_and_Cosmology_of_G ravity-Like_Fields/links/55e358d108aede0b5733bf06/Introduction-to-the-Physics-Astrophysics-and-Cosmology-of-Gravity-Like-Fields.pdf.

Dumas, Stephane. 2016. *Catalogue of SETI Publications.* ResearchGate. https://www. researchgate.net/publication/288993481_Catalogue_of_SETI_publications.

Duncan, Charles. 2020. "What does the military know about UFOs? Senate committee wants public report." Impact 2020, June 25, 2020. https://www .mcclatchydc.com/news/nation-world/national/article243768802.html.

Dunphy, Seamus. 2014. "UFO Fever in America's Historical Newspapers: The Mysterious Airships of 1896-87." Readex Blog. *Readex: A Division of NewsBank*, September 12, 2014. https://www.readex.com/blog/ufo-fever-americas-historical-newspapers-mysterious-airships-1896-97.

Dürr, Detlef, Sheldon Goldstein, and Nino Zanghi. 1996. "Bohmian Mechanics as the Foundation of Quantum Mechanics." In *Bohmian Mechanics and Quantum Theory: An Appraisal*, edited by James T. Cushing, Arthur Fine and Sheldon Goldstein, 21-44. Boston Studies in the Philosophy of Science (BSPS) book series, volume 184. Dordrecht: Kluwer Academic Publishers. https://link.springer.com/chapter/10.1007/978-94-015-8715-0_2.

———. 2003 <1992>. "Quantum Equilibrium and the Origin of Absolute Uncertainty." *Journal of Statistical Physics* 67: 843-907. arXiv, submitted August 6, 2003 (v1). https://arxiv.org/abs/quant-ph/0308039.

Duve, Christian de. 1995. *Vital Dust: Life as a Cosmic Imperative*. New York: Basic Books.

DW (Deutsche Welle). 2022. "Stakes are high over Ukraine crisis, but Putin is 'not a gambler.'" Interview with Fyodor Lukyanov. *DW (Deutsche Welle)*, February 3, 2022. https://www.dw.com/en/stakes-are-high-over-ukraine-crisis-but-putin-is-not-a-gambler/av-60642464.

Eberhart, George M. 2022. *UFOs and Intelligence: A Timeline*. Academia.edu. https://www.academia.edu/43868466/UFOs_and_Intelligence_A_Timeline_By_George_M_Eberhart.

EE (Extraterrestrials and The Environment: Nuclear Power Plants, Nuclear weapons and ETs (1944-2007)). n.d. ExopoliticsRadio. https://www.bibliotecapleyades.net/exopolitica/esp_exopolitics_ZZZZY.htm.

Egan, Matt, Allison Morrow, and David Goldman. 2023. "First Republic secures $30 billion rescue from large banks." *CNN*, March 17, 2023. https://www.cnn.com/2023/03/16/investing/first-republic-bank/index.html.

Elliott, Josh K. 2021. "The CIA released thousands of UFO documents online. Here's how to read them." *Global News*, January 13, 2021. https://globalnews.ca/news/7573277/cia-ufo-documents-declassified-online/.

ENS (European Nuclear Society). 2022. "Nuclear Power Plants in Europe." ENS (European Nuclear Society), December 1, 2022. https://www.euronuclear.org/glossary/nuclear-power-plants-in-europe/.

EPA (United States Environmental Protection Agency). 2023. "Radioactive Fallout From Nuclear Weapons Testing." United States Environmental Protection Agency, last updated February 13, 2023. https://www.epa.gov/radtown/radioactive-fallout-nuclear-weapons-testing.

EURACTIV (EURACTIV.com with Reuters). 2022. "Belarus says Russian nuclear-capable Iskander missile systems ready for use." *EURACTIV.com with Reuters*, December 26, 2022. https://www.euractiv.com/section/global-europe/news/belarus-says-russian-nuclear-capable-iskander-missile-systems-ready-for-use/.

Evans, Gareth, and George Wright. 2023. "Mystery surrounds objects shot down by US military." *BBC News*, February 13, 2023. https://www.bbc.com/news/world-us-canada-64620064.

Exoplanet.eu. n.d. "Catalog." Exoplanet.eu. http://exoplanet.eu/catalog/.

Explore Aztec. n.d. "Aztec UFO Crash Site." Explore Aztec, March 25, 2007. http://www.aztecnm.com/aztec/ufocrashsite.html.

Exoplanet.eu. n.d. "Catalog." Exoplanet.eu. http://exoplanet.eu/catalog/.

Falconbridge (Falconbridge, ON). n.d. "Falconbridge, ON." Military Communications & Electronics Museum. http://www.c-and-e-museum.org/Pinetreeline/other/other15/other15e.html.

Fandom. n.d.-a. "Kardashev Scale. Type IV." Fandom. https://kardashev.fandom.com/wiki/Type_IV.

———. n.d.-b. "Kardeshev Scale. Type V." Fandom. https://kardashev.fandom.com/wiki/Type_V.

———. n.d.-c. Kardeshev Scale. Type VI." Fandom. https://kardashev.fandom.com/wiki/Type_VI.

FAS (Federation of American Scientists). n.d. *The External Referral Working Group: Its Origins and Development*. FAS (Federation of American Scientists) Project on Government Secrecy. https://sgp.fas.org/advisory/erwg.html.

Fasan, Ernst. 1970. *Relations with Alien Intelligences; The Scientific Basis of Metalaw*. Berlin: Verlag.

FBI (Federal Bureau of Investigation). 1950. "Guy Hottel Part 1 of 1." FBI Records: The Vault. https://vault.fbi.gov/hottel_guy/Guy%20Hottel%20Part%201%20of%201/view.

———. 2013. "UFOs And The Guy Hottel Memo." FBI.gov, March 25, 2013. https://www.fbi.gov/news/stories/ufos-and-the-guy-hottel-memo.

Fein, Esther B. 1989. "U.F.O. Landing Is Fact, Not Fantasy, the Russians Insist." *The New York Times*, October 11, 1989. https://www.nytimes.com/1989/10/11/world/ufo-landing-is-fact-not-fantasy-the-russians-insist.html.

Ferreira, Becky. 2023. "Government Scientists Discover Entirely New Kind of Quantum Entanglement in Breakthrough." *VICE*, January 4, 2023. https://www.vice.com/en/article/88qj3z/government-scientists-discover-entirely-new-kind-of-quantum-entanglement-in-breakthrough.

Fihn, Beatrice, editor. 2013. *Unspeakable suffering – the humanitarian impact of nuclear weapons*. Genève: Reaching Critical Will – a programme of Women's International League for Peace and Freedom. February 1, 2013. https://d3n8a8pro7vhmx.cloudfront.net/ican/pages/1043/attach-ments/original/1620205155/UnspeakableSuffering-web.pdf?1620205155.

Finney, Ben. 1990. "The impact of contact." *Acta Astronautica* 21, no. 2: 117-21.

Finnsson, Luna. 2009. "Danish UFO files now open to public." *IceNews: News from the Nordics*, February 6, 2009. https://www.icenews.is/2009/02/06/danish-ufo-files-now-open-to-public/.

Fiscaletti, Davide. 2018. *The Geometry of Quantum Potential: Entropic Information of the Vacuum*. Singapore, Hackensack, NJ, and London: World Scientific. Chapter 1 is available at https://www.worldscientific.com/doi/pdf/10.1142/9789813227989_0001.

Foltynova, Kristyna. 2022. "Russia's Stranglehold On The World's Power Cycle." *RadioFreeEurope/RadioLiberty*, September 1, 2022. https://www.rferl.org/a/russia-nuclear-power-industry-graphics/32014247.html.

Foust, Jeff. 2022. "Rogozin removed as head of Roscosmos as seat barter agreement signed." *SpaceNews*, July 15, 2022. https://spacenews.com/rogozin-removed-as-head-of-roscosmos-as-seat-barter-agreement-signed/.

Fox, James, director. 2004. *Out of the Blue: The Definitive Investigation of the UFO Phenomenon*. UFOTV. https://www.microsoft.com/en-us/p/ufotv-presents-out-of-the-blue-the-definitive-investigation-on-the-ufo-phenomenon/8d6kgwzl5tz7?activetab=pivot%3aoverviewtab.

Freeman, Ben, and William D. Hartung. 2023. "Unwarranted Influence, Twenty-First-Century-Style: Not Your Grandfather's Military-Industrial Complex." *TomDispatch*, May 4, 2023. https://tomdispatch.com/unwarranted-influence-twenty-first-century-style/.

Fuerza Aérea Uruguaya. n.d. "Comisión Receptora e Investigadora de Denuncias de Objetos Voladores No Identificados (CRIDOVNI)." Fuerza Aérea Uruguaya. https://www.fau.mil.uy/es/articulos/182-comision-receptora-e-investigadora-de-denuncias-de-objetos-voladores-no-identificados-cridovni.html.

FUFOF (French UFO Files). 2015. "French UFO Files." Internet Archive, December 27, 2015. https://archive.org/details/FrenchUFOFiles/00137R/.

Garamone, Jim. 2023. "Air Force Shoots Down 'High-Altitude Object' off Alaskan Coast." U.S. Department of Defense, February. 10, 2023. https://www.defense.gov/News/News-Stories/Article/Article/3295813/air-force-shoots-down-high-altitude-object-off-alaskan-coast/.

GCWUFO (Grant Cameron Whitehouse UFO). 2023a. "GRANT CAMERON Canadian Government Study, Sky Canada Project 1/2." March 10, 2023. https://www.youtube.com/watch?v=rHXFuHUo2eQ.

———. 2023b. "GRANT CAMERON Sky Project UFO Canadian Study part 3." March 17, 2023. https://www.youtube.com/watch?v=qmMEynh78xE.

———. 2023c. "GRANT CAMERON Part 2 of Sky Canada Project, Canadian Government UFO Study." March 14, 2023. https://www.youtube.com/watch?v=c_q7RE_tjvU&t=3s.

geipan. n.d. "Les cas du GEIPAN." geipan (Groupe d'Études et d'Informations sur les Phénomènes Aérospatiaux Non Identifiés. https://www.cnes-geipan.fr/.

Gestin, Pierre. 1973. *Phénomènes Spatiaux*. July 1973 (February 1961). Loqueffret, France.

Gevaerd, Ademar José. 2005. "Brazil Releases Classified Data Recognises UFO Research Confirmation of Great News in Brazilian Ufology." *rense.com*, May 23, 2005. https://rense.com/general65/braz.htm.

Gill, Alfred. 1978. *The Gill Sightings, June 1959: Anglican Missionary, Father W. Gill Describes His Experiences in New Guinea during July, 1959*. Moorabbin, Victoria: Victorian UFO Research Society. https://catalogue.nla.gov.au/ Record/2951975.

Gitle Hauge, Bjørn, Anna-Lena Kjøniksen, and Erling Petter Strand. 2017. "Optical luminosity of the transient luminous phenomena in Hessdalen, Norway." 19th EGU (European Geosciences Union) General Assembly, EGU2017, proceedings from the conference held April 23-28, 2017, Vienna, Austria, 8468. Publication date April 2017. https://ui.adsabs.harvard.edu/abs/2017EGUGA..19.8468G/abstract.

Gitle Hauge, Bjørn, Anna-Lena Kjøniksen, Erling Petter Strand, Jacques Zlotnicki, and George Vargemezis. 2016. "Magnetic influence on the unidentified luminous phenomena in Hessdalen, Norway." EGU General Assembly 2016, April 17-22, 2016 in Vienna Austria. Publication date April 2016. https://ui.adsabs.harvard.edu/abs/2016EGUGA..18.7638G/abstract.

Glette-Iversen, Ingrid, and Terje Aven. 2021. "On the meaning of and relationship between dragon-kings, black swans and related concepts." *Reliability*

Engineering and System Safety 211 (July), no. 107625: 1-13. https://reader .elsevier.com/reader/sd/pii/S0951832021001678?token=B21B5BE82A029CC8E00 7D7C441E1277A8C06B2AF76B5C9155667F5CFC5A7C237B7CE47B79AE60F8F7 00E559BE91D4F77&originRegion=us-east-1&originCreation=20230225145322.

Goldstein, Sheldon. 1998a. "Quantum Theory without Observers—Part One." *Physics Today* 51, no. 3: 42.

———. 1998b. "Quantum Theory without Observers—Part Two." *Physics Today* 51, no. 4: 38.

Good, Timothy. 1988 <1987>. *Above Top Secret: The Worldwide UFO Cover-Up.* New York: William Morrow and Company, Inc.

———. 2007 <2006>. *Need to Know: UFOs, the Military and Intelligence.* New York: Pegasus Books.

Gosson, Maurice A. de. 2001. *The Principles of Newtonian and Quantum Mechanics.* London: Imperial College Press.

GOV.UK. 2021. "Integrated Review of Security, Defence, Development and Foreign Policy 2021: nuclear deterrent." GOV.UK, March 17, 2021. [Additional information added to page on March 16, 2023.] https://www.gov.uk/ guidance/integrated-review-of-security-defence-development-and-foreign-policy-2021-nuclear-deterrent.

———. 2023. "New nuclear fuel agreement alongside G7 seeks to isolate Putin's Russia." GOV.UK, April 16, 2023. https://www.gov.uk/government/news/new-nuclear-fuel-agreement-alongside-g7-seeks-to-isolate-putins-russia.

GPM (Given Place Media). n.d. "The Mysterious Battle of Los Angeles, 1942." Los Angeles Almanac. http://www.laalmanac.com/history/hi07s.php.

Graham, R.D., Robert Hurst, Robert J. Thirkettle, Clive Rowe, and Philip H. Butler. 2007. "Experiment to Detect Frame Dragging in a Lead Super-conductor." *Physica C*, July 6, 2007 [preprint]. https://www.academia/edu/ 30289810/Experiment_to_detect_frame_dragging_in_a_lead_superconductor.

Grebennikova, T.V., A.V. Syroeshkin, E.V. Shubralova, O.V. Eliseeva, L.V. Kostina, N.Y Kulikova, O.E. Latyshev, M.A. Morozova, A.G. Yuzhakov, I.A. Zlatsky, M.A. Chichaeva, and O.S. Tsygankov. 2018. "The DNA of Bacteria of the World Ocean and the Earth in Cosmic Dust at the International Space Station." *The Scientific World Journal* 2018 (April 18): 7 pages. Article ID 7360147. https://hindawi.com/journals/tswj/2018/7360147/.

Greenewald, John. "UFOs: The Central Intelligence Agency (CIA) Collection." The Black Vault, January 7, 2021. https://www.theblackvault.com/documentarchive/ufos-the-central-intelligence-agency-cia-collection/.

Greer, Steven M. 2017. "2001 National Press Club Event (Presented by Dr. Steven Greer)." April 24, 2017. https://www.youtube.com/watch?v=4DrcG7VGgQU&t=0s.

Grieco, Kelly A., Christopher Preble, James Siebens, Elias Yousif, Rachel Stohl, Marla Keenan, Samantha Turner, Evan Cooper, Frank O'Donnell, and Martyn Williams. 2022. "Experts React: The Biden Administration's National Defense Strategy." Stimson, November 2, 2022. https://www.stimson.org/2022/experts-react-the-biden-administrations-national-defense-strategy/.

Gross, Daniel. 2022. "Are Unidentified Aerial Phenomena (UAP) Produced by Advanced Extraterrestrial Intelligences (ETIs)? A View of the Future of Humanity as a Model for the Emergence of Extraterrestrial Intelligences." In *Extraterrestrial Intelligence: Academic and Societal Implications*, edited by Jensine Andresen and Octavio A. Chon Torres, 112-23. Newcastle upon Tyne, U.K.: Cambridge Scholars Publishing.

Gross, Judah Ari. 2021. "Israel receives 3 more F-35 fighter jets." *The Times of Israel*, April 25, 2021. https://www.timesofisrael.com/liveblog_entry/israel-receives-3-more-f-35-fighter-jets/.

Gross, Loren E. 1988 <1972>. *UFO'S A History: 1946: The Ghost Rockets*. First Edition 1972, Enlarged Second Edition 1982, Enlarged Third Edition 1988. Fremont, CA: Privately Published.

Guthke, Karl S. 1990. *The Last Frontier: Imagining Other Worlds from the Copernican Revolution to Modern Science Fiction*. Ithaca, NY: Cornell University Press.

Haines, Avril. 2021. "Our Future in Space: Ignatius Forum." Washington National Cathedral, November 10, 2021. https://www.youtube.com/watch?v=UwyPk_f8aAA.

Haines, Gerald K. 1997 <1995>. "CIA's Role in the Study of UFOs, 1947-90." *Studies in Intelligence* 40, no. 5: 67-84. https://www.cia.gov/static/105bd8290b90de13ee136fecc9fe863f/cia-role-study-UFOs.pdf. [The 1997 version of this article also as published the same years as *Studies in Intelligence*, Semiannual Edition. no. 1, 67-84. The 1995 version of this article was published as "The CIA's Role in the Study of UFOs, 1947-90." *Studies in Intelligence* 39, no. 4: 75-92.]

Hajnoczi, Amb. Thomas. 2021. "Implications of Germany's accession to the Treaty on the Prohibition of Nuclear Weapons." September 23, 2021. https://d3n8a8pro7vhmx.cloudfront.net/ican/pages/2306/attachments/original/1632406535/I mplications_of_Germanys_accession_to_the_TPNW.pdf?1632406535. [German version first published by Greenpeace Germany is available at https://www. greenpeace.de/sites/default/files/publications/gpde_22072021_atomwaffenverb otsvertrag.pdf.]

Hall, Richard H. 2001. *The UFO Evidence: A thirty-Year Report*. Volume 2. Lanham, MD: Scarecrow Press.

Hambling, David. 2020. "The EmDrive Just Won't Die." *Popular Mechanics*, September 11, 2020. https://www.popularmechanics.com/space/rockets /a33917439/emdrive-wont-die/.

Harrington, Anne. 1996. *Reenchanted Science: Holism in German Culture from Wilhelm II to Hitler*. Princeton, NJ: Princeton University Press.

Harris, Melanie J., Nalin Chandra Wickramasinghe, David Lloyd, Jayant V. Narlikar, P. Rajaratnam, Michael P. Turner, Shirwan Al-Mufti, Max K. Wallis, S. Ramadurai, and Fred Hoyle. 2002. "Detection of living cells in stratospheric samples." *Proceedings of the International Society for Optics and Photonics* 4495: 192-98. https://spie.org/Publications/Proceedings/Paper/10.1117/12.454758?SSO=1.

Hartle, James B. 1992. "Spacetime quantum mechanics and the quantum mechanics of spacetime." Lectures given at Les Houches École d'été, Gravitation et quantifications, July 9-17, 1992. [These lectures were delivered as part of Les Houches Summer School on Gravitation, Session 57, July 5-August 1, 1992.]

Harvey, John. n.d. "Dr. John Harvey, Principal Deputy, Assistant to the Secretary of Defense for Nuclear and Chemical and Biological Defense Programs." CSIS (Center for Strategic & International Studies). https://csis-website-prod.s3.amazonaws.com/s3fs-public/legacy_files/files/attachments/091113 _John_Harvey_Bio.pdf.

Hastings, Robert L. 2023. "AARO Interviews 'UFOs and Nukes' Witnesses." UFOs & Nukes, February 22, 2023. https://www.ufohastings.com/articles /aaro-interviews-ufos-and-nukes-witnesses.

———. director. 2016. *UFOs and Nukes: The Secret Link Revealed*. Verifiable Pictures. [Topics discussed in this film also are discussed in Hastings 2017<2008>.]

———. 2017 <2008>. *UFOs and Nukes: Extraordinary Encounters at Nuclear Weapons Sites*. Second edition, revised and updated. [A publisher is not listed for the second edition, but AuthorHouse published the first edition in 2008.]

———. 2006. "UFO sightings at ICBM sites and nuclear Weapons Storage Areas." NICAP (National Investigations Committee on Aerial Phenomena). https://www.nicap.org/babylon/missile_incidents.htm.

Häuser, Jochem, and Walter Dröscher. 2015. "Elementary Primer of Field Propulsion Physics." *Frontiers in Space Propulsion*, edited by Takaaki Musha. Nova Science Publishers, Inc. [chapter 7]. ResearchGate. https://www.research gate.net/publication/273381056_Elementary_primer_of_field_propulsion _physics. [Although the author's surname is Häuser, with the diacritic, this document uses 'Hauser,' without the diacritic.]

———. 2017. "Gravity beyond Einstein? Part I: Physics and the Trouble with Experiments." *Z. Naturforsch* 72, no. 6 (April 29)a: 493-525. https://www. degruyter.com/document/doi/10.1515/zna-2016-0479/html?lang=de. [Although the author's surname is Häuser, with the diacritic, this document uses 'Hauser,' without the diacritic.]

———. 2019. "Gravity beyond Einstein? Part II: Fundamental Physical Principles, Number Systems, Novel Groups, Dark Energy, and Dark Matter, MOND." *Z. Naturforsch* 74, no. 5 (April 10)a: 387-446. https://www.degruyter.com/ document/doi/10.1515/zna-2018-0559/html?lang=en. [Although the author's surname is Häuser, with the diacritic, this document uses 'Hauser,' without the diacritic.]

Heidmann, Jean. 1995. *Extraterrestrial Intelligence.* Cambridge, U.K.: Cambridge University Press.

Heisenberg, Werner. 1949 <1930>. *The Physical Principles of the Quantum Theory.* Translated by Carl Eckart and Frank C. Hoyt. New York: Dover Publications, Inc.

———. 1959 <1958>. *Physics and Philosophy: The Revolution in Modern Science.* London: George Allen and Unwin.

Henry, Richard C. 1988. "UFOs and NASA." Invited Essay. *Journal of Scientific Exploration* 2, no. 2: 93-142. https://henry.pha.jhu.edu/ufosNASA.pdf.

Hernández, Gabriela Rosa, and Daryl G. Kimball. 2022. "Russia Blocks NPT Conference Consensus Over Ukraine." Arms Control Association, September 2022. https://www.armscontrol.org/act/2022-09/news/russia-blocks-npt-conference -consensus-over-ukraine#:~:text=Russia%20blocked%20the%202022% 20nuclear,nuclear%20power%20plant%20in%20Ukraine.

Hersh, Seymour. 2023a. "From The Gulf of Tonkin To The Baltic Sea: The secret and incomplete history of US-Norway collaboration in covert operations." Substack, February 22, 2023. https://seymourhersh.substack.com/p/from-the-gulf-of-tonkin-to-the-baltic?utm_source=%2Fsearch%2FRobert%2520Baer&utm_medium=reader2

———. 2023b. "How America Took Out The Nord Stream Pipeline." Substack, February 8, 2023. https://seymourhersh.substack.com/p/how-america-took-out-the-nord-stream

Hicks, Kathleen H. 2021. "Establishment of the Airborne Object Identification and Management Synchronization Group." Memorandum for Senior Pentagon Leadership, Commanders of the Combatant Commands, Defense Agency and Field Activity Directors. United States Department of Defense, November 23, 2021. https://www.airforcemag.com/app/uploads/2021/11/ESTABLISHMENT-OF-THE-AIRBORNE-OBJECT-IDENTIFICATION-AND-MANAGEMENT-SYNCHRONIZATION-GROUP.pdf.

Hiley, Basil J. 1999. "Active Information and Teleportation." In *Epistemological and Experimental Perspectives on Quantum Physics*, edited by Daniel Greenberger, W.L. Reiter, and Anton Zeilinger, 113-26. Dordrecht: Kluwer Academic Publishers.

———. 1995. "The Conceptual Structure of the Bohm Interpretation of Quantum Mechanics." In *Symposium on the Foundation of Modern Physics: 70 Years of Matter Waves*, edited by Kalervo Vihtori Laurikainen, Claus Montonen and K. Sunnarborg, 99-118. Editions Frontiers, Gif-sur-Yvette.

———. 2002. *Proceedings of the Conference "Quantum Theory, Reconsideration of Foundations,"* edited by Andrei Khrennikov, 141-62. Sweden: Växjö University Press. Conference June 17-21, 2011. [Page numbers refer to the PDF available at http://www7.bbk.ac.uk/tpru/BasilHiley/Vexjo2001W.pdf.]

———. 2001. *Non-Commutative Geometry, the Bohm Interpretation and the Mind-Matter Relationship.* Computing Anticipatory Systems: CASYS 2000-Fourth International Conference Proceedings, edited by Daniel M. Dubois, Vol. 573: 77-88. New York: AIP Press.

Hiley, Basil J., Robert E. Callaghan, and Owen Maroney. 2001. "Quantum Trajectories, Real, Surreal or an Approximation to a Deeper Process?" http://www7.bbk.ac.uk/tpru/RecentPublications/QTrajs.pdf.

Hiley, Basil J., and F. David Peat, editors. 1987. *Quantum Implications*. London and New York: Routledge & Kegan Paul.

Hiley, Basil J., and Paavo Pylkkänen. 2005. "Can Mind Affect Matter Via Active Information?" *Mind & Matter* 3, no. 2: 7-27.

Himmelstein, Sue. 2022. "The global status of nuclear power." Engineering360, December 23, 2022. https://insights.globalspec.com/article/19674/the-global-status-of-nuclear-power#:~:text=According%20to%20the%20World%20Nuclear,26%20fewer%20than%20in%202011.

History.com. n.d. "USS Roosevelt 'Gimbal' UFO: Declassified Video." History.com. https://www.history.com/topics/paranormal/uss-roosevelt-gimbal-ufo-declassified-video.

Howard, Paul, director. 2020. *Infinite Potential: The Life & Ideas of David Bohm*. Imagine Films. https://www.infinitepotential.com/.

Howell, Elizabeth. 2021. "Fermi Paradox: Where are the aliens?" *Space.com*, December 17, 2021. https://www.space.com/25325-fermi-paradox.html.

———. 2020. "Russian cosmonaut spots 'space guests' amid dazzling auroras in video. They're not aliens." *Space.com*, August 22, 2020. https://www.space.com/russian-cosmonaut-space-guests-video-from-station.html.

Hoyle, Fred, and Nalin Chandra Wickramasinghe. 1982. *Evolution from Space*. London: J.M. Dent & Sons Limited.

———, editors. 2000. *Astronomical Origins of Life*. Dordrecht / Boston / London: Kluwer Academic Publishers.

Hoyle, Fred, Geoffrey Burbidge, and Jayant V. Narlikar. 2000. *A Different View of Cosmology: From a static universe through the big bang towards reality*. Cambridge: Cambridge University Press.

HPSCI (House Permanent Select Committee on Intelligence). 2022a. "Chairman Carson Delivers Opening Statement at Counterterrorism, Counterintelligence, Counterproliferation Subcommittee Hearing on Unidentified Aerial Phenomena." Press Releases. U.S. House of Representatives Permanent Select Committee on Intelligence, Washington, D.C., May 17, 2022. https://intelligence.house.gov/news/documentsingle.aspx?DocumentID=1197.

———. 2022b. "Open C3 Subcommittee Hearing on Unidentified Aerial Phenomena." May 17, 2022. https://www.youtube.com/watch?v=aSDweUbGBow.

Hudson, Kate. 2021a. "It's Time to Ban the Bomb." *Tribune*, January 22, 2021. https://tribunemag.co.uk/2021/01/banning-the-bomb.

———. 2023. "Depleted Uranium: the last thing Ukrainians need." Campaign for Nuclear Disarmament (CND), March 30, 2023. https://cnduk.org/depleted-uranium-the-last-thing-ukrainians-need/.

———. 2021b. "Why Boris Loves the Bomb." *Tribune*, March 17, 2021. https://tribunemag.co.uk/2021/03/why-boris-loves-the-bomb.

Hudson, Kate. 2023. "Depleted Uranium: the last thing Ukrainians need." Campaign for Nuclear Disarmament (CND), March 30, 2023. https://cnduk.org/depleted-uranium-the-last-thing-ukrainians-need/.

Huyghe, Patrick. 1979. "U.F.O. Files: The Untold Story." *The New York Times*, October 14, 1979, 106. https://www.nytimes.com/1979/10/14/archives/ufo-files-the-untold-story.html.

IAA (The International Academy of Astronautics). 2000. *A Decision Process for Examining the Possibility of Sending Communications to Extraterrestrial Civilizations: A Proposal*. Paris: International Academy of Astronautics. https://iaaspace.org/wp-content/uploads/iaa/Studies/seti.pdf.

IAEA (International Atomic Energy Agency Board of Governors). 2021. *Verification and monitoring in the Islamic Republic of Iran in light of United Nations Security Council resolution 2231 (2015)*. Report by the Director General, September 7, 2021. Derestricted September 15, 2021. https://www.iaea.org/sites/default/files/21/09/gov2021-39.pdf.

ICAN (International Campaign to Abolish Nuclear Weapons). 2021. *Complicit: 2020 Global Nuclear Weapons Spending*. ICAN (International Campaign to Abolish Nuclear Weapons), June 7, 2021. https://d3n8a8pro7vhmx.cloudfront.net/ican/pages/2161/attachments/original/1622825593/Spending_Report_Web.pdf?1622825593.

———. 2020. *Enough is Enough: 2019 Global Nuclear Weapons Spending*. ICAN (International Campaign to Abolish Nuclear Weapons), May 13, 2020. https://d3n8a8pro7vhmx.cloudfront.net/ican/pages/1549/attachments/original/1589365383/ICAN-Enough-is-Enough-Global-Nuclear-Weapons-Spending-2020-published-13052020.pdf?1589365383.

———. n.d.-a. "Full text of the Treaty on the Prohibition of Nuclear Weapons." ICAN (International Campaign to Abolish Nuclear Weapons). https://www.icanw.org/tpnw_full_text.

———. n.d.-b. "The Human Cost of Nuclear Testing." ICAN (International Campaign to Abolish Nuclear Weapons). https://www.icanw.org/nuclear_tests.

———. 2022a. *No Place To Hide: Nuclear Weapons and the Collapse of Health Care Systems*. ICAN (International Campaign to Abolish Nuclear Weapons), February 10, 2022. https://d3n8a8pro7vhmx.cloudfront.net/ican/pages/2544/attachments/original/1644334250/NoPlacetoHide-ICAN-Report-Feb2022-web.pdf?1644334250.

———. 2018. *Nuclear Weapons Ban Monitor 2018*. ICAN (International Campaign to Abolish Nuclear Weapons), October 1, 2018. https://www.icanw.org/nuclear_weapons_ban_monitor_2018.

———. 2019a. *Nuclear Weapons Ban Monitor 2019: Tracking Progress towards a World Free of Nuclear Weapons*. ICAN (International Campaign to Abolish Nuclear Weapons), October 1, 2019. https://www.icanw.org/nuclear_ban_monitor_2019.

———. 2011. *Nuclear weapons spending: a theft of public resources*. ICAN (International Campaign to Abolish Nuclear Weapons), September 1, 2011. https://d3n8a8pro7vhmx.cloudfront.net/ican/pages/1044/attachments/original/1576780954/Screen_Shot_2019-12-19_at_17.59.34.png?1576780954

———. 2015. *A Pledge to Fill the Legal Gap: Vienna Conference 2014*. ICAN (International Campaign to Abolish Nuclear Weapons), February 28, 2015. https://d3n8a8pro7vhmx.cloudfront.net/ican/pages/743/attach-ments/original/1620207351/ViennaReport.pdf?1620207351.

———. 2019b. *Schools of Mass Destruction: American Universities in the U.S. Nuclear Weapons Complex*. ICAN (International Campaign to Abolish Nuclear Weapons), November 13, 2019. https://d3n8a8pro7vhmx.cloudfront.net/ican/pages/430/attachments/original/1574113227/ICAN-Schools-of-mass-Destruction_nov2019.pdf?1574113227.

———. 2022b. *Squandered: 2021 Global Nuclear Weapons Spending*. ICAN (International Campaign to Abolish Nuclear Weapons), June 14, 2022. https://assets.nationbuilder.com/ican/pages/2873/attachments/original/1655145777/Spending_Report_2022_web.pdf?1655145777.

———. n.d.-c. "Why a ban?" ICAN (International Campaign to Abolish Nuclear Weapons). https://www.icanw.org/why_a_ban.

iceage2012. 2021. "UAP at Spain's La Palma volcano." Imgur, October 11, 2021. https://imgur.com/gallery/cXeraw8.

IfA (Institute for Astronomy, University of Hawaii. 2018. "Earth's First Know Interstellar Visitor Unmasked." Institute for Astronomy Press Release. https://www.ifa.hawaii.edu/info/press-releases/Oumuamua/.

IPFM (International Panel on Fissile Materials). 2022a. "Fissile material stocks." International Panel on Fissile Materials, May 2, 2022. https://fissile materials.org/.

———. 2022b. "International Panel on Fissile Materials: About IPFM." International Panel on Fissile Materials, July 23, 2022. https://fissilematerials.org /ipfm/about.html.

———. 2015. *Global Fissile Material Report 2015: Nuclear Weapon and Fissile Material Stockpiles and Production.* Eighth annual report of the International Panel on Fissile Materials. https://fissilematerials.org/library/gfmr15.pdf.

Iran International. 2021. "Iran Can Have Enough Uranium For A Nuke In One Month – Report." Iran International, September 14, 2021. https://old.iran intl.com/en/world/iran-can-have-enough-uranium-nuke-one-month-report.

Isgleas, Daniel. 2009. "Hay aún 40 casos de ovnis sin explicación." *El Pais*, June 7, 2009. https://historico.elpais.com.uy/090607/pnacio-421863/nacional/Hay-aun-40-casos-de-ovnis-sin-explicacion/.

Jackson, Herb. 2009. "North Bergen man is homeland security assistant for President Obama." *NorthJersey.com*, December 5, 2009. https://web.archive .org/web/20110507023129/http://www.northjersey.com/news/120509_North_Be rgen_man_is_homeland_security_assistant_for_President_Obama.html.

JD (Jonathan Dimbleby). 2002. Jonathan Dimbleby. *ITV (Independent Television)*, March 24, 2002.

Jeffrey Sachs. 2023. Jeffrey Sacs on the Nord Stream Investigation |UN Security council|. February 23, 2023. https://www.youtube.com/watch?v=OZyjFUXX GBA.

Jiang, Linhua, Nobunari Kashikawa, Shu Wang, et al. 2020. "Evidence for GN-z11 as a luminous galaxy at redshift 10.957." *Nature Astronomy*, December 14, 2020. https://www.nature.com/articles/s41550-020-01275-y.

Johnson, Donald A. 2002. "Do Nuclear Facilities Attract UFOs?" CUFON.org (The Computer UFO Network), July 14, 2002. http://www.cufon.org/contributors/DJ/Do%20Nuclear%20Facilities%20Attract%20UFOs.pdf.

John, Tara, and Bethlehem Feleke. 2020. "A looming plague: The fight to contain a locust invasion that could push millions into hunger." *CNN*, May 28, 2020, https://edition.cnn.com/interactive/2020/05/africa/locusts/.

Johnson, David W. 2022. *Aligning Megalithic Sites of Southern England and Carnac, France with Groundwater Features: Aligning the Three Worlds*. Rhinebeck, NY: Monkfish Book Publishing Company.

Johnson, Douglas Dean. 2023. "Crash Story: The Trinity UFO Crash Hoax." Mirador, May 1, 2023. https://douglasjohnson.ghost.io/crash-story-the-trinity-ufo-crash-hoax/.

Johnson, Jesse. 2020. "Japanese Defense Ministry unveils protocol for encountering UFOs." *The Japan Times*, September 14, 2020. https://www.japantimes.co.jp/news/2020/09/14/national/japan-defense-ministry-ufo/.

Jones, Eric M. 1985. *"Where is Everybody?" An Account of Fermi's Question*. Los Alamos: Los Alamos National Laboratory.

Jordans, Frank. 2023. "Over and out: Germany switched off its last nuclear power plants." *AP (Associated Press) News*, April 15, 2023. https://apnews.com/article/germany-nuclear-power-plants-shut-energy-76dfaa223f88fedff138b9a63a6f0da.

Jozuka, Emiko. 2015. "How Volcanic Eruptions Threaten Nuclear Power Plants." *Vice*, November 14, 2015. https://www.vice.com/en/article/pgaj9b/how-volcanic-eruptions-threaten-nuclear-power-plants.

Judd, Donald J. 2023. "National security adviser will lead new 'interagency team' to study unidentified aerial objects." *CNN*, February 13, 2023. Part of "February 13, 2023 Latest on the unidentified objects shot down over North America" by Aditi Sangal, Elise Hammond, Maureen Chowdhury, and Mike Hayes, Updated 9:56 p.m. ET, February 13, 2023. https://www.cnn.com/politics/live-news/unidentified-objects-us-airspace-white-house-02-13-2023/h_7b05b16b3c7977c11cb19d7db7d0f903.

Jung, Carl G. 1969. "On the Psychology of the Trickster-Figure." In *The Archetypes and the Collective Unconscious*. Second edition. Translated by Richard Francis Carrington Hull. Princeton: Princeton University Press. Volume 9 (Part I):

Archetypes and the Collective Unconscious of *The Collected Works of C. G. Jung*, 255-72. Princeton: Princeton University Press (Bollingen Series).

Kalantarova, Olena. 2021. Methodological Pluralism Through the Lens of the Buddhist Doctrine of Time Kālacakra: An Interview with Dr. Jensine Andresen." *Філософська думка (Philosophical Thought)* 2: 165-83.

Kaleka, Amardeep S., director. 2013. *Sirius*. Bayview Films.

Kauffman, George B., and Isaac Mayo. 1997. "The Story of Nitinol: The Serendipitous Discovery of the Memory Metal and Its Applications." *The Chemical Educator* 2 (June): 1-21.

Kaufmann, Pierre, and Peter A. Sturrock. 2004. "On Events Possibly Related to the Brazil Magnesium." *Journal of Scientific Exploration* 18, no. 2: 283-91.

Keane, Daniel. 2021. "Nuclear-powered submarines have 'long history of accidents', Adelaide environmentalist warns." *ABC (Australian Broadcasting Corporation) News*, September 17, 2021. https://www.abc.net.au/news/2021-09-17/nuclear-submarines-prompt-environmental-and-conflict-concern/100470362.

Keel, John A. 1970-1971. "Mystery Aeroplanes of the 1930s." *Flying Saucer Review*, a four-part article: Part 1: vol. 16, no. 3 (May/June 1970): 10-13; Part 2: vol. 16, no. 4 (July/August 1970): 9-14; Part 3: vol. 17, no 4 (July/August 1971): 17-19; Part 4: vol. 17, no. 5 (September/October 1971): 20-22, 28. https://www.scribd.com /document/21787053/Mystery-Aeroplanes-of-the-1930-s-by-John-A-Keel.

———. 1969. "The Principle of Transmogrification." *Flying Saucer Review* 15, no. 4 (August): 27-8, 31. https://www.scribd.com/document/385489728/THE-PRINCIPLE-OF-TRANSMOGRIFICATION-by-John-A-Keel.

Kehoe, Adam, and Marc Cecotti. 2021. "Multiple Destroyers Were Swarmed By Mysterious 'Drones' Off California Over Numerous Nights." *The Drive*, March 23, 2021. https://www.thedrive.com/the-war-zone/39913/multiple-destroyers-were-swarmed-by-mysterious-drones-off-california-over-numerous-nights.

Kessler, Glenn. 2022. "Does Russia sell nearly $1 billion in uranium to the U.S. a year?" *The Washington Post*, April 20, 2022. https://www.washing-tonpost .com/politics/2022/04/20/does-russia-sell-nearly-1-billion-uranium-us-year/.

Khrennikov, Andrei Yu. 2000. "Classical and quantum mechanics on *p*-adic trees of ideas." *Biosystems* 56, nos. 2-3 (April-May): 95-120.

KI (Kim Iversen). 2023. "Conversation With Economist Richard Werner | The Plandemic Was Used To Usher In TOTAL CONTROL." Originally aired

February 24, 2023 on Rumble. https://www.youtube.com/watch?v=FFAbK20r Xa0&t=46s.

Kimball, Daryl. 2018. "Fissile Material Cut-off Treaty (FMCT) at a Glance." Arms Control Association, June 2018. https://www.armscontrol.org/factsheets /fmct.

———. 2022a. "The Nuclear Nonproliferation Treaty (NPT) at a Glance." Arms Control Association, August 2022. https://www.armscontrol.org/factsheets /nptfact.

———. 2022b. "Timeline of the Nuclear Nonproliferation Treaty (NPT)." Arms Control Association, August 2022. https://www.armscontrol.org/factsheets /Timeline-of-the-Treaty-on-the-Non-Proliferation-of-Nuclear-Weapons-NPT.

King, Darryn. 2020 <2018>. "Why Mysterious Green Fireballs Worried the U.S. Government in 1948." History Stories, August 17, 2018, updated January 16, 2020. https://www.history.com/news/ufos-green-fireballs-nuclear-facilities-new-mexico.

Kirby, Paul. 2022. "Russia-Ukraine war: Top official Rogozin wounded in Ukrainian shelling." *BBC*, December 22, 2022. https://www.bbc.com /news/world-europe-64063046.

Kirkpatrick, Seán. 2023. *The Defense Department's UAP Mission & Civil Aviation*. Presentation by the Director of the All-Domain Anomaly Resolution Office to the National Academies of Sciences, Engineering, and Medicine's Transportation Research Board. US Department of Defense & Office of the Director of National Intelligence, January 11, 2023. https://drive .google.com/file/d/1Lln8JFxbtKRw8U5KjBiLIfFOOf45EAta/view

Klass, Philip J. 1983. *UFOs: The Public Deceived*. New York: Prometheus Books.

Knorr, Wolfgang. 2020. "The age of stability is over, and coronavirus is just the beginning." The Conversation, April 17, 2020. https://theconver-sation.com/the-age-of-stability-is-over-and-coronavirus-is-just-the-beginning-136380.

Koerner, David, and Simon LeVay. 2000. *Here Be Dragons: The Scientific Quest for Extraterrestrial Life*. Oxford: Oxford University Press.

Korpela, Eric J. 2019. "SETI: Its Goals and Accomplishments." In *Handbook of Astrobiology*, edited by Vera M. Kolb, 727-40. Boca Raton, FL: CRC Press (Taylor & Francis Group).

Krepon, Michael. 2021a. "How to Avoid Nuclear War." *War on the Rocks*, November 8, 2021. https://warontherocks.com/2021/11/how-to-avoid-nuclear-war/.

———. 2021b. "The rise and demise of arms control." *Bulletin of the Atomic Scientists*, November 15, 2021. https://thebulletin.org/2021/11/the-rise-and-demise-of-arms-control/.

———. 2021c. *Winning and Losing the Nuclear Peace: The Rise, Demise, and Revival of Arms Control*. Stanford: Stanford University Press.

Kristensen, Hans M. 2019. "Urgent: Move US Nuclear Weapons Out Of Turkey." Federation of American Scientists (FAS), October 16, 2019. https://fas.org/blogs/security/2019/10/nukes-out-of-turkey/.

Kristensen, Hans M., and Matt Korda. 2022a. "The 2022 Nuclear Posture Review: Arms Control Subdued By Military Rivalry." Federation of American Scientists (FAS), October 27, 2022. https://fas.org/blogs/security/2022/10/2022-nuclear-posture-review/.

———. 2022b. "The Nuclear Notebook: Israeli nuclear weapons, 2021." *Bulletin of the Atomic Scientists* 78, no. 1: 38-50. *Conflict in space* (Special Issue, January 17). https://www.tandfonline.com/doi/full/10.1080/00963402.2021.2014239.

———. 2019. "United States nuclear forces, 2019." *Bulletin of the Atomic Scientists* 75, no. 3: 122-134. https://www.tandfonline.com/doi/epdf/10.1080/00963402.2019.1606503?needAccess=true&role=button.

Kristensen, Hans M., Matt Korda, Moritz Kütt, Zia Mian, and Pavel Podvig. 2021. "10. World nuclear forces." In *SIPRI [Stockholm International Peace Research Institute] Yearbook 2021: Armaments, Disarmament and International Security*. Oxford University Press. [Page numbers refer to the PDF available at https://www.sipri.org/yearbook/2021/10.]

Kuhn, Thomas. 1962. *The Structure of Scientific Revolutions*. Chicago: The University of Chicago Press.

Kuiper, Thomas B. H., and Mark Morris. 1977. "Searching for Extraterrestrial Civilizations." *Science* 196: 616.

Lennane, Richard, and Tim Wright. 2021. *A Non-Nuclear Alliance: Why NATO Members Should Join the UN Ban on Nuclear Weapons*. ICAN (International Campaign to Abolish Nuclear Weapons), June 10, 2021. https://d3n8a8pro7vhmx.cloudfront.net/ican/pages/2165/attachments/original/1623235224/ICAN-NATO-report-final.pdf?1623235224.

Lewis-Kraus, Gideon. 2021. "How the Pentagon Started Taking U.F.O.s Seriously." *The New Yorker*, April 30, 2021. https://www.newyorker.com/magazine /2021/05/10/how-the-pentagon-started-taking-ufos-seriously.

Li, Wanyee. 2020. "Secret UFO Files? In Canada the truth is out there – online and searchable." *The Toronto Star*, July 28, 2020. https://www.thestar.com/news /canada/2020/07/27/secret-ufo-files-in-canada-the-truth-is-out-there-online- and-searchable.html.

Library and Archives Canada. 2007 <2005>. "Canada's UFOs: The Search for the Unknown." Library and Archives Canada, created August 11, 2005, updated December 14, 2007. https://web.archive.org/web/20160116131034/http:www. collectionscanada.gc.ca/ufo/#.

Liljegren, Anders, and Clas Svahn. 1987. "The Ghost Rockets." In *UFOs: 1947-1987, The 40-Year Search for An Explanation*, edited by Hilary Evans with John Spencer for The British UFO Research Association, 32-38. London: Fortean Tomes.

Lindley, David. 2005. "The Birth of Wormholes." *Physical Review Focus* 15, no. 11 (March 25). https://physics.aps.org/story/v15/st11.

Lock, Samantha. 2022. "Ukraine crisis live news: Russia open to 'diplomacy', says Putin, amid fresh shelling in east." *The Guardian*, February 23, 2022, blog post entry 1:50 a.m., "Putin ready for 'diplomatic solutions' but Russia's interests non-negotiable.'" https://www.theguardian.com/world/live/2022/feb/23/ukraine -russia-news-crisis-latest-live-updates-putin-biden-europe-sanctions-russian- invasion-border-troops.

Lopez, C. Todd. 2022. "DOD Office Moving Ahead in Mission to Identify 'Anomalous Phenomena.'" U.S. Department of Defense, December 17, 2022. https://www.defense.gov/News/News-Stories/Article/Article/324 9317/dod-office-moving-ahead-in-mission-to-identify-anomalous-phenomena/.

Lucas, Greg. n.d. "The Battle of Los Angeles." CAL@170, By the California State Library: 170 Stories Celebrating the State of California's First 170 Years. https://cal170.library.ca.gov/february-24-1942-the-battle-of-los-angeles-2/.

Maccabee, Bruce. n.d. "The Fantastic Flight of JAL1628." http://brumac .mysite.com/JAL1628/JL1628.html.

———. 1985. *The Government UFO Connection: A Collection of UFO Documents from the Governments of the U.S.A. and Canada*. Mount Rainier, MD: Fund for UFO Research (FUFOR).

Maccabee, Bruce, and Brad Sparks. n.d. "Analysis of the Photos of an Unidentified Object Observed by the Astronauts of SKYLAB 3." National Investigations

Committee On Aerial Phenomena (NICAP). http://www.nicap.org/reports/730920skylabIII_report2.htm.

Macrakis, Kristie. 1998. "Reenchanted Science: Holism in German Culture from Wilhelm II to Hitler." Book review. *German Politics and Society* 16, no. 3 (Fall): 150–53. https://www.doi.org/10.1353/bhm.1997.0149.

Madrak, Susie. 2019. "What About That UFO Over The White House That Just Prompted A Lockdown?" *Crooks and Liars*, November 27, 2019. https://crooksandliars.com/2019/11/what-about-ufo-over-white-house-prompted-0.

Magramo, Kathleen, Eliza Mackintosh, Aditi Sangal, Mike Hayes, and Leinz Vales. 2023. "Xi and Putin meet in Moscow as Russia's war in Ukraine continues." *CNN*, March 20, 2023. https://edition.cnn.com/europe/live-news/russia-ukraine-war-news-03-20-23/h_31e9cc5712f2a58064b1377034395978.

Mallove, Eugene F., Mary M. Connors, Robert L. Forward, and Zbigniew Paprotny. 1978. *A Bibliography on the Search for Extraterrestrial Intelligence*. NASA Reference Publication 1021. NASA (National Aeronautics and Space Administration), Scientific and Technical Information Office, March 1978. https://ntrs.nasa.gov/api/citations/19780013076/downloads/19780013076.pdf.

Malmstrom. n.d. "History of Malmstrom Air Force Base." Malmstrom Air Force Base. https://www.malmstrom.af.mil/About-Us/History/Malm-strom-History/.

Mark, Julian. 2023. "Senate panel: Credit Suisse helped ultra-rich Americans dodge taxes." *The Washington Post*, March 29, 2023. https://www.washingtonpost.com/business/2023/03/29/credit-suisse-tax-evasion-senate-report/.

Marker, Jonathan. 2019. "Public Interest in UFOs Persists 50 Years After Project Blue Book Termination." *National Archives News*, December 5, 2019. https://www.archives.gov/news/articles/project-blue-book-50th-anniversary.

Marlow, Iain. 2023. "US Fears a War-Weary World May Embrace China's Ukraine peace Bid." *Bloomberg*, March 22, 2023. https://www.bloomberg.com/news/articles/2023-03-23/us-fears-a-war-weary-world-may-embrace-china-s-ukraine-peace-bid?leadSource=uverify%20wall#xj4y7vzkg.

Maroney, Owen, and Basil J. Hiley. 1999. "Quantum State Teleportation Understood Through the Bohm Interpretation." *Foundation of Physics* 29: 1403-15.

Masetti, Maggie. 2015. "How Many Stars in the Milky Way?" National Aeronautics and Space Administration (NASA), Goddard Space Flight Center, July 22, 2015.

https://asd.gsfc.nasa.gov/blueshift/index.php/2015/07/22/how-many-stars-in-the-milky-way/.

Massar, Serge, and Stefano Pironio. 2012. "A Closer Connection Between Entanglement and Nonlocality." *Physics* 5, no. 56 (May 14). https://physics.aps.org/articles/v5/56.

Mazzola, Michael, director. 2017. *Unacknowledged*. Auroris Media.

———. 2021. "The Cosmic Hoax – An Exposé. NTLGNT media." July 24, 2021. https://www.youtube.com/watch?v=Ecj87JoUyY4.

MCAA (Marie Curie Alumni Association). 2015. "Scientific Investigation on Unidentified Anomalous Phenomena." *6th MCAA Newsletter*. https://www.mariecuriealumni.eu/newsletter/scientific-investigation-unidentified-anomalous-phenomena-uap-0.

McDonald, James E. 1969. "Science in Default: Twenty-Two Years of Inadequate UFO Investigations." Presentation at the American Association for the Advancement of Science, 134th Meeting, General Symposium, Unidentified Flying Objects, Boston, December 27, 1969.

———. 1968. "Statement on Unidentified Flying Objects," submitted to the House Committee on Science and Astronautics, Rayburn Bldg., Washington, D.C. 29 July 1968. http://kirkmcd.princeton.edu/JEMcDonald/mcdonald_hcsa_68.pdf.

McFall, Caitlin. 2022. "Putin says one missile will trigger 'hundreds' of warheads in stark message on nuclear deterrence." *Fox News*, December 9, 2022. https://www.foxnews.com/world/putin-says-one-missile-trigger-hundreds-warheads-stark-message-nuclear-deterrence.

Meech, Karen. 2018. "The Alien Protocols." Season 13, Episode 3 of *Ancient Aliens*. History Channel, May 11, 2018. https://www.history.com/shows/ancient-aliens/season-13/episode-3.

Meessen, A., 2012. "Evidence of Very Strong Low Frequency Magnetic Fields." *PIERS [Progress In Electromagnetics Research Symposium] Proceedings*, Moscow, Russia, August 19-23, 2012. http://www.cobeps.org/pdf/meessen_evidence.pdf.

MEM (Middle East Monitor). 2023. "BRICS discussing decision on Saudi Arabia, Iran memberships this year." *MEM (Middle East Monitor)*, February 16, 2023. https://www.middleeastmonitor.com/20230216-brics-discussing-decision-on-saudi-arabia-iran-memberships-this-year/.

Messmer, Marion. 2023. "Opinion: The other nuclear threat you might have missed from Putin's speech." *CNN Opinion*, March 2, 2023. https://www.cnn.com /2023/03/02/opinions/russia-nuclear-test-putin-messmer/index.html.

Merton, Robert K. 1973. *The Sociology of Science: Theoretical and Empirical Investigations.* Chicago: University of Chicago Press.

Michael, Donald N. 1960. *Proposed Studies on the Implications of Peaceful Space Activities for Human Affairs.* A Report Prepared for the Committee on Long-Range Studies of the National Aeronautics and Space Administration [NASA] by the Brookings Institution. December 1960. Washington, D.C.: Brookings Institution. https://ntrs.nasa.gov/search.jsp?R=19640053196.

Miller, Gordon L. 1987. *Resonance, Information and the Primacy of Process: Ancient Light on Modern Information and Communication Theory.* Ph.D. Dissertation, Rutgers University.

Miller, Warner A., and John A. Wheeler. 1984. "Delayed-Choice Experiments and Bohr's Elementary Quantum Phenomenon." In *Foundations of Quantum Mechanics in the Light of New Technology*, edited by Susumu Kamefuchi, Hajime Ezawa, Yoshimasa Murayama, Shinichi Nomura, and Mikio Namiki, 140-157. Central Research Laboratory, Hitachi, Ltd., Kokubunji, Tokyo, Japan, August 29-31, 1983. Tokyo: Physical Society of Japan.

Minton, Leslie. 2018. "The future of humans' relationship with space and alien life." *ASU (Arizona State University) News*, February 16, 2018. https://news .asu.edu/20180216-discoveries-asu-aaas-meeting-human-space-future.

MIT CIS (Massachusetts Institute of Technology Center for International Studies). n.d. "History." MIT Center for International Studies. http://cis.mit.edu/history.

Mitra, Anwesha. 2023. "BRICS alliance working to create its own currency, says Russian official." *Mint*, April 2, 2023. https://www.live-mint.com/news /world/brics-alliance-working-to-create-its-own-currency-says-russian-official-11680445842563.html.

MOD ([U.K.] Ministry of Defence). 1951. Directorate of Scientific Intelligence and Joint Technical Intelligence Committee. "Unidentified Flying Objects." Report by the "Flying Saucer" Working Party, June 1951. http://www.ianridpath.com /ufo/flying%20saucer%20working%20party.pdf.

Monk, Nicholas A. M., and Basil J. Hiley. 1998. "A Unified Algebraic Approach to Quantum Theory." *Foundations of Physics Letters* 11: 371-77.

Moniz, Ernest J. 2020. "The risk of a nuclear attack is back at historic levels, 75 years after Hiroshima." *NBC News*, August 6, 2020. https://www.nbcnews .com/think/opinion/risk-nuclear-attack-back-historic-levels-75-years-after-hiroshima-ncna1235925.

Montesdeoca, Juan Fernando. 2019. "La Fuerza Aérea de Perú confirma avistamiento de ovnis cerca del aeropuerto de Lima." *CNN Latinoamérica*, June 18, 2019. https://cnnespanol.cnn.com/2019/06/18/la-fuerza-aerea-de-peru-confirma-avistamiento-de-ovnis-cerca-del-aeropuerto-de-lima/.

Moret, Leuren. 2004. "Depleted Uranium: The Trojan Horse of Nuclear War." *World Affairs: The Journal of International Issues* 8, no. 2 (April-June): 101-18. http://www.leurenmoret.info/archive/du-trojan-horse.html.

Morris, Michael S., Kip S. Thorne, and Ulvi Yurtsever. 1988. "Wormholes, Time Machines, and the Weak Energy Condition." *Physical Review Letters* 61, no. 13 (September 26): 1446-49.

Morrison, Philip, and William H. McNeill. 1973. "The Consequences of Contact." In *Communication with Extraterrestrial Intelligence: CETI*, edited by Carl Sagan, 333-49. Cambridge, MA: MIT Press.

Morrison, Philip, John Billingham, and John. Wolfe, editors. 1977. *The Search/or Extraterrestrial Intelligence*. NASA, SP-419. Washington, D.C.: National Aeronautics and Space Administration, NASA History Office.

Mosher, Dave. 2017. "An alien hunter explains why extraterrestrial visitors are unlikely—despite the US government's UFO evidence," *Business Insider*, December 21, 2017. https://www.businessinsider.com/pentagon-ufo-alien -research-program-footage-nytimes-2017-12.

Moskvitch, Katia. 2013. "A Link Between Wormholes and Quantum Entanglement." *Science*, December 2, 2013. https://www.sciencemag.org/news/2013/12/link -between-wormholes-and-quantum-entanglement.

Mossige-Norheim, Thea, and Lars-Olof Strömberg. 2022. "Polisen: Drönare övar kärnkraftverk i natt." *Expressen*, January 15, 2022. https://www. expressen.se/nyheter/uppgift-till-polisen-foremal-over-karnkraftverken/.

Mount, Adam. 2021. "What Is the Sole Purpose of U.S. Nuclear Weapons?" Federation of American Scientists (FAS), September 16, 2021. https://fas.org/pub-reports/sole-purpose/.

MSNBC. 2023. "White House announces team to address 'unidentified aerial objects.'" Statement by National Security official John Kirby. *MSNBC*, February 13, 2023. https://www.youtube.com/watch?v=FYfr8daqNVk.

Muraviev, Alexey. 2019. "Russia's nuclear submarine disaster will test President Vladimir Putin and his navy." *ABC (Australian Broadcasting Corporation) News*, July 3, 2019. https://www.abc.net.au/news/2019-07-03/russias-nuclear-submarine-disaster-test-vladimir-putin-navy/11274964.

MWL (Martin Willis Live Shows). 2021. "09-14-21 Robert Salas, Martin Keller, UAPs/UFOs & Nukes, and Mainstream Media." September 14, 2021. https://www.youtube.com/watch?v=GOW8QNpWJao.

———. 2020. "10-31-20 Paul Stonehill, RUSSIA'S USO SECRETS and UFOs Around the World." October 31, 2020. https://www.youtube.com/watch?v=7 CAFzyaTphI.

———. 2016. "Interviiew [*sic*] with Paul Stonehill, Russia's USO Secrets & Ufos, 08-24-2016." August 24, 2016. https://www.youtube.com/watch?v=SjQONKw-zG8.

NAA (National Archives of Australia). 1957-1971. "Scientific Intelligence – General – Unidentified Flying Objects." NAA: A13693, 309/2/000. https://record search.naa.gov.au/SearchNRetrieve/Interface/ViewImage.aspx?B=30030606&S= 7&R=0.

Nanu, Maighna. 2023. "Watch: Vladimir Putin deploys nuclear submarines in surprise Pacific drills." *The Telegraph*, April 18, 2023. https://www.telegraph .co.uk/world-news/2023/04/18/watch-vladimir-putin-nuclear-submarines-pacific/.

Narayanan, Nirmal. 2019. "Five UFOs Spotted Above Nuclear Power Plant In Fukushima, Conspiracy Theorists Suspect Alien Visit." *International Business Times (IBT)*, August 24, 2019. https://www.ibtimes.co.in/five-ufos-spotted-above-nuclear-power-plant-fukushima-conspiracy-theorists-suspect-alien-visit-804142.

NARCAP (National Aviation Reporting Center On Anomolous Phenomena). n.d. "International Technical Reports." National Aviation Reporting Center On Anomolous Phenomena. https://www.narcap.org/international-reports.

NAS (National Academy of Sciences). 1962. *A Review of Space Research*. The Report of the Summer Study Conducted under the Auspices of the Space Science Board of the National Academy of Sciences at the State University of Iowa, Iowa City, Iowa, June 17-August 10, 1962. Washington, D.C.: National Research Council. https://nap.national-academies.org/catalog/12421/a-review-of-space-research.

NASA (National Aeronautics and Space Administration). 2022. "NASA to Set Up Independent Study on Unidentified Aerial Phenomena." NASA.gov, June 9, 2022. Updated December 22, 2022 to reflect a change in terminology for UAP from Unidentified Aerial Phenomena to Unidentified Anomalous Phenomena. https://www.nasa.gov/feature/nasa-to-set-up-independent-study-on-unidentified-aerial-phenomena/.

―――. n.d. "Pioneers – Hermann Oberth and Wernher von Braun." NASA .gov. https://www.nasa.gov/audience/foreducators/rocketry/image-gallery/rp_ OberthAward.jpg.html#.Y16TFnbMJD8.

NASA (National Aeronautics and Space Administration) History Office. 2003. "SETI: The Search/or Extraterrestrial Intelligence." National Aeronautics and Space Administration, NASA History Program Office. https://history.nasa .gov/seti.html.

NASIC (National Air and Space Intelligence Center). n.d. "National Air and Space Intelligence Center Heritage." National Air and Space Intelligence Center. https://www.nasic.af.mil/About-Us/Fact-Sheets/Article/611728/national-air-and-space-intelligence-center-heritage/.

The National Archives [U.K.]. 2011. "UAP in the UK Air Defence Region: Volume 3." Created May 15, 2006, last updated July 26, 2011. https://webarchive .nationalarchives.gov.uk/ukgwa/20121110115232/http://www.mod.uk/DefenceIn ternet/FreedomOfInformation/PublicationScheme/SearchPublicationScheme/Ua pInTheUkAirDefenceRegionVolume3.htm.

―――. 2013. "UFOs: Newly released UFO files from the UK government." The National Archives [U.K.], files released in June 2013, http://www.national archives.gov.uk/ufos/.

National Archives [U.S.]. 2019. UFO Exhibit. National Archives [U.S.]. December 6, 2019-January 8, 2020.

National Security Archive. 2016. "Unilateral U.S. nuclear pullback in 1991 matched by rapid Soviet cuts." National Security Archive, September 30, 2016. https://nsarchive.gwu.edu/briefing-book/nuclear-vault-russia-programs/2016-09-30/unilateral-us-nuclear-pullback-1991-matched.

NATO (North Atlantic Treaty Organization). 2023. "NATO's nuclear deterrence policy and forces." North Atlantic Treaty Organization (NATO), April 11, 2023. https://www.nato.int/cps/en/natohq/topics_50068.htm.

————. 1968. A Report by the Military Committee to the Defence Planning Committee on Overall Strategic Concept for the Defense of the North Atlantic Treaty Organization Area." Final Decision on MC 14/3. North American Military Committee, January 16, 1968. NATO Strategy Documents *1949-1969*. https://www.nato.int/docu/stratdoc/eng/a680116a.pdf.

NCTV (New China TV). 2023. "GLOBALink | UN Security Council probe into Nord Stream explosions is global priority: U.S. scholar." *New China TV*, February 23, 2023, https://www.youtube.com/watch?v=1g3UM5VK4qs.

NDAA FY22 (National Defense Authorization Act for Fiscal Year 2022). 2021. Public Law 117-81, 135 Stat. 1541-2450. Section 1683, 578-83. https://www.congress.gov /117/plaws/publ81/PLAW-117publ81.pdf. [Because the NDAA for FY22 was finalized in 2021, bibliographic consistency would render this citation 'NDAA FY22 2021, page number.' For readability and to prevent confusion, I have rendered it as 'NDAA FY22, page number.']

NDAA FY23 (National Defense Authorization Act for Fiscal Year 2023). 2022. Rules Committee Print 117-70, Text of the House Amendment to the Senate Amendment to H.R. 7776 [Showing the text of the James M. Inhofe National Defense Authorization Act for Fiscal Year 2023]. Section 1673, 565-68. https://www.congress.gov/117/bills/hr7776/BILLS-117hr7776enr.pdf. [Because the NDAA for FY23 was finalized in 2022, bibliographic consistency would render this citation 'NDAA FY23 2022, page number.' For readability and to prevent confusion, I have rendered it as 'NDAA FY23, page number.']

Nelson, Clarence William Nelson II (a.k.a., Bill Nelson). 2021. "Interview with Bill Nelson October 19, 2021." October 19, 2021. https://www.youtube.com/watch? v=9hH1XEqKlTs.

New Scientist, and Afp. 2007. "France opens up its UFO files." *New Scientist*, March 22, 2007. https://www.newscientist.com/article/dn11443-france-opens-up-its-ufo-files/.

News18 India. 2021. "Pentagon Just Admitted to 'Testing' UFO Wreckage. But What Did They Discover?" *News18 India*, February 16, 2021. https://www.news18.com/news/buzz/the-pentagon-just-admitted-to-testing-ufo-wreckage-heres-what-they-have-discovered-3436577.html.

NIA (Nuclear Industry Association). 2023. "New Nuclear Fuel Agreement Alongside G7 Seeks to Isolate Putin's Russia." Nuclear Industry Association, April 17, 2023. https://www.niauk.org/new-nuclear-fuel-agreement-alongside-

g7-seeks-to-isolate-putins-russia/#:~:text=An%20alliance%20between%20
the%20UK,Russia's%20grip%20on%20supply%20chains.

NICAP (National Investigations Committee On Aerial Phenomena). n.d.-a.
"Category 3: Electromagnetic Effects." NICAP (National Investigations
Committee On Aerial Phenomena), Years 1945 to 1963, https://
nicap.org/cat3-to1964.htm. Years 1964 to 1992, https://nicap.org/cat3-
1964on.htm.

———. 1960. *Electro-Magnetic Effects Associated with Unidentified Flying Objects
(UFOs)*. Washington, D.C.: Subcommittee of the National Investigations
Committee on Aerial Phenomena (NICAP). https://majesticdocuments.com
/pdf/nicap-electromagneticeffects_june60.pdf.

———. n.d.-b. "UFO Auto-Tracked, Sends Phony IFF." RADCAT Case Directory
Category 9, RADAR. National Investigations Committee On Aerial Phenomena
(NICAP), May 15, 1964. http://www.nicap.org/640515whitesands_dir.htm.

The Nimitz Encounters. 2022. "2004 USS Ronald Reagan UFO Encounter Witness –
Patrick Gokey." July 5, 2022. https://www.youtube.com/watch?v=O5KMG
_WYwH0.

nippon (nippon.com). 2021. "Japan's Nuclear Power Plants in 2021." nippon.com,
March 31, 2021. https://www.nippon.com/en/japan-data/h00967/.

Northrop Grumman. 2022. "Sentinel – The Ground Based Strategic Deterrent."
Northrop Grumman. https://www.northropgrumman.com/space/sentinel/.

Norton-Taylor, Richard. 2013. "Alien nation: MoD releases final UFO files." *The
Guardian*, June 21, 2013. https://www.theguardian.com/uk/2013/jun/21/last-
release-mod-ufo-files.

———. 2023. "Did the UK Deploy a Nuclear-Armed Submarine to the Falklands
Conflict?" *Declassified UK*, April 6, 2023. https://declassifieduk.org/did-uk-
deploy-nuclear-armed-submarine-to-falklands-conflict/.

NPAIP (Norwegian People's Aid, and ICAN Partner). 2022. *Nuclear Weapons Ban
Monitor 2021: Tracking progress towards a world without nuclear weapons*. March
2022. Oslo: Norwegian People's Aid and ICAN Partner. https://banmonitor.org
/files/Nuclear-Weapons-Ban-Monitor/TNWBM_2021_290322_pages_small_
web-2.pdf.

NTI (Nuclear Threat Initiative). 2023a. "Comprehensive Nuclear Test Ban Treaty
(CTBT)." NTI (Nuclear Threat Initiative). https://www.nti.org/education-
center/treaties-and-regimes/comprehensive-nuclear-test-ban-treaty-ctbt/.

———. 2023b. "FMCT: Proposed Fissile Material (Cut-Off) Treaty (FMCT)." https://www.nti.org/education-center/treaties-and-regimes/proposed-fissile-material-cut-off-treaty/.

NWC (U.S. Naval War College). 2022. "Intelligence Studies: Types of Intelligence Collection." U.S. Naval War College Library, September 29, 2022. https://usnwc.libguides.com/c.php?g=494120&p=3381426.

NYP (New York Post). 2022. "NEW! UFOs, Werewolves & Ghosts | Shocking truth of Pentagon AAWSAP program | The Basement Office." May 12, 2022. https://www.youtube.com/watch?v=6XD4gQS_-qY.

NYT (The New York Times). 1963. "Extraterrestrial Life." *The New York Times*, January 10, 1963. https://www.nytimes.com/1963/01/10/archives/extraterrestrial-life.html.

———. 1979. "C.I.A. Papers Detail U.F.O. Surveillance." *The New York Times*, January 14, 1979. https://www.nytimes.com/1979/01/14/archives/cia-papers-detail-ufo-surveillance-agencys-secret-studies-convince.html.

NZDF (New Zealand Defence Force). 1981. *Unidentified Flying Objects (UFO) Files Copies for Release to the Public*. Internet Archive, uploaded December 11, 2015 [declassified December 2010]. https://archive.org/details/NewZealandUFO/AIR-1080-6-897-Volume-1-1978-1981/mode/2up.

Obama, Barack Hussein II. 2021. "Reggie Watts to Barack Obama: What's w/Dem Aliens?" The Late Show with James Corden. https://www.youtube.com/watch?v=xp6Ph5iTIgc.

Öberg, Karin I., Viviana V. Guzmán, Catherine Walsh, et al. 2021. "Molecules with ALMA at Planet-forming Scales (MAPS). I. Program Overview and Highlights. *The Astrophysical Journal Supplement Series* 257, no. 1 (November): (29 pages). https://doi.org/10.3847/1538-4365/ac1432.

Oberhaus, Daniel. 2019. A Mythical Form of Space Propulsion Finally Gets a Real Test. *Wired*, June 5, 2019. https://www.wired.com/story/a-mythical-form-of-space-propulsion-finally-gets-a-real-test/.

———. 2020. NASA's EmDrive Leader Has a New Interstellar Project. *Wired*, May 8, 2020. https://www.wired.com/story/nasas-emdrive-leader-has-a-new-interstellar-project/.

———. 2018. "The Pentagon Released New Documents About the 'Tic-Tac UFO.'" *Vice*, May 24, 2018. https://www.vice.com/en_us/article/9k8kaz/the-pentagon -released-new-documents-about-the-tic-tac-ufo.

Oberth, Hermann. 1954. "Lecture Notes For Lecture About Flying Saucers 1954." https://drive.google.com/file/d/1JskZ-ljVL2Uz0CtxcBSb7Q6_w6UThwyB/view.

OCSA (Office of the Chief Science Advisor of Canada). 2023. "OCSA's Sky Canada Project." March 14, 2023. https://www.youtube.com/watch?v=c_q7RE_tjvU.

ODNI (Office of the Director of National Intelligence). 2022. *2022 Annual Report on Unidentified Aerial Phenomena*. McLean, VA: Office of the Director of National Intelligence, January 12, 2023. https://www.dni.gov/files/ODNI/documents/ assessments/Unclassified-2022-Annual-Report-UAP.pdf. [Although the unclassified version of this report was released in 2023, I am citing it under 2022 since reports are expected every year.]

———. 2021. *Preliminary Assessment: Unidentified Aerial Phenomena*. McLean, VA: Office of the Director of National Intelligence, June 25, 2021. https://www.dni.gov/files/ODNI/documents/assessments/Prelimary- Assessment-UAP-20210625.pdf.

———. n.d.-a. *U.S. Intelligence Community Budget*. McLean, VA: Office of the Director of National Intelligence. https://www.dni.gov/index.php/what-we-do /ic-budget.

———. n.d.-b. *Members of the IC*. McLean, VA: Office of the Director of National Intelligence. https://www.dni.gov/index.php/what-we-do/members-of-the-ic.

———. n.d.-c. "National Intelligence Council – Who We Are." McLean, VA: Office of the Director of National Intelligence. https://www.dni.gov/index.php /who-we-are/organizations/mission-integration/nic/nic-who-we-are#:~:text=The %20NIC's%20National%20Intelligence%20Officers,of%20regional%20and%20 functional%20issues.

Oliver, Bernard M., and John Billingham, editors. 1971. *Project Cyclops: A Design Study of a System for Detecting Extraterrestrial Intelligent Life*. Prepared under Stanford / NASA / Ames Research Center 1971 Summer Faculty Fellowship Program in Engineering Systems Design. https://ntrs.nasa.gov/api/ citations/19730010095/downloads/19730010095.pdf.

Oswald, Michael, director. 2014. *Princes of the Yen: Central Banks and the Transformation of the Economy*. Written by Michael Oswald. https:// www.imdb.com/title/tt4172710/. [Also uploaded on Independent POV as

"Princes of the Yen | Documentary Film." November 4, 2014. https://www.youtube.com/watch?v=p5Ac7ap_MAY.]

Otis, Daniel. 2021. "Credible UFO Reports Are Being Ignored, Declassified Canadian Government Documents Reveal." *Vice*, November 29, 2021, https://www.vice.com/en/article/5dggbx/canada-military-rcaf-ufos-sightings.

Page, Thornton. 1969. "Scientific Study of Unidentified Flying Objects: Final Report of Research Conducted by the University of Colorado for the Air Force Office of Scientific Research under the Direction of Edward U. Condon." Book review. *American Journal of Physics* 37, no. 10: 1071-72. https://pubs.aip.org/aapt/ajp /article/37/10/1071/1048124/Scientific-Study-of-Unidentified-Flying-Objects.

Panda, Ankit, and Vipin Narang. 2021. "Sole Purpose Is Not No First Use: Nuclear Weapons and Declaratory Policy." *War on the Rocks*, February 22, 2021. https://warontherocks.com/2021/02/sole-purpose-is-not-no-first-use-nuclear-weapons-and-declaratory-policy/.

Parfitt, Tom. 2009. "Action man Vladimir Putin turns submariner at Lake Baikal." *The Guardian*, August 2, 2009. https://www.theguardian.com/world/2009/ aug/02/russia-vladimir-putin-lake-baikal.

Parks, Miles, Deepa Shivaram, and Greg Myre. 2023. "More UFOs Shot Down By US Air Force." *NPR (National Public Radio)*, February 13, 2023. https://www .npr.org/2023/02/13/1156606425/more-ufos-shot-down-by-us-air-force.

Pasetta, Martin, director. 1988. *UFO Cover-Up?: Live!* Written by Barry Taff and Tracy Tormé. https://www.imdb.com/title/tt0846738/.

Pasternack, Alex. 2016. "The Moon-Walking, Alien-Hunting, Psychic Astronaut Who Got Sued By NASA." *Vice*, May 14, 2016. https://www.vice.com/en /article/aek7ez/astronaut-edgar-mitchell-outer-space-inner-space-and-aliens.

Paul, Deanna. 2019. "How angry pilots got the Navy to stop dismissing UFO sightings." *The Washington Post*, April 25, 2019. https://www.washington post.com/national-security/2019/04/24/how-angry-pilots-got-navy-stop-dismissing-ufo-sightings/.

PBS (Public Broadcasting Service). 2011. "The Fabric of the Cosmos: Quantum Leap." Season 13, Episode 18 of *NOVA*, November 15, 2011. https://www.pbs.org/video/nova-the-fabric-of-the-cosmos-quantum-leap/.

PDIWG (Planetary Defense Interagency Working Group). 2023. *National Preparedness Strategy & Action Plan for Near-Earth Object Hazards and Planetary*

Defense. A Product of the Planetary Defense Interagency Working Group of the National Science & Technology Council. April 2023. https://www.whitehouse .gov/wp-content/uploads/2023/04/2023-NSTC-National-Preparedness-Strategy -and-Action-Plan-for-Near-Earth-Object-Hazards-and-Planetary-Defense.pdf.

The Peak. 2023. "Science Advisor looks skyward." The Peak: Proudly Canadian, March 3, 2023. https://www.readthepeak.com/stories/03-23-science-advisor-looks-skyward.

Pearce, Fred. 2015. "Global Extinction Rates: Why Do Estimates Vary So Wildly?" Yale Environment 360, August 17, 2015. https://e360.yale.edu/features/global_ extinction_rates_why_do_estimates_vary_so_wildly#:~:text=Convention%20on %20Biological%20Diversity%20concluded,as%2010%20percent%20a%20decade.

Perez, Zamone. 2023. "Congress calls for more funding of Pentagon UFO office." *Military Times*, March 31, 2023. https://www.militarytimes.com/news/your-military/2023/03/31/congress-requests-more-funds-for-pentagon-ufo-office-in-budget-request/.

Perkins, Richard H., Michelle T. Bensi, Jacob Philip, and Selim Sancaktar. 2011. *Screening Analysis Report for the Proposed Generic Issue on Flooding of Nuclear Power Plant Sites Following Upstream Dam Failures*. U.S. Nuclear Regulatory Commission, Office of Nuclear Regulatory Research, Division of Risk Analysis. https://big.assets.huffingtonpost.com/flooding.pdf.

Phelan, Matthew. 2019. "Navy Pilot Who Filmed the 'Tic Tac' UFO Speaks: 'It Wasn't Behaving by the Normal Laws of Physics.'" *Intelligencer – New York Magazine*, December 19, 2019. https://nymag.com/intelligencer /2019/12/tic-tac-ufo-video-q-and-a-with-navy-pilot-chad-underwood.html.

Picheta, Rob. 2020. "Aliens definitely exist and they could be living among us on Earth, says Britain's first astronaut." *CNN*, January 6, https://edition.cnn .com/2020/01/06/uk/helen-sharman-aliens-exist-scli-scn-gbr-intl/index.html.

Pike, John. 2000. "CVN-68 Nimitz-Class." FAS (Federation of American Scientists) Military Analysis Network, maintained by Robert Sherman, updated January 8, 2000. https://man.fas.org/dod-101/sys/ship/cvn-68.htm.

———. 1996. "Central Intelligence Agency." FAS (Federation of American Scientists), Intelligence Resource Program, maintained by Steven Aftergood, updated September 23, 1996. https://irp.fas.org/cia/ciahist.htm.

Pilger, John, director. 2016. *The Coming War on China*. Written by John Pilger. https://www.imdb.com/title/tt6197028/. [Also uploaded on True Story Documentary Channel as "Nuclear war is not only imaginable, but planned –

True Story Documentary Channel." January 12, 2020. https://www.youtube.com/watch?v=vAfeYMONj9E.]

Polansky, Anne (with the assistance of Lawrence Criscione). 2018. "As Fossil Fuels Melt the Planet, Could Climate Change Cause a Nuclear Meltdown?" Government Accountability Project, April 9, 2018. https://whistle blower.org/general/climate-science-watch/2018-04-09-as-fossil-fuels-melt-the-planet-could-climate-change-cause-a-nuclear-meltdown/.

Pollina, Richard. 2023. "Lawmakers call for more funding of Pentagon UFO office following Biden budget request." *The New York Post*, April 4, 2023. https://nypost.com/2023/04/04/lawmakers-call-for-more-funding-of-pentagon-ufo-office/.

Pompeo, Michael R. 2022. "Nuclear Weapons, China, and a Strategic Defense Initiative for this Century." The National Interest, January 18, 2022. https://nationalinterest.org/feature/nuclear-weapons-china-and-strategic-defense-initiative-century-199549.

Pozsar, Zoltan. 2023. "Great power conflict puts the dollar's exorbitant privilege under threat." Opinion. *Financial Times*, January 20, 2023. https://www.ft.com/content/3e05b491-d781-4865-b0f7-777bc95ebf71.

President of Russia. 2022. "Joint Statement of the Russian Federation and the People's Republic of China on the International Relations Entering a New Era and the Global Sustainable Development." http://en.kremlin.ru/supplement/5770.

Princeton University. 2023. "Frank von Hippel (1937-present). Princeton University." https://nuclearprinceton.princeton.edu/people/frank-von-hippel.

PRPS (Paranormal Research Paul Stonehill). 2021. "The UFO That Rattled Moscow and Soviet Strategic Missile Forces … ." May 31, 2021. https://www.youtube.com/watch?app=desktop&v=Vt-WtxbcC_E.

———. 2022. "UNDERWATER MYSTERIES OF LAKE BAIKAL: SHAMAN, UFOS, ALIENS." November 18, 2022. https://www.youtube.com/watch?v=2n NzF21tPRE.

Pylkkänen, Paavo T. I. 2007. *Mind, Matter and the Implicate Order*. Berlin/Heidelberg/New York: Springer-Verlag.

Quest TV. 2020. "Did The Soviet Union Discover Aliens In The Deepest Lake In The World? | UFOs: The Lost Evidence." September 21, 2020. https://www.youtube.com/watch?v=DVq7gBH70WE.

Ramírez Muro, Verónica. 2014. "Por qué Perú reabrió la oficina de búsqeda de ovnis." *BBC*, February 5, 2014. https://www.bbc.com/mundo/noticias/2014/02/140205_peru_ovni_gobierno_busqueda_vh.

Randall, Audrey. 2014. "Earthquakes and Nuclear Power Plants." *Science On a Sphere*, August 1, 2014. https://sos.noaa.gov/catalog/datasets/earthquakes-and-nuclear-power-plants/.

Randerson, James. 2007. "MoD opens its files on UFO sightings to public." *The Guardian*, May 3, 2007. https://www.theguardian.com/politics/2007/may/03/spaceexploration.military.

Randle, Kevin D. 2022. *The Washington Nationals: Flying Saucers over the Capital*. Pontefract, West Yorkshire, UK: Flying Disk Press

Ray, Deepak K. 2022. "Even a small nuclear war threatens food security." *Nature Food* 3 (August 15): 567-68. https://www.nature.com/articles/s43016-022-00575-y.

Regan, Helen, Julia Kesaieva, Mariya Knight, Katharina Krebs, Kostan Nechyporenko, and Stephanie Halasz. 2023. "Evacuations from Zaporizhzhia renew concerns for nuclear power plant safety." *CNN*, May 8, 2023. https://www.cnn.com/2023/05/08/europe/zaporizhzhia-evacuations-ukraine-russia-intl-hnk/index.html.

Regis, Jr., Edward, editor. 1985. *Extraterrestrials: Science and alien intelligence,* edited by Edward Regis, Jr. Cambridge, U.K.: Cambridge University Press.

Revell, Eric. 2023. "Feds may throw struggling First Republic Bank a lifeline by expanding emergency lending program." *Fox Business*, March 26, 2023. https://www.foxbusiness.com/economy/feds-may-throw-struggling-first-republic-bank-lifeline-expanding-emergency-lending-program.

Ridge, Francis L. 2005a. "The Camp Hood Sightings." The Nuclear Connection Project, updated September 22, 2005. https://www.nicap.org/texas/texassightings.htm.

———. 2005b. "The New Mexico Sightings." The Nuclear Connection Project, updated September 3, 2005. https://www.nicap.org/nmexico/newmexicosightings.htm.

———. 2016 <2003>. "Nuclear Connection Project (NCP)." Established July 1, 2003, updated December 23, 2016. http://www.nicap.org/ncp.htm.

Rising, David. 2023. "UN nuclear watchdog's safety concerns rise as fighting intensifies near Ukraine plant." *PBS (Public Broadcasting Service)*, May 7, 2023. https://www.pbs.org/newshour/world/un-nuclear-watchdogs-safety-concerns-rise-as-fighting-intensifies-near-ukraine-plant.

Rivas, Ray, director. 1974. *UFOs: Past, Present, and Future*. Written by Robert Emenegger. https://www.imdb.com/title/tt0365874/.

Rogan, Tom. 2019a. "Birds, balloons, or a UFO: What violated DC airspace?" *Washington Examiner*, November 27, 2019. https://www.washington examiner.com/opinion/birds-balloons-or-a-ufo-what-violated-dc-airspace.

———. 2022. "Unidentified drones sighted flying over US government nuclear labs." *Washington Examiner*, December 22, 2022. https://www.washington examiner.com/opinion/unidentified-drones-sighted-flying-over-us-government-nuclear-labs.

———. 2019b. "US Army signs contract to study UFO material and make better weapons." *Washington Examiner*, October 18, 2019. https://www.washington examiner.com/opinion/us-army-signs-contract-to-study-ufo-material-and-make-better-weapons.

Rogers, Paul. 2023a. "Putin's nuclear threat and NATO nuclear strategy." Culturico, February 18, 2023. https://culturico.com/2023/02/18/putins-nuclear-threat-and-nato-nuclear-strategy/.

———. 2023b. "The risk of nuclear war over Ukraine is real. We need diplomacy now." *openDemocracy*, April 14, 2023. https://www.opendemocracy.net/en/russia-ukraine-nuclear-war-threat-crisis-diplomacy-resolve/.

———. 1996. *Sub-strategic Trident: A slow-burning fuse*. University of London: Centre for Defence Studies.

Rogoway, Tyler, and Joseph Trevithick. 2020a. "Mysterious Drone Incursions Have Occurred Over U.S. THAAD Anti-Ballistic Missile Battery In Guam." *The Drive*, September 14, 2020. https://www.thedrive.com/the-war-zone/36085/troubling-drone-incursions-have-occurred-over-guams-thaad-anti-ballistic-missile-battery.

———. 2020b. "The Night A Mysterious Drone Swarm Descended On Palo Verde Nuclear Power Plant." *The Drive*, July 29, 2020. https://www.thedrive.com/the-war-zone/34800/the-night-a-drone-swarm-descended-on-palo-verde-nuclear-power-plant.

———. 2019. "The SR-71 Blackbird's Predecessor Created 'Plasma Stealth' By Burning Cesium-Laced Fuel." *The Drive*, September 12, 2019. https://www.thedrive.com/the-war-zone/29787/the-sr-71-blackbirds-predecessor-created-plasma-stealth-by-burning-cesium-laced-fuel.

Rubio, Marco @marcorubio. 2023. "Why is the White House creating a new 'interagency team' to monitor,investigate & report on unidentified aerial objects when we already have @DoD_AARO which we helped created over two years ago?" February 13, 2023, 3:02 PM. Tweet.

Rudd, Kevin. 2021. "How to keep US-China rivalry from starting a nuclear arms race." *South China Morning Post*, December 19, 2021. https://www.scmp.com/comment/opinion/article/3159963/how-keep-us-china-rivalry-starting-nuclear-arms-race.

Russian Roswell (Russian Roswell: Inside Kapustin Yar). n.d. *Russian Roswell: Inside Kapustin Yar*. History Channel. https://www.history.com/videos/russian-roswell-inside-kapustin-yar?cmpid=MRSS_tvguide_HIS.

Sagan, Carl, editor. 1973. *Communication with Extraterrestrial Intelligence: CETI*. Cambridge, MA: MIT Press.

———, editor. 1971. *Soviet-American Conference on the Problems of Communication with Extraterrestrial Intelligence*. Proceedings from conference held on September 5-11, 1971 at the Byurakan Astrophysical Observatory, U.S.S.R. Conference was organized jointly by the U.S. National Academy of Sciences (with assistance from the U.S. National Science Foundation) and the U.S.S.R. Academy of Sciences (Akademiiā̄nauk SSSR).

Sagan, Carl, and Frank Drake. 1975. "The Search for Extraterrestrial Intelligence." *Scientific American* 232, no. 5 (May): 80-89. https://www.jstor.org/stable/24949801.

Sanderson, Ivan T. 1970. *Invisible Residents: A Disquisition upon Certain Matters Maritime, and the Possibility of Intelligent Life under the Waters of This Earth*. Cleveland, OH: The World Publishing Co. [Reprinted in 2005 as *Invisible Residents: The Reality of Underwater UFOs*. Kempton, IL: Adventures Unlimited Press.]

Sanger, David E. 2021. "Hundreds of Scientists Ask Biden to Cut the U.S. Nuclear Arsenal." *The New York Times*, December 16, 2021. https://www.nytimes.com/2021/12/16/us/politics/scientists-letter-nuclear-arsenal.html.

SASC (United States Senate Committee on Armed Services). 2023. Subcommittee on Emerging Threats and Capabilities. "Open/Closed: To Receive Testimony on the Mission, Activities, Oversight, and Budget of the All-Domain Anomaly Resolution Office." April 19, 2023. https://www.armed-services.senate .gov/hearings/to-receive-testimony-on-the-mission-activities-oversight-and-budget-of-the-all-domain-anomaly-resolution-office. [The hearing begins at timestamp 1:53:27.]

Scahill, Jeremy. 2023. "Russia Calls for U.N. Investigation of Nord Stream Attack, as Hersh Accuses White House of False Flag." *The Intercept*, March 25, 2023. https://theintercept.com/2023/03/25/nord-stream-russia-investigation-seymour-hersh/.

Scaruffi, Piero. 2000. *David Bohm: Wholeness and the Implicate Order (Routledge, 1980)*. Book review. https://www.scaruffi.com/mind/bohm.html#:~:text=Bohm%20sug gested%20that%20the%20implicate,and%20ultimately%20space%20and%20time.

Schaffer, Jonathan. 2003. "Is there a fundamental level?" *Nous* 37, no. 3: 498-517. http://jonathanschaffer.org/fundamental.pdf.

Schaper, Annette. 2013. *Highly Enriched Uranium, a Dangerous Substance that Should Be Eliminated*. PRIF-Report No. 124. Frankfurt am Main, Germany: Peace Research Institute Frankfurt (PRIF). https://www.hsfk.de/fileadmin/HSFK/ hsfk_downloads/prif124.pdf.

Schenkel, Peter. 1988. *ETI: A challenge for change*. New York: Vantage Press.

Schneider, Mycle. 2022. *The World Nuclear Industry Status Report 2022*. Paris: A Mycle Schneider Consulting Project. October 2022. https://www.world nuclearreport.org/IMG/pdf/wnisr2022-v3-hr.pdf.

Schrödinger, Erwin. 1926. "Quantisierung als Eigenwertproblem (trans. Quantization as an Eigenvalue Problem)." *Annalen der Physik* 79: 361-76. https://uni-tuebingen.de/fileadmin/Uni_Tuebingen/Fakultaeten/MathePhysik/ Institute/IAP/Forschung/MOettel/Geburt_QM/schrodinger_AnnPhys_385_437 _1926.pdf.

———. 1944. *What Is Life? The Physical Aspect of the Living Cell*. Cambridge: Cambridge University Press.

Scoles, Sarah. 2019. "It's Sentient." *The Verge*, July 31, 2019. https://www.theverge .com/2019/7/31/20746926/sentient-national-reconnaissance-office-spy-satellites-artificial-intelligence-ai.

Scully, Marlan O. 1998. "Do Bohm trajectories always provide a trustworthy physical picture of particle motion?" *Physica Scripta* 1998, no. T76: 41-46. https://iopscience.iop.org/article/10.1238/Physica.Topical.076a00041/meta.

seniorsam. 2015. "COVER UP AT The Vandenberg UFO Shot Down a Missile THERE !!!." June 12, 2015. https://www.youtube.com/watch?v=i1LF6u7jgi0.

SETI@home. n.d. https://setiathome.berkeley.edu/.

SETI Institute. 2004. "The Center for SETI Research." *SETI Institute*. [This version of the website is no longer available.]

Sguazzin, Antony. 2023. "BRICS Debates Expansion as Iran, Saudi Arabia Seek Entry." *Bloomberg*, February 15, 2023. https://www.bloomberg.com/news/articles/2023-02-15/brics-debates-expansion-as-iran-saudi-arabia-seek-entry?leadSource=uverify%20wall.

Sheldrake, Rupert. 1981. *A New Science of Life: The Hypothesis of Formative Causation*. Los Angeles: J.P. Tarcher.

———. 1995 <1981>. *A New Science of Life: The Hypothesis of Morphic Resonance*. Rochester, VT: Park Street Press.

———. 1995 <1988>. *The Presence of the Past: Morphic Resonance and the Habits of Nature*. Rochester, VT: Park Street Press.

Shiva, Vandana. 1999. *Biopiracy: The Plunder of Nature and Knowledge*. Boston: South End Press.

Siberian Times. 2015. "Aliens and UFOs at the world's deepest lake." *The Siberian Times*, March 30, 2015. https://siberiantimes.com/other/others/features/f0077-aliens-and-ufos-at-worlds-deepest-lake/.

Siegel, Ethan. 2022. "There are more galaxies in the Universe than even Carl Sagan ever imagined." Big Think, June 22, 2022. https://bigthink.com/starts-with-a-bang/galaxies-in-universe/.

Sigalos, MacKenzie. 2023. "What the failures of Signature, SVB and Silvergate mean for the crypto sector." *CNBC*, March 12, 2023.

Simmons, Ann M., and Austin Ramzy. 2023. "Russia-China Summit Showcases Challenge to the West." *The Wall Street Journal*, March 21, 2023. https://www.wsj.com/articles/china-xi-jinping-vladimir-putin-meet-in-russia-400d39e1.

Sinclair, Ward, and Art Harris. 1979. "What Were Those Mysterious Craft?" *The Washington Post*, January 19, 1979. https://www.washingtonpost.com/

archive/politics/1979/01/19/what-were-those-mysterious-craft/1b9d1f3d-dddb-
4a92-87b3-0143aa5d7a3e/.

Siripala, Thisanka. 2020. "Japan's Defense Ministry Launches Protocol for UFO
Sightings." *The Diplomat*, September 18, 2020. https://thediplomat.com
/2020/09/japans-defense-ministry-launches-protocol-for-ufo-sightings/.

Slijepcevic, Predrag, and Chandra Wickramasinghe. 2021. "Reconfiguring SETI in
the microbial context: Panspermia as a solution to Fermi's paradox." *BioSystems*
206, no. 104441. https://doi.org/10.1016/j.biosystems.2021.10441.

Smith, R. Jeffrey. 2021. "The US Nuclear Arsenal is Becoming More Destructive and
Possibly More Risky." *The Center for Public Integrity*, October 29, 2021.
https://publicintegrity.org/national-security/future-of-warfare/nuclear-
weapon-arsenal-more-destructive-risky/.

Snyder, Susi. 2021. *Perilous Profiteering: The companies building nuclear arsenals and their
financial backers.* November 11, 2012. Utrecht: PAX and ICAN.
https://d3n8a8pro7vhmx.cloudfront.net/ican/pages/2331/attachments/original/
1637141262/2021_Perilous_Profiteering_Final.pdf?1637141262.

———. 2022. *Rejecting Risk: 101 Policies against nuclear weapons.* January 19, 2022.
Utrecht: PAX and ICAN. https://d3n8a8pro7vhmx.cloudfront.net/ican/
pages/2490/attachments/original/1642593421/RejectingRisk-eb.pdf?1642593421.

Sokolski, Henry, and Andrea Stricker. 2023. "Biden is letting American help fund
Russia's nuclear-weapon complex." *New York Post*, April 6, 2023.
https://nypost.com/2023/04/06/biden-is-letting-america-help-fund-russias-
nuclear-weapon-complex/. [This article also is available on the Foundation for
Defense of Democracies (FDD) website, https://www.fdd.org/analysis/2023/
04/06/biden-is-letting-america-help-fund-russias-nuclear-weapon-complex/.]

Sorkin, Andrew Ross, Ravi Mattu, Bernhard Warner, Sarah Kessler, Michael J. de la
Merced, Lauren Hirsch, and Ephrat Livni. 2023. "Davos Worries About a
'Polycrisis.'" *The New York Times*, January 17, 2023. https://www.nytimes
.com/2023/01/17/business/dealbook/davos-world-economic-forum-
polycrisis.html.

Sornette, Didier, and Guy Ouillon. 2012. "Dragon-kings: Mechanisms, statistical
methods and empirical evidence." *The European Physical Journal Special Topics*
205, no. 1: 1-26.

SPIEGEL. 2012. "Israel's Deployment of Nuclear Missiles on Subs from Germany."
SPIEGEL International, June 4, 2012. https://www.spiegel.de/international

/world/israel-deploys-nuclear-weapons-on-german-built-submarines-a-
836784.html.

Stanley, Jason. 2021. "America is now in fascism's legal phase." *The Guardian*,
December 22, 2021. https://www.theguardian.com/world/2021/dec/22/america
-fascism-legal-phase.

Stanley, Robert. 2006. *Close Encounters on Capitol Hill*. Providence, RI: Ūnicŭs Press.

———. 2011. *Covert Encounters in Washington, D.C.* Providence, RI: Ūnicŭs Press.

———. 2005. "UFOs on Capitol Hill." *Nexus Magazine*, Part 1: August-September
2005, 55-58; Part 2: October-November 2005, 45-51, 77.

Starr, Barbara. 2022. "Test rocket carrying component for future nuclear armed
ICBM explodes after takeoff." *CNN*, July 8, 2022. https://
www.cnn.com/2022/07/08/politics/test-rocket-explodes/index.html.

Steele, Edward J., Shirwan Al-Mufti, Kenneth A. Augustyn, et al. 2018. "Cause of
Cambrian Explosion - Terrestrial or Cosmic?" *Progress in Biophysics and
Molecular Biology* 136: 3-23. https://doi.org/10.1016/j.pbiomolbio.2018.03.004.

Steele, Edward J., Reginald M. Gorczynski, Robyn A. Lindley, et al. 2019. "Lamarck
and panspermia - On the Efficient Spread of Living Systems Throughout the
Cosmos." *Biophysics and Molecular Biology* 149: 10-32.
https://doi.org/10.1016/j.pbiomolbio.2019.08.010.

Stewart, Will. 2023. "Putin's nuclear show of strength to the West: Russia holds
surprise drills involving its nuke bombers." *Daily Mail*, April 18, 2023.
https://www.dailymail.co.uk/news/article-11984653/Russia-holds-surprise-
drills-involving-nuke-bombers.html.

Stonehill, Paul. 1998. *The Soviet UFO Files: Paranormal Encounters Behind the Iron
Curtain*. New York: Quadrillion Publishing.

Stonehill, Paul, and Philip Mantle. 2017. *Russia's Roswell Incident: And Other Amazing
UFO Cases From The Former Soviet Union*. Pontefract, West Yorkshire, UK: Flying
Disk Press.

———. 2020 <2016>. *Russia's USO Secrets: Unidentified Submersible Objects in Russian
and International Waters*. Pontefract, West Yorkshire, UK: Flying Disk Press.

———. 2010. *UFO Case Files of Russia*. Niagara Region, Ontario, Canada: 11th
Dimension Press.

Stride, Scot. 2005. "Interview by Dr. David Livingston." *The Space Show*, Broadcast 359 (Special Edition, July 11). https://www.thespaceshow.com/show/11-jul-2005/broadcast-359-special-edition.

Sturrock, Peter A. 1987. "An Analysis of the Condon Report on the Colorado UFO Project." *Journal of Scientific Exploration* 1, no. 1: 75-100. http://www.scientificexploration.org/journal/volume-1-number-1-1987

———. 2001. "Composition Analysis of the Brazil Magnesium." *Journal of Scientific Exploration* 15, no. 1: 69-95.

———. 1994a. "Report on a Survey of the Membership of the American Astronomical Society Concerning the UFO Problem: Part 1." *Journal of Scientific Exploration* 8, no. 1: 1-45.

———. 1994b. "Report on a Survey of the Membership of the American Astronomical Society Concerning the UFO Problem: Part 2." *Journal of Scientific Exploration* 8, no. 2: 153-95.

———. 1994c. "Report on a Survey of the Membership of the American Astronomical Society Concerning the UFO Problem: Part 3." *Journal of Scientific Exploration* 8, no. 3: 309-46.

———. 2004. "Time-Series Analysis of a Catalog of UFO Events: Evidence of a Local-Sidereal-Time Modulation." *Journal of Scientific Exploration* 18, no. 3: 399–419. http://www.scientificexploration.org/journal/volume-18-number-3-2004.

———. 1999. *The UFO Enigma: A New Review of the Physical Evidence*. New York: Warner Books, Inc.

———. 1974. "UFO Reports from AIAA Members." *Aeronautics and Astronautics* 12, no. 5 (May): 60-64.

Sturrock, Peter A., Von R. Eshleman, Thomas E. Holzer, et al. 1998. "Physical Evidence Related to UFO Reports: The Proceedings of a Workshop Held at the Pocantico Conference Center, Tarrytown, New York, September 29-October 4, 1997." *Journal of Scientific Exploration* 12, no. 2: 179-229. https://www.research gate.net/publication/224791605_Physical_Evidence_Related_to_UFO_Reports_ The_Proceedings_of_a_Workshop_Held_at_the_Pocantico_Conference_Center _Tarrytown_New_York_September_29_-_October_4_1997.

Svahn, Clas. 2022-2023. Emails to author, December 24, 2022 and April 6, 2023.

Swarbrick, Susan. 2020. "UFO special: The seven strangest unexplained sightings in Scotland's skies." *The Herald*, May 9, 2020. https://www.heraldscotland .com/life_style/18436635.ufo-special-seven-strangest-unexplained-sightings-scotlands-skies/.

Swords, Michael D. 2006. "We know where you live." *International UFO Reporter (IUR)* 30, no. 2: 7-12.

Tajmar, Martin, and Clovis J. de Matos. 2006. "Towards a new test of general relativity?" European Space Agency, March 23, 2006. https://www.esa.int/ Enabling_Support/Preparing_for_the_Future/Discovery_and_Preparation/Tow ards_a_new_test_of_general_relativity.

Talbert, A.E. 1955a. "Conquest of gravity aim of top scientists in U.S." *New York Herald-Tribune*, November 30, 1955: 1, 36.

———. 1955b. "Scientists taking first steps in assault on gravity barrier." *The Miami Herald*, November 30, 1955: 1, 2-A.

Taleb, Nassim Nicholas. 2012. *Antifragile: Things That Gain From Disorder*. New York: Random House.

Tannenwald, Nina. 2022. "'Limited' Tactical Nuclear Weapons Would Be Catastrophic." *Scientific American*, March 10, 2022. https://www.scientific american.com/article/limited-tactical-nuclear-weapons-would-be-catastrophic/.

Tarter, Jill, and Michael Michaud, editors. 1990. *SETI Post-Detection Protocol. Acta Astronautica* 21, no. 2 (Special Issue, February): 69-154.

TASS (Telegrafnoye Agentstvo Sovetskogo Soyuza, Russian Телеграфное агентство Советского Союза). 2023a. "No plans to use nuclear weapons in Ukraine – Russia's UN mission." *TASS (Russian News Agency)*, March 1/2, 2023. https://tass.com/politics/1583503.

———. 2022a. "Risk of nuclear war is increasing – Putin." *TASS (Russian News Agency)*, *TASS (Russian News Agency)*, December 7, 2022. https://tass.com/politics/1547371.

———. 2023b. "Russia, China convinced that nuclear war must never be unleashed – joint statement." *TASS (Russian News Agency)*, March 21/22, 2023. https://tass.com/russia/1592589.

———. 2023c. "Russia ready to defend itself with any weapon, including nuclear – Medvedev." *TASS (Russian News Agency)*, February 22, 2023. https://tass.com/defense/1580211.

———. 2022b. "Russian scientists carry out research on so-called UFOs, Roscosmos chief says." *TASS (Russian News Agency)*, June 11, 2022. https://tass.com/russia/1463895.

———. 2020. "UFO video footage captured by Russian cosmonaut sent for analysis – Roscosmos." *TASS (Russian News Agency)*, August 19, 2020. https://tass.com/science/1191653.

———. 2023d. "Ukraine's attempt to take Crimea reason enough for Russia to use any weapons – Medvedev." *TASS (Russian News Agency)*, March 24, 2023. https://tass.com/politics/1593963.

Tavares, Frank. 2020. "About Half of Sun-Like Stars Could Host Rocky, Potentially Habitable Planets." NASA Ames, October 29, 2020. https://www.nasa.gov/feature/ames/kepler-occurrence-rate.

Temple, Robert. 2007. "The prehistory of panspermia: Astrophysical or metaphysical." *International Journal of Astrobiology* 6: 169-80.

Thayer, Gordon D. 1971. "UFO encounter II - The Lakenheath England, Radar-Visual UFO case, August 13-14, 1956." *Aeronautics and Astronautics* 9, no. 9 (September): 60-64.

Theophrastus. 1999. *Enquiry into Plants*. Books 1-5. Translated by Arthur Hort. Cambridge, MA: Harvard University Press.

Thériault, Annie, and Florence Ogola. 2023. "Richest 1% bag nearly twice as much wealth as the rest of the world put together over the past two years." Oxfam International, January 16, 2023. https://www.oxfam.org/en/press -releases/richest-1-bag-nearly-twice-much-wealth-rest-world-put-together- over-past-two-years.

Thom, René. 1975 <1972>. *Structural Stability and Morphogenesis*. Reading, MA: W.A. Benjamin, Inc. Originally published in 1972 as *Stabilité structurelle et morphogénèse* by W.A. Benjamin, Inc.]

Thompson, Mark. 2023. "UBS is buying Credit Suisse in bid to halt banking crisis." *CNN*, March 19, 2023. https://www.cnn.com/2023/03/19/business/credit-suisse -ubs-rescue/index.html.

Tingley, Brett. 2022a. "Mysterious Drone Incursions Confirmed Over Sweden's Nuclear Facilities." *The Drive*, January 15, 2022. https://www.thedrive.com/the- war-zone/43901/mysterious-drone-incursions-confirmed-over-swedens- nuclear-facilities-this-weekend.

————. 2022b. "Russia dismisses space agency chief in wake of international controversies." *Space.com*, July 15, 2022. https://www.space.com/roscosmos-russia-space-agency-director-rogozin-out.

Tirone, Jonathan. 2023. "Nuclear Powers Pledge to Push Putin Out of Uranium Markets." *Bloomberg*, April 17, 2023. https://www.bloomberg.com/news/articles/2023-04-17/nuclear-powers-pledge-to-push-putin-out-of-uranium-markets#xj4y7vzkg.

Tórrez, Robert J. 2004. *UFOs over Galisteo and other stories of New Mexico's history*. Albuquerque: University of New Mexico Press. https://academic.oup.com/mind/article/LIX/236/433/986238.

Totten, Bill. 2003. "Pulling away the curtains from the 'Princes of the Yen.'" *The Japan Times*, August 10, 2003. https://www.japantimes.co.jp/culture/2003/08/10/books/book-reviews/pulling-away-the-curtains-from-the-princes-of-the-yen/.

Tough, Allen. 1991. *Crucial Questions About the Future*. Lanham, MD: University Press of America. [Quotation is from chapter entitled, "Intelligent Life in the Universe: What Role Will It Play in Our Future?"] http://members.aol.com/AllenTough/bok.html.

Trevithick, Joseph. 2023a. "Broken AARO? Pentagon UAP Office's Role Questioned Following Shootdowns." *The Drive*, February 15, 2023. https://www.the-drive.com/the-war-zone/broken-aaro-pentagon-uap-offices-role-questioned-following-shootdowns.

————. 2019. "Did Israel Just Conduct A Ballistic Missile Test From a Base On Its Mediterranean Coast?" *The Drive*, December 6, 2019. https://www.thedrive.com/the-war-zone/31358/did-israel-just-conduct-a-ballistic-missile-test-from-a-base-on-its-mediterranean-coast.

————. 2022. "Swedish Security Agency Declares A National Event As Drone Incursions Over Nuclear Sites Grow." *The Drive*, January 17, 2022. https://www.thedrive.com/the-war-zone/43905/swedish-security-agency-declares-a-national-event-as-drone-incursions-over-nuclear-sites-grow.

————. 2023b. "U.S. Scrambles To Get A Handle On Mysterious Aerial Incursions (Updated)." *The Drive*, February 13, 2023. https://www.thedrive.com/the-war-zone/u-s-scrambles-to-get-a-handle-on-mysterious-aerial-incursions.

Tritten, Travis J. 2019a. Email exchange with author.

———. 2019b. "UFO Group Sharing Exotic Materials With Army for Combat Vehicles." *Bloomberg Government*, October 21, 2019, https://about.bgov.com /news/ufo-group-sharing-exotic-materials-with-army-for-combat-vehicles/.

TT (Toftenes, Terje, and Truls Toftenes, directors). 2010. *The Day Before Disclosure*. Written by Terje Toftenes and Truls Toftenes. https://top-documentary films.com/day-before-disclosure/?__cf_chl_f_tk=0JrpLcXL8fSqoMujNnB2jAIyy VR2pcABb8VH0JNf.cs-1642594230-0-gaNycGzNCKU.

Turing, Alan M. 1950. "Computing Machinery and Intelligence." *Mind* LIX (59), no. 236 (October): 433-60. https://phil415.pbworks.com/f/TuringComputing.pdf.

Twining, Nathan F. 1947. "AMC Opinion Concerning 'Flying Discs.'" Secret Memo. September 23, 1947. http://www.nicap.org/twining_letter_docs.htm.

UAF (Unsealed Alien Files). 2015. "The 1952 Wave." Season 4, Episode 2 of *Unsealed Alien Files*. April 24, 2015. https://m.imdb.com/title/tt4743504/?ref_=nm_flmg _slf_41.

uaptheory. 2022. "On Our New Picture of the Galaxy & the Likelihood of Alien Life." https://www.uaptheory.com/habitable-worlds/.

UCS (Union of Concerned Scientists). 2015. "Frequently Asked Questions about Taking Nuclear Weapons Off Hair-Trigger Alert." Union of Concerned Scientists (UCS), January 2015. https://www.ucsusa.org/sites/default/ files/attach/2015/01/Hair-Trigger%20FAQ.pdf.

Uda, Také. 2010. "Declassified U.S. Government Documents on the UFO-Nukes Connection: A Representative Cross-Section." https://s3.documentcloud .org/documents/9330/declassified-u-s-government-documents-on-the-ufo- nuclear-weapons-connection.pdf.
[These declassified U.S. government documents were released by UFO researcher Robert L. Hastings at the September 27, 2010 National Press Club press conference in Washington, D.C.]

UFOs (UFOs: The White House Files). 2019. "UFOs: The White House Files." Season 1, Episode 1 of *UFOs: The White House Files*. History Channel, December 6, 2019. https://www.xfinity.com/stream/entity /6345212213226581112.

UFOs & Nukes. n.d. "Documents." *UFOs & Nukes*. https://www.ufohastings .com/documents.

UIA (Union of International Associations). 2022. "Irresponsible genetic manipulation." UIA: The Encyclopedia of World Problems & Human Potential, March 21, 2022. http://encyclopedia.uia.org/en/problem/139036.

UN (United Nations). 2022. "Non-Proliferation Treaty Conference Ends without Adopting Substantive Outcome Document Due to Opposition by One Member State." United Nations | Meetings Coverage and Press Releases, August 26, 2022. https://press.un.org/en/2022/dc3850.doc.htm.

————. 2000. "Report of the Committee on the Peaceful Uses of Outer Space." In *General Assembly Official Records, Fifty-fifth Session, Supplement No. 20 (A/55/20)*. June 26, 2000. New York: United Nations. https://www.unoosa .org/pdf/gadocs/A_55_20E.pdf.

————. 2023. "Security Council Rejects Draft Resolution Establishing Commission to Investigate Sabotage of Nord Stream Pipeline." United Nations | Meetings Coverage and Press Releases, March 27, 2023. https://press.un.org /en/2023/sc15243.doc.htm.

————. 2017. "United Nations Conference to Negotiate a Legally Binding Instrument to Prohibit Nuclear Weapons, Leading Towards their Total Elimination | 16 February, 27 – 31 March, 15 June – 7 July 2017." New York: United Nations. https://www.un.org/disarmament/tpnw/index.html.

UNODA (United Nations Office for Disarmament Affairs). n.d.-a. "Comprehensive Nuclear-Test-Ban Treaty (CTBT)." United Nations Office for Disarmament Affairs (UNODA). https://www.un.org/disarmament/wmd/nuclear/ctbt/.

————. n.d.-b. "Conference on Disarmament." United Nations Office for Disarmament Affairs (UNODA). https://www.un.org/disarmament /conference-on-disarmament/.

————. n.d.-c. "Treaty on the prohibition of nuclear weapons." United Nations Office for Disarmament Affairs (UNODA). https://www.un.org /disarmament/wmd/nuclear/tpnw/.

UNOOSA (United Nations Office for Outer Space Affairs). 2023. "Committee on the Peaceful Uses of Outer Space." United Nations Office for Outer Space Affairs (UNOOSA). https://www.unoosa.org/oosa/en/ourwork/copuos/index.html.

Urton, James. 2020. "Earth's cousins: Upcoming missions to look for 'biosignatures' in the atmospheres of nearby worlds." *UW [University of Washington] News*, February 14, 2020. https://www.washington.edu/news/2020/02/14/exoplanet -atmospheres-biosignatures/.

USAFE (United States Air Forces in Europe). 1948. "TOP SECRET." November 4, 1948. http://www.nicap.org/docs/481104usafedoc.pdf.

USAWC (United States Army War College). 2019. *Implications of Climate Change for the U.S. Army*. United States Army War College. https://climateandsecurity .files.wordpress.com/2019/07/implications-of-climate-change-for-us-army_ army-war-college_2019.pdf.

USCENTCOM (U.S. Central Command). n.d. "Area of Responsibility." U.S. Central Command. https://www.centcom.mil/AREA-OF-RESPONSIBILITY/.

U.S. DOD (U.S. Department of Defense). 2022a. "2022 Nuclear Posture Review." U.S. Department of Defense, October 27, 2022. https://s3.amazonaws.com /uploads.fas.org/2022/10/27113658/2022-Nuclear-Posture-Review.pdf.

———. 2022b. "Department of Defense Releases its 2022 Strategic Reviews – National Defense Strategy, Nuclear Posture Review, and Missile Defense Review." U.S. Department of Defense, October 27, 2022. https:// www.defense.gov/News/Releases/Release/Article/3201683/department-of- defense-releases-its-2022-strategic-reviews-national-defense-stra/.

———. 2021. "DoD Announces the Establishment of the Airborne Object Identification and Management Synchronization Group (AOIMSG)." U.S. Department of Defense, November 23, 2021. https://www.defense.gov /News/Releases/Release/Article/2853121/dod-announces-the-establishment-of- the-airborne-object-identification-and-manag/.

———. 2022c. "DoD Announces the Establishment of the All-domain Anomaly Resolution Office." U.S. Department of Defense, July 20, 2022. https://www.defense.gov/News/Releases/Release/Article/3100053/dod- announces-the-establishment-of-the-all-domain-anomaly-resolution-office/.

———. 2020a. "Establishment of Unidentified Aerial Phenomena Task Force." U.S. Department of Defense, August 14, 2020. https://www.defense.gov/News /Releases/Release/Article/2314065/establishment-of-unidentified-aerial- phenomena-task-force/.

———. 2020b. "Statement by the Department of Defense on the Release of Historical Navy Videos." U.S. Department of Defense, April 27, 2020. https://www.defense.gov/News/Releases/Release/Article/2165713/statement- by-the-department-of-defense-on-the-release-of-historical-navy-videos/.

———. 2023. "Statement From Secretary of Defense Lloyd J. Austin III." U.S. Department of Defense, February 4, 2023. https://www.defense.gov /News/Releases/Release/Article/3288535/statement-from-secretary-of-defense-lloyd-j-austin-iii/.

———. 2022d. "USD(I&S) Ronald Moultrie and Dr. Sean Kirkpatrick Media Roundtable on the All-domain Anomaly Resolution Office." U.S. Department of Defense, December 16, 2022. https://www.defense.gov/News/Transcripts /Transcript/Article/3249303/usdis-ronald-moultrie-and-dr-sean-kirkpatrick-media-roundtable-on-the-all-domai/.

U.S. DOS (U.S. Department of State). n.d. "New START Treaty." U.S. Department of State. https://www.state.gov/new-start/#:~:text=The%20United%20States%20 and%20the,force%20on%20February%205%2C%202011.

USS (United States Senate). 2023. "Letter from Members of the United States Senate." February 16, 2023. https://www.gillibrand.senate.gov/wp-content/ uploads/2023/02/Sen.-Gillibrand-Rubio-AARO-Funding-Request-Letter-FINAL-2.17.23.pdf.

U.S. SSCI (United States Senate Select Committee on Intelligence). 2022. Intelligence Authorization Act for Fiscal Year 2023. REPORT together with ADDITIONAL VIEWS. 117th Congress, 2nd Session, Report 117-132, July 20, 2022. https://www.congress.gov/117/crpt/srpt132/CRPT-117 srpt132.pdf.

van Brugen, Isabel. 2022. "Russian State TV Warns of Nuclear War That Only 'Mutants' Will Survive." *Newsweek*, June 7, 2022. https://www.newsweek .com/russian-state-tv-vladimir-solovyov-nuclear-war-nato-support-mutants-survive-ukraine-1713352.

Vergun, David. 2023. "Efforts Underway to Recover Object Downed Over Lake Huron." U.S. Department of Defense, February 13, 2023. https://www .defense.gov/News/News-Stories/Article/Article/3296905/efforts-underway-to -recover-object-downed-over-lake-huron/.

Visit Hessdalen. n.d. *Project Hessdalen*. http://www.hessdalen.org/.

Visser, Matt. 1996. *Lorentzian Wormholes: From Einstein to Hawking*. New York: Springer-Verlag.

von Ludwiger, Illobrand. n.d. *The New Worldview of the Physicist Burkhard Heim*. http://heim-theory.com/wp-content/uploads/2016/03/I-v-Ludwiger -The-New-Worldview-of-the-Physicist-Burkhard-Heim-160321.pdf.

Waddington, Conrad H. 1956. *Principles of Embryology*. London: Allen and Unwin.

Wainwright, Milton, and Fawaz Alshammari. 2010. "The forgotten history of panspermia and theories of life from space." *Journal of Cosmology* 7: 1771-76.

Watson, Nigel. 2007. "Official UFO Files Are Online." PRLOG (Press Release Log), April 20, 2007. https://www.prlog.org/10014365-official-ufo-files-are-online.html.

Weber, Renée. 1987. "Meaning as being in the implicate order philosophy of David Bohm: a conversation." In *Quantum Implications: Essays in Honour of David Bohm*, edited by Basil J. Hiley and F. David Peat, 436-50. Abingdon, U.K.: Routledge. [No page numbers are cited because I use the online version of this chapter, "Bohm: A Change of Meaning is a Change of Being, 'Meaning as being in the implicate order,'" available at https://ontoscopy.net/extras/bohm-a-change-of-meaning-is-a-change-of-being.]

———, editor. 1986. *Dialogues with Scientists and Sages: The Search for Unity*. London and New York: Routledge & Kegan Paul.

Weinstein, Dominique. 2001. *Unidentified Aerial Phenomena – Eighty Years of Pilot Sightings: A Catalog of Military, Airliner, and Private Pilots sightings from 1916 to 2000*. February 2001. https://static1.squarespace.com/static/5cf80ff422b5a9000 1351e31/t/5d02eb46935aac0001690f62/1560472408972/narcap_revised_tr-4.pdf.

Wells, Llyd E., John C. Armstrong, and Guillermo Gonzalez. 2003. "Reseeding of early earth by impacts of returning ejecta during the late heavy bombardment." *Icarus* 162, no. 1 (March): 38-46.

Werner, Richard A. 2003. *Princes of the Yen: Central Bankers and the Transformation of the Economy*. London: M.E. Sharpe.

Wesoff, Eric. 2022. "How did the US nuclear industry fare in 2022?" *Canary Media*, December 28, 2022. https://www.canarymedia.com/articles/nuclear/how-did -the-us-nuclear-industry-fare-in-2022.

WH (The White House). 2023. "NSTC: National Preparedness Strategy and Action Plan for Near-Earth Object Hazards and Planetary Defense." The White House, April 3, 2023. https://www.whitehouse.gov/ostp/news-updates/2023/04/03/ nstc-national-preparedness-strategy-and-action-plan-for-near-earth-object-hazards-and-planetary-defense/.

Wheeler, John A. 1992. *At Home in the Universe*. American Institute of Physics Masters of Modern Physics Series. Woodbury, NY: American Institute of Physics (AIP).

———. 1962. *Geometrodynamics*. New York: Academic Press.

Wickramasinghe, Janaki, (Nalin) Chandra Wickramasinghe, and William Napier. 2010. *Comets and the Origin of Life*. Singapore: World Scientific Publishing Co.

Wickramasinghe, Nalin Chandra. 2012. "DNA sequencing and predictions of the cosmic theory of life." *Astrophysics and Space Science* 343: 1-5, published September 7, 2012, issued January 2013. https://link.springer.com /article/10.1007/s10509-012-1227-y.

———. 2022a. "Giant Comet C/2014 UN271 (Bernardinelli-Bernstein) Provides New Evidence for Cometary Panspermia." *International Journal of Astronomy and Astrophysics* 12: 1-16. https://www.scirp.org/journal/ijaa.

———. 2022b. "Panspermia versus Abiogenesis: A Clash of Cultures." *Journal of Scientific Exploration* 36, no. 1 (Spring 2022 [May 22, 2022]): 121-29. https://doi.org/10.31275/20222199.

———. 2014. *The Search for our Cosmic Ancestry*. Singapore: World Scientific Publishing Co.

Wickramasinghe, Nalin Chandra, Fred Hoyle, and David Lloyd. 1996. "Eruptions of comet Hale-Bopp at 6.5 AU." *Astrophysics and Space Science* 240: 161-65. https://link.springer.com/article/10.1007/BF00640204.

———. 1999. "Eruptions of Comet Hale-Bopp at 6.5 AU." *Astrophysics and Space Science* 268: 373-78. https://link.springer.com/article/10.1023/A:1002409423674.

Wickramasinghe, Nalin Chandra, and Gensuke Tokoro. 2014a. "Life as a Cosmic Phenomenon: The Socio-Economic Control of a Scientific Paradigm." *Journal of Astrobiology & Outreach* 2, no. 2: 1-7. https://academia.edu/48090249/Life_as_a _Cosmic_Phenomenon_The_Socio-Economic_Control_of_a_Scientific_Paradigm.

———. 2014b. "Life as a cosmic phenomenon 2: The panspermia trajectory of Homo sapiens." *Journal of Astrobiology & Outreach* 2, no. 2: 115. https://www.research-gate.net/publication/270869470_Life_as_a_Cosmic_Phenomenon_2The_Panspe rmia_Trajectory_of_Homo_sapiens.

Wickramasinghe, Nalin Chandra, Gensuke Tokoro, and Milton Wainwright. 2015. "The transition from earth-centered biology to cosmic life." *Journal of Astrobiology & Outreach* 3, no. 1: 1000122. https://www.researchgate.net /publication/277573132_The_Transition_from_Earth-centred_Biology_to_ Cosmic_Life.

Wickramasinghe, Nalin Chandra, Dayal T. Wickramasinghe, Christopher A. Tout, John C. Lattanzio, and Edward J. Steele. 2019. *Astrophysics and Space Science* 364. Article number 205 (November 20). https://doi.org/10.1007/s10509-019-3698-6.

Wilpert, Gregory. 2021. "Reconstructing US-China Relations." Institute for New Economic Thinking. July 22, 2021. https://www.ineteconomics.org/perspectives/blog/reconstructing-us-china-relations.

Wilson, Edward O. n.d. "E.O. Wilson: What Does E.T. Really Look Like?" Big Think. https://bigthink.com/videos/what-does-et-really-look-like-with-eo-wilson/.

Woolsey, R. James, interview. 2021. "Amb. R. James Woolsey on the Cold War, JFK Assassination & UFOs." April 2, 2021. https://www.youtube.com/watch?v=_vQF4QzUX9M&t=3s.

Woolsey, R. James, William R. Graham, Henry F. Cooper, Fritz Ermarth, and Peter Vincent Pry. 2021. "Iran Probably Already Has the Bomb. Here's What to Do about It." *National Review*, March 19, 2021. https://www.national review.com/2021/03/iran-probably-already-has-the-bomb-heres-what-to-do-about-it/.

———. 2016. "Underestimating Nuclear Missile Threats from North Korea and Iran." *National Review*, February 12, 20216. https://www.national review.com/2016/02/iran-north-korea-nuclear/.

WPR (World Population Review). 2022. "Failed States 2022." World Population Review. https://worldpopulationreview.com/country-rankings/failed-states.

———. 2023. "Nuclear Weapons by Country 2023." World Population Review. https://worldpopulationreview.com/country-rankings/nuclear -weapons-by-country.

Wright, Dan. 2019. *The CIA UFO Papers: 50 Years of Government Secrets and Cover-ups.* Newburyport, MA: MUFON [Mutual UFO Network].

Wright, Jason T. 2017. "Prior Indigenous Technological Species." Version 2 submitted April 30, 2017. arXiv, submitted April 24, 2017 (v1), last revised April 30, 2017 (v2). http://arxiv.org/abs/1704.07263v2.

Xia, Lili, Alan Robock, Kim Scherrer, Cheryl S. Harrison, Benjamin Leon Bodirsky, Isabelle Weindl, Jonas Jägermeyr, Charles G. Bardeen, Owen B. Toon, and Ryan Heneghan. 2022. "Global food insecurity and famine from reduced crop, marine fishery and livestock production due to climate disruption from nuclear war soot injection." *Nature Food* 3: 586-96. file:///C:/Users/jensi/Downloads/s43016-022-00573-0.pdf.

Youssef, Nancy A., and Lindsay Wise. 2023. "Pentagon's Unidentified-Object Office Is Underfunded, Senators Say." *The Wall Street Journal*, February 14, 2023. https://www.wsj.com/articles/pentagons-unidenti-fied-object-office-is-underfunded-senators-say-b435af26.

Yturria, Santiago. 2004. "Mexican DoD Acknowledges UFOs In Mexico." The Light Party, May 11, 2004. https://lightparty.com/Spirituality/MexicanUFOsReal.htm.

Zalasiewicz, Jan, Mark Williams, Alan Haywood, and Michael Ellis. 2011. The "Anthropocene: a new epoch of geological time?" *Philosophical Transactions of the Royal Society A* 369: 835-41. https://royalsocietypublishing.org/doi/10.1098/rsta.2010.0339#:~:text=Anthropogenic%20changes%20to%20the%20Earth's,is%20widely%20and%20seriously%20debated.

Zeh, H. Dieter. 1999. "Why Bohm's Quantum Theory." *Foundation of Physics Letters* 12: 197-200, submitted December 21, 1998 (v1), last revised March 27, 1999 (v2). https://arxiv.org/pdf/quant-ph/9812059.pdf.

About the Author

Jensine Andresen

Jensine Andresen (Ph.D. Harvard University) holds a B.S.E. in Civil Engineering from Princeton University, where she also earned a Certificate from the School of Public and International Affairs. She completed an M.A. degree at Columbia University in Social Anthropology with a focus on China. She also earned A.M. (master's) and Ph.D. degrees at Harvard University from the Committee on the Study of Religion with a focus on Indo-Tibetan Buddhism. Dr. Andresen' translation of a renowned Sanskrit text and Tibetan annotations is forthcoming as *The Kālacakra Tantra: The Initiation Chapter with the Vimalaprabhā Commentary*.

Dr. Andresen served as a Visiting Assistant Professor at the University of Vermont, where she taught Science and Religion and world religions. She also was an Assistant Professor at Boston University in the interdisciplinary doctoral program on Science, Philosophy, and Religion. Dr. Andresen also held two academic appointments as a Visiting Scholar at Columbia University, and later she was appointed as an Officer of Research, Associate Research Scholar at Columbia.

Dr. Andresen edited *Religion in Mind: Cognitive Perspectives on Religious Belief, Ritual, and Experience* (Cambridge University Press 2001) and she is a co-editor of *Cognitive Models and Spiritual Maps: Interdisciplinary Explorations of Religious Experience* (Imprint Academic 2000) and of Extraterrestrial Intelligence: Academic and Societal Implications (Cambridge Scholars Publishing 2022). Dr. Andresen's recent chapter, "Two Elephants in the Room of Astrobiology," published in *Astrobiology: Science, Ethics, and Public Policy* (Wiley/Scrivener 2021) examines Unidentified Aerial Phenomena (UAP) in the context of the militarization and weaponization of space, which Dr. Andresen opposes. Her single-authored monograph, *Safe Space* (Tuthi 2023), is forthcoming, as is her invited chapter "Which Science?" in *Astroanthropology: Science, Ethics, and Religion*, edited by Arvin Gouw, Brian

Patrick Green, Junghyung Kim, and Ted Peters (forthcoming 2024). Her single-authored monograph, *Extraterrestrial Mind*, is in preparation.

Dr. Andresen has published multiple entries in *Encyclopedia of Science and Religion* (Macmillan, 2003), a chapter in *Fifty Years in Science and Religion: Ian G. Barbour and his Legacy* (Ashgate, 2004), and articles in many peer-reviewed journals and other publications, including *The International Journal for the Psychology of Religion, Harvard Theological Review, Dreaming, The American Society of International Law, Proceedings of the 96th Annual Meeting, March 13-16, 2002, Washington, DC: The Legalization of International Relations/The Internationalize of Legal Relations, Journal of Cultural Diversity, Journal of Sleep Research, The Journal of Religion, Zygon: Journal of Religion and Science, Religion and Education, Isis (Supplement, Catching up with the Vision: Essays on the Occasion of the 75th Anniversary of the Founding of the History of Society Society)*, and Boston University's *Focus*. Dr. Andresen is one of the co-authors of *Report on Ecumenical Faith and Genetics Working Group*, which was written as part of work with the Episcopal Diocese of Massachusetts Faith and Genetics Working Group. She also created a six-part videotape series, *Bioethics and Society: Scientific, Ethical, Legal, and Religious Perspectives on Genetic Technologies*.

In addition to her work in academia, Dr. Andresen has held various positions in finance, business, and government.